PLASTICITY OF THE NEUROMUSCULAR SYSTEM

The Ciba Foundation is an international scientific and educational charity. It was established in 1947 by the Swiss chemical and pharmaceutical company of CIBA Limited—now CIBA-GEIGY Limited. The Foundation operates independently in London under English trust law.

The Ciba Foundation exists to promote international cooperation in biological, medical and chemical research. It organizes about eight international multidisciplinary symposia each year on topics that seem ready for discussion by a small group of research workers. The papers and discussions are published in the Ciba Foundation symposium series. The Foundation also holds many shorter meetings (not published), organized by the Foundation itself or by outside scientific organizations. The staff always welcome suggestions for future meetings.

The Foundation's house at 41 Portland Place, London, W1N 4BN, provides facilities for meetings of all kinds. Its Media Resource Service supplies information to journalists on all scientific and technological topics. The library, open seven days a week to any graduate in science or medicine, also provides information on scientific meetings throughout the world and answers general enquiries on biomedical and chemical subjects. Scientists from any part of the world may stay in the house during working visits to London.

Ciba Foundation Symposium 138

PLASTICITY OF THE NEUROMUSCULAR SYSTEM

A Wiley – Interscience Publication

1988

JOHN WILEY & SONS

Chichester · New York · Brisbane · Toronto · Singapore

Published in 1988 by John Wiley & Sons Ltd, Chichester, UK.

Suggested series entry for library catalogues:
Ciba Foundation Symposia

Ciba Foundation Symposium 138
x + 273 pages, 61 figures, 6 tables

Library of Congress Cataloging-in-Publication Data

Plasticity of the neuromuscular system.
p. cm. — (Ciba Foundation symposium : 138)
'Symposium of Plasticity of the Neuromuscular System, held at the Ciba Foundation, London, 19–21 January 1988'—Contents p.
Edited by David Evered and Julie Whelan.
'A Wiley–Interscience publication.'
Includes bibliographies and indexes.
ISBN 0 471 91902 0
1. Muscles—Growth—Congresses. 2. Neuroplasticity—Congresses. 3. Neuromuscular transmission—Congresses. I. Evered, David. II. Whelan, Julie. III. Symposium on Plasticity of the Neuromuscular System (1988 : Ciba Foundation) IV. Series.
[DNLM: 1. Motor Neurons—physiology—congresses. 2. Muscles—growth & development—congresses. 3. Neuromuscular Junction—growth & development—congresses. 4. Neuronal Plasticity—congresses. W3 C161F v. 138 / WL 102.9 P715 1988]
QP321.P555 1988
591.1′852—dc19
DNLM/DLC
for Library of Congress 88-20780
CIP

British Library Cataloguing in Publication Data

Plasticity of the neuromuscular system.
1. Man. Musculoskeletal system
I. Evered, David II. Whelan, Julie
III. Series
612′.7

ISBN 0 471 91902 0

Typeset by Inforum Ltd, Portsmouth
Printed and bound in Great Britain at The Bath Press, Avon

Contents

Symposium on Plasticity of the neuromuscular system, held at the Ciba Foundation, London, 19–21 January 1988

The topic for this symposium was proposed by Professor Gerta Vrbová

Editors: David Evered (Organizer) and Julie Whelan

Participants

M.C. Brown University Laboratory of Physiology, University of Oxford, Parks Road, Oxford OX1 3PT, UK

A.J. Buller (*Chairman*) Lockhall Cottage, Cow Lane, Steeple Aston, Oxford OX5 3SG, UK

B.M. Carlson Department of Anatomy & Cell Biology, Medical Science Building II, University of Michigan Medical School, Ann Arbor, Michigan 48109, USA

C.M. Crowder Department of Pharmacology, Washington University School of Medicine, Box 8103, 660 South Euclid Avenue, St Louis, Missouri 63110, USA

M.R. Dimitrijevic Division of Restorative Neurology & Human Neurobiology, Baylor College of Medicine, 7000 Fannin, Suite 2140, Houston, Texas 77030, USA

V. Dubowitz Department of Paediatrics & Neonatal Medicine, Royal Postgraduate Medical School, Hammersmith Hospital, Du Cane Road, London W12 0HS, UK

B. Eisenberg Department of Physiology, Rush Medical College, Rush-Presbyterian-St Luke's Medical Center, 1750 West Harrison Street, Chicago, Illinois 60612, USA

T. Gordon Department of Pharmacology, Faculty of Medicine, University of Alberta, 9-70 Medical Sciences Building, Edmonton, Alberta, Canada T6G 2H7

C.E. Henderson* Institut Pasteur, Neurobiologie Moléculaire, 28 rue du Docteur Roux, F-75724 Paris cédex 15, France

---

* *Present address*: Biochimie CNRS-INSERM, B.P.5051, 34033 Montpellier Cedex, France.

P.N. Hoffman Departments of Ophthalmology & Neurology, 5-167 Meyer Building, Johns Hopkins Hospital, 600 N. Wolfe St, 601 North Broadway, Baltimore, Maryland 21205, USA

N. Holder Department of Anatomy and Human Biology, King's College London, Strand, London WC2R 2LS, UK

D. Kernell Department of Neurophysiology, University of Amsterdam, AMC, Meibergdreef 15, 1105 AZ Amsterdam, The Netherlands

C. Lance-Jones School of Medicine, Department of Neurobiology, Anatomy & Cell Science, University of Pittsburgh, Pittsburgh, Pennsylvania 15261, USA

M.B. Lowrie Department of Anatomy & Embryology, University College London, Gower Street, London WC1E 6BT, UK

A.W. Mudge MRC Developmental Neurobiology Programme, Department of Biology, Medawar Building, University College London, Gower Street, London WC1E 6BT, UK

B. Nadal-Ginard Department of Pediatrics, Harvard Medical School, Children's Hospital, 300 Longwood Avenue, Boston, Massachusetts 02115, USA

R.W. Oppenheim Department of Anatomy, The Bowman Gray School of Medicine, Wake Forest University, 300 South Hawthorne Road, Winston-Salem, North Carolina 27103, USA

D. Pette Fakultät für Biologie, Universität Konstanz, Postfach 5560, D-7750 Konstanz 1, Federal Republic of Germany

R.R. Ribchester Department of Physiology, University Medical School, Teviot Place, Edinburgh EH8 9AG, UK

F. Rieger† Department of Molecular & Developmental Biology, The Rockefeller University, 1230 York Avenue, New York, NY 10021-6399, USA

N. Rubinstein Department of Anatomy, University of Pennsylvania School of Medicine, Philadelphia, Pennsylvania 19104, USA

† *Present address*: Groupe de Recherches INSERM-CNRS, Biologie et Pathologie Neuromusculaires, 17 rue du Fer-à-Moulin, F-75005 Paris, France.

D. Sanes Departments of Otolaryngology and Physiology & Biophysics, New York University Medical Center, 550 First Avenue, New York, NY 10016, USA

J. Szczepanowska (*Ciba Foundation Bursar)* Laboratory of Protein Metabolism, Nencki Institute of Experimental Biology, 3 Pasteur Str, 02-093 Warszawa, Poland

S. Thesleff Department of Pharmacology, University of Lund, Sölvegatan 10, S-223 62 Lund, Sweden

D.C. Van Essen Division of Biology 216-76, California Institute of Technology, Pasadena, California 91125, USA

G. Vrbová Department of Anatomy & Embryology, University College London, Gower Street, London WC1E 6BT, UK

F.S. Walsh Department of Neurochemistry, Institute of Neurology, The National Hospital, Queen Square, London WC1 3BG, UK

R. Zak Department of Medicine, University of Chicago, 5841 S Maryland Avenue, Box 360, Chicago, Illinois 60637, USA

Introduction

A.J. Buller

Lockhall Cottage, Cow Lane, Steeple Aston, Oxfordshire OX5 3SG, UK

It is now nine years since the last major meeting on plasticity in the neuromuscular system was held. That meeting took place in Konstanz in 1979 and was organized by Dirk Pette, who is with us today. I acted as honorary chairman to that meeting, and would like to quote from my Introduction to the proceedings which were subsequently published (1980). I wrote 'It cannot be pretended that the meeting at Konstanz solved all our problems or differences, it would have been unthinkable that it could. What it did was to provide an opportunity for free, unbridled discussion (with more than a trace of humour) of those remaining enigmas which beset this fascinating field. Such opportunities are the lifeblood of modern scientific exchange. The written paper has its place — including within a volume such as this — but the face-to-face exchange with scientific friends and colleagues from other parts of the world is the essence of the small symposium. The meeting at Konstanz was a model of its kind. It sent us back to our respective corners of the globe refreshed, excited and greatly appreciative of our sponsors' and host's generosity'. Through the generosity of the Ciba Foundation we now have the opportunity to sort out some of the problems that remained after Konstanz. Again I doubt that we expect to draw matters to a final conclusion, but we can certainly make progress, to judge from the abstracts of the formal papers to be presented.

Unfortunately, some old friends and some new colleagues that I was personally looking forward to meeting cannot be with us for personal and family reasons — among them, Professor Mu-Ming Poo of Yale University School of Medicine, Dr Terje Lømo of the University of Oslo, and Dr Marilyn Duxson of the University of Otago. Nevertheless, I believe that we have more than a quorum of experts drawn from many disciplines (including biochemistry, physiology, developmental neurobiology, molecular biology pathology and medicine) who will, I am certain, produce a very lively symposium on this medically and scientifically important subject.

Reference

Pette D (ed) 1980 Plasticity of muscle. Walter de Gruyter, Berlin/New York

Physiological factors influencing the growth of skeletal muscle

Brenda R. Eisenberg, David J. Dix and John M. Kennedy*

Departments of Physiology, Rush Medical College and **University of Illinois, Chicago, Illinois 60612, USA*

Abstract. The growth of muscle can be regulated by developmental changes or by alterations in hormone levels or in the rate or amount of work demanded. The mechanisms and structures involved in growth processes can be studied by controlling these factors. The models used are chicken anterior latissimus dorsi (ALD) muscle under the influence of overloading and rabbit tibialis anterior (TA) muscle under the influence of chronic nerve stimulation. Both models involve changes in the isoform of myosin that is expressed. Methods of study include quantitative ultrastructural analysis, immunofluorescence and *in situ* mRNA hybridization. In overloaded chick ALD fibres polysomes are non-uniformly distributed between the myofibrils and in a peripheral annulus even though subcellular concentrations of the new isoform are not found. In normal rabbit muscle the highest concentration of myosin mRNA detected by *in situ* hybridization is found in the subsarcolemmal zone. In stimulated TA polysomes are found between myofibrils. It appears that the myosin mRNA accumulates at specific cell locations before translation; then diffusion of isomyosin and rapid exchange into myofibrils follows. Therefore, regulation of growth may be possible at the transcriptional, translational and assembly stages.

1988 Plasticity of the neuromuscular system. Wiley, Chichester (Ciba Foundation Symposium 138) p 3–21

Muscle cell organelles show a strong periodic repeat and offer a superb example of subcellular organization. The sizes and distribution of sarcoplasmic reticulum, mitochondria and contractile proteins are well known (reviewed by Eisenberg 1983). However, these structures are not static throughout life and striated muscle cells can adapt to different work outputs. Increased work produces an increased total muscle mass (hypertrophy) and disuse leads to loss (atrophy). This adaptive growth occurs in a controlled way to maintain a well-defined internal composition. Some of the mechanisms that might be involved in regulating and directing the growth of muscle are discussed in this paper. Particular attention is paid to the structural distribution of mRNA and polysomes and the effect on regional myofibrillar assembly.

Models of adaptive growth

Several experimental models exist for increasing the size of a muscle fibre by increasing the work. We have used a weight applied to a chicken wing to produce rapid hypertrophy of the anterior latissimus dorsi (ALD) and an acceleration of the normal developmental programme in isoform switching (Kennedy et al 1986, 1988). Isoforms are slightly modified forms of proteins. They differ in their properties and are expressed when functional demands on the muscle are drastically changed, often in response to an alteration in the firing pattern of the α-motor neuron. Many experimental situations induce such changes (e.g. exercise, spinal section, tenotomy and cross-innervation) but the exact nature of the work done by the muscle fibre is seldom known. The simplest model for isoform switching is provided by driving the nerve activity by a stimulator (Pette & Vrbová 1985, Salmons & Henriksson 1981). Initial findings showed that maintained low frequency (10 Hz) stimulation characteristic of a slow motoneuron nerve makes a fast muscle become slow, and maintained high frequency (40 Hz) stimulation characteristic of a fast nerve makes a denervated slow muscle become fast. We have used a continuous 10 Hz stimulation to transform the fast myosin heavy chain (MHC) of the rabbit tibialis anterior (TA) into a slow MHC isoform (Eisenberg & Salmons 1981).

The relationships between fibre size, isoform switching and activation pattern are not understood. In most mammals the largest fibres are fast-glycolytic (type IIB) and rapidly produce large forces in response to a voluntary burst of high frequency activity. The smallest fibres are slow-oxidative (type I) and usually sustain posture by reflex control of maintained low frequency activity. It is not commonly realized that under many chronic stimulation protocols muscles atrophy severely (Eisenberg & Salmons 1981). However, weight loss can be avoided if rests are given during the day or if the frequency is parcelled into intermittent bursts (Pette & Vrbová 1985). Burst patterns can also override the frequency-dependent isoform switching. These complex effects may be correlated with the load actually moved during the contraction (Fitts et al 1986).

Cell and molecular biological processes

The biological processes involved in muscle growth are the same as for any eukaryotic cell. Transcription of the DNA for numerous muscle genes occurs in the nucleus such that the amounts and kinds of RNA are coordinated and regulated as needed. The RNA is processed into mRNA and exported to the cytoplasm where distribution and translation follow. Cytoplasmic processes can also be regulated and modified. The newly translated proteins are assembled into the respective organelles. The assembly process and monomer pool may themselves provide feedback regulation on translation.

mRNA distribution

Some information about the cellular location of actively translated message in muscle can be obtained by observing the anatomical distribution of polyribosomes. In striated muscle, polysomes are usually thought to be found between the myofibrils (Eisenberg 1983) where cytoskeletal filaments and other organelles are located.

Study of ribosome distribution fails to distinguish myosin mRNA from other messenger RNA and free mRNA goes undetected. mRNA has been visualized by labelling complementary RNA probes and hybridizing them *in situ* on cultured cells or sections. It is significant that the first studies using probes for specific mRNAs have shown highly specific intracellular distributions. In cultures from embryonic chick skeletal muscle the MHC mRNA is first seen in the cytoplasm of adjacent myoblasts on fusion (John et al 1977). In cultured fibroblasts the intracellular distribution of actin mRNA is peripherally located near stress fibres, whereas vimentin and tubulin are near the nucleus (Lawrence & Singer 1986).

Myosin heavy chain expression

One of the most abundant contractile proteins is the myosin heavy chain, and the expression and regulation of its genes have been studied in detail. DNA sequencing and hybridization techniques suggest the existence of at least eleven MHC genes in rat (Wydro et al 1983). Sequence comparison of rabbit MHC isoforms displays their homologies. Rabbit α- and β-cardiac MHC cDNAs in the S2 region are more than 95% similar (Sinha et al 1982) and both are 75% similar to the nucleotide sequence of a fast skeletal cDNA (Maeda et al 1987).

Methods

The animal models used are described in detail elsewhere: overloading of a slow-tonic chick muscle (Kennedy et al 1986, 1988) and chronic stimulation of a fast-twitch rabbit muscle (Eisenberg & Salmons 1981).

Anatomical techniques included stereological morphometry (Eisenberg 1983) and immunofluorescence with a slow-specific antibody (HPM7, a gift from Dr R. Zak) (Eisenberg et al 1985). Endogenous mRNA was detected in sections by *in situ* hybridization with a biotinylated riboprobe (Dix & Eisenberg 1988a,b). The probe was transcribed from the 1.1 kb *Sac*I fragment of pMHC-81, encoding the C-terminal end of the myosin light meromyosin

(Sinha et al 1982). RNA probes (riboprobes) have some advantages in that either the hybridizing (cRNA) or the non-hybridizing (mRNA) riboprobe can be transcribed. The hybridizing cRNA probe bound tightly to the natural mRNA and stayed within the section. The non-hybridizing mRNA riboprobe did not hybridize but might bind non-specifically to tissue elements and remain as a single-stranded RNA until removal by ribonuclease and washing. Frozen sections used for *in situ* hybridization of RNA probes had the additional advantage of allowing immunofluorescence on serial sections. Fixation of the frozen section retained structural detail, allowing fine resolution of the cellular localization of specific mRNAs. In summary, frozen sections were fixed, prehybridized and carried through the hybridization, wash, and enzymic detection steps.

Results

Overloaded chick muscle

The application of a weight overload to the humerus of chickens induced hypertrophy of ALD muscle fibres; the fibre diameter increased and the muscle weight almost doubled after 14 days of overloading (Kennedy et al 1986). This growth was accompanied by a rapid and almost complete replacement of slow-tonic myosin isoform. SM1, by another slow-tonic isoform SM2. A microscopic evaluation of these rapidly growing fibres showed a marked change in a subsarcolemmal annulus seen as a halo in cross-section or a rim in longitudinal section (Fig. 1A). Ultrastructural examination revealed that the subsarcolemmal annulus was rich in polyribosomes and cytoplasm but relatively devoid of myofibrils (Fig. 1B). Some mechanism must be responsible for this remarkable accumulation of actively translating mRNA.

Since subjective observation suggested that numerous changes in morphology were occurring, some findings were quantified by using stereological analysis of randomly sampled groups of electron micrographs. The myofibres showed a decrease in myofibrillar content (from 72% to 58%) with the polysomal fraction increasing. The polyribosomal region of the subsarcolemmal shell almost doubled and there was also a marked increase of polysomes between the myofibrils (Fig. 2A).

Normal and stimulated rabbit muscle

In normal rabbit skeletal muscle polysomes are only rarely identified in thin sections at the ultrastructural level. However, polysomes have been noted in experimental situations of increased use. In a re-examination of the electron micrographs used in the analysis of the chronically stimulated TA (Eisenberg & Salmons 1981), we noted an increase in polysomes. They were most

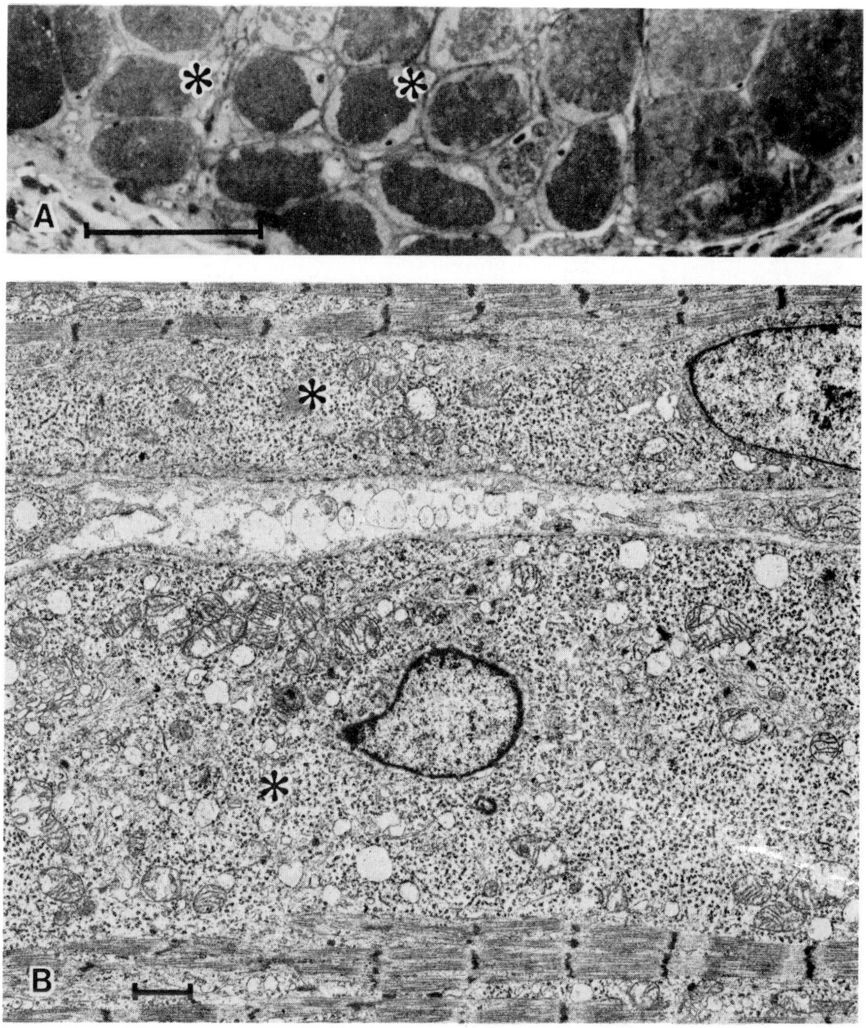

FIG. 1. Overloaded chick ALD muscle showing subsarcolemmal annulus rich in polysomes at asterisks. (A) Light micrograph after five days of overload, cut 1μm thick in cross-section with annulus. Bar, 50 μm. (B) Electron micrograph after three days of overload, cut in longitudinal section. Bar, 1 μm. (From Kennedy et al 1988, with permission.)

abundant nine to 11 days after the initiation of the 10 Hz stimulation (Fig. 2B). The polysomes were usually found in a coiled configuration at the I band between the myofibrils. Occasionally, a straight polysome was seen lying transversely or longitudinally oriented in the I band as if being aligned by a cytoskeletal filament (Fig. 2B).

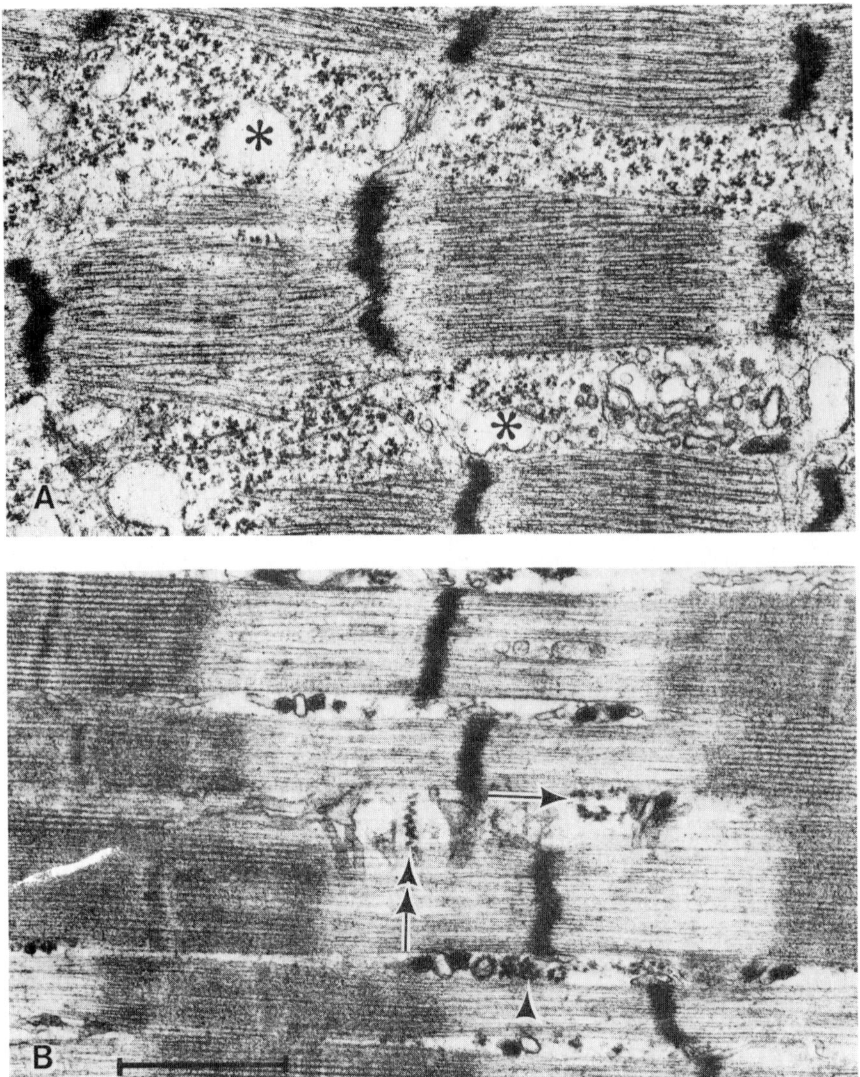

FIG. 2. (A) Electron micrograph of longitudinally sectioned chick ALD muscle showing inter-myofibrillar spaces rich in polysomes at asterisks, after two weeks of overload. Bar, 1μm. (B) Electron micrograph of longitudinally sectioned rabbit TA muscle after 11 days of chronic stimulation. Note that polysomes lie between fibrils in coils (arrowhead), longitudinally (single arrow), or transversely (double arrow). Bar 1 μm.

Adult rabbit skeletal muscle has been successfully frozen, fixed, hybridized, and studied by enzymic detection for the distribution of myosin heavy chain mRNA within the cells. The pattern of MHC mRNA distribution

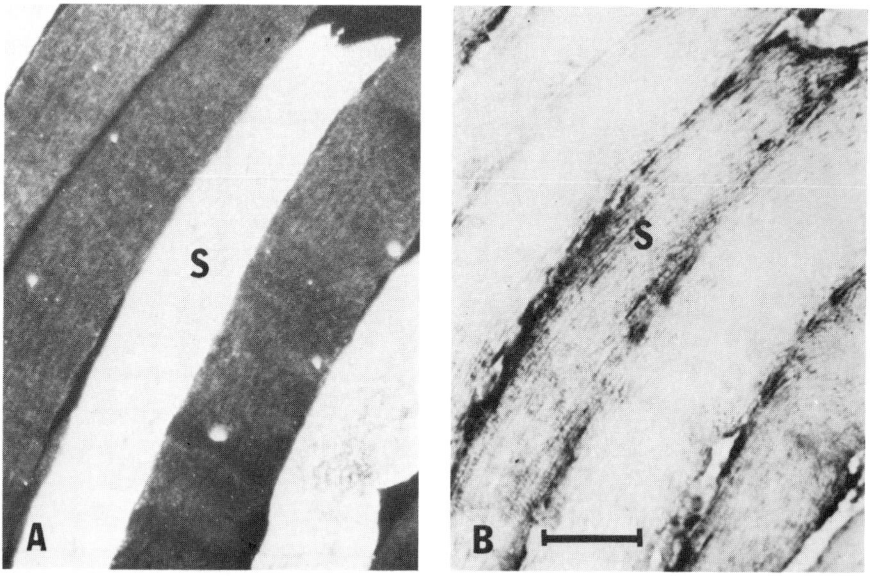

FIG. 3. Serial frozen sections of normal rabbit medial gastrocnemius muscle cut longitudinally. (A) Immunofluorescent detection of an anti-slow muscle monoclonal antibody (HPM7, kindly provided by Dr R. Zak) which identifies the slow-twitch fibres (S). (B) *In situ* hybridization to myosin heavy chain mRNA using an α-cardiac biotinylated riboprobe detected enzymically with streptavidin phosphatase. Bar, 25 μm.

within the cell was seen at subcellular resolutions with light microscopy. The biotinylated cRNA probe complementary to some forms of MHC mRNA hybridized cellular message (Fib. 3B). The slow-twitch fibres, classified immunologically (Fig. 3A), hybridized the α-cardiac myosin mRNA probe while the fast IIB fibres did not (Fig. 3B) (Dix & Eisenberg 1988a,b). The mRNA was concentrated in the subsarcolemmal and inter-myofibrillar regions of a fibre, indicating that mechanisms for subcellular concentration of mRNA were operational.

Discussion

Transcriptional regulation

Muscle gene expression is subject to numerous regulatory factors which are integrated in the nucleus in ways which are not yet well understood. The transcriptional regulation is affected by muscle growth factors and physiological 'activity' factors (Fig. 4). The muscle growth factors might include developmental factors, nutritional status, hormone levels (thyroid, insulin,

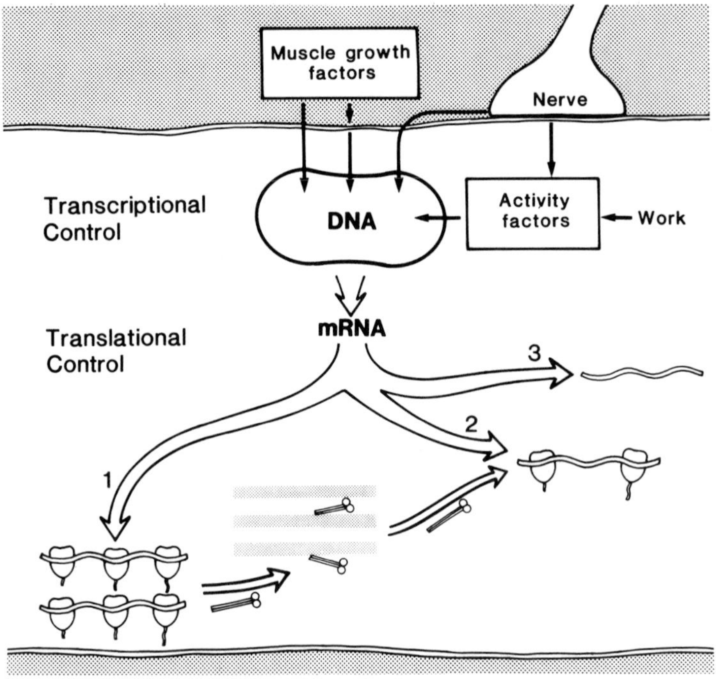

FIG. 4. Diagram of potential pathways involved in the transcriptional and translational regulation of myosin expression in skeletal muscle. The possible routes taken by MHC mRNA are indicated. (1) mRNA travels to the subsarcolemmal region where it is actively translated into myosin. The new double-headed myosin molecules build new filaments and are rapidly exchanged throughout existing myofibrils. (2) A lesser amount of mRNA travels to the inter-myofibrillar spaces where it is translated less efficiently because of autoregulation by excess or degraded myosin. (3) Some mRNA is not translated and is non-polyribosomal.

testosterone), and perhaps neurotrophic agents (Zak 1984). The physiological work of the fibre is also important and it appears that the load against which the muscle contracts is a most significant element (Fitts et al 1986). However, the actual mechanisms are not yet known.

Examples of macroscopic differences in the transcriptional regulation of skeletal muscle nuclei are known. The neuromuscular junction region has the highest levels of mRNA for the acetylcholine receptor subunits (Merlie & Sanes 1985), suggesting that these nuclei are independently regulated. In ectopically stimulated muscle fibres the excited zone accumulates different MHC isoforms from the remaining fibre (Salviati et al 1986). Segmental domains of polymorphic expression of myosin light and heavy chains occur in fibres during chronic stimulation (Staron et al 1987). Nuclear domains in the myotendon region might also exist, because the greatest concentrations of vinculin and talin (proteins associated with mechanical adhesion) are present here.

Translational control

The distribution and interaction of mRNA with other cellular components is important to the post-transcriptional regulation of gene expression in the cytoplasm. Considerable attempts have been made using biochemical extractions to assign structural compartments for the free and polysomal mRNAs. The distribution of mRNA may well change when it goes from a non-translated to a translated state. In L6 myoblasts and myotubes, 85% of mRNA is found in polysomes which are bound to the cytoskeleton (Pramanik et al 1986). Translation of MHC mRNA in unfused L6 myoblasts is blocked, whereas the same mRNA becomes polysomal and is translated in post-fusion L6 myotubes (Endo & Nadal-Ginard 1987). In unfused myoblasts the MHC mRNAs bind a different set of ribonuclear proteins (mRNPs) in polysomal or free mRNP fractions (Ruzdijic et al 1985).

Autoregulation

mRNA has been shown to be concentrated close to the site where newly translated cytoskeletal proteins are inserted into the existing framework (Fulton et al 1980). Close spatial associations of mRNA with an organelle could create a microenvironment for the autoregulation of the translation of message by its product. Autoregulation of tubulin is by excess of monomeric tubulin, leading to increased degradation of β-tubulin mRNA by an interaction with the first 16 translated codons (Gay et al 1987). These results suggest that MHC mRNA, if actively translated, could be found in association with the filaments. Co-regulation between the number of thin and thick filaments in the sarcomere might favour the possibility of interaction between MHC mRNA and actin filaments. Actin stress fibre-like structures appear to direct myofibrillogenesis in cultured chick cardiac myocytes (Dlugosz et al 1984). Thin filaments may act as a template and may concentrate myosin in the appropriate area of the sarcomere to enhance thick filament assembly and alignment. Once all the available template sites were filled, monomeric myosin concentration would increase. This excess could link myosin assembly and the autoregulation of mRNA: unassembled myosin would stop further translation.

Assembly

After translation the newly synthesized isomyosin diffuses and exchanges for existing isomyosin in the myofibrils (Wenderoth & Eisenberg 1987). This process has been confirmed by the uniform distribution throughout the fibre seen in several cases where transition between two isoforms is occurring — for example, in chick, where the early slow myosin, SM1, is being replaced by the later form, SM2 (Kennedy et al 1986). Regional accumulations of a newly synthesized isoform of myosin have never been found at the submicroscopic level.

Summary

The growth of muscle is regulated by developmental changes, or by alterations in hormone levels or in the rate or amount of work demanded. Regulation by these factors occurs at the transcriptional level in individual nuclei. A hypothesis for translational control in the cytoplasm has been discussed, namely that MHC mRNA associates with the myofibrils and that redistribution of message may occur during rapid changes in synthesis rates and regional growth (Fig. 4). In normal muscle fibres the subcellular location of MHC mRNA appears to be subsarcolemmal and between the myofibrils, perhaps confirming these notions. Extreme usage of muscle produces an unusual rearrangement of ribosomes. For example, in overloaded slow-tonic fibres of chicken ALD, dense ribosomal accumulations are seen in a subsarcolemmal annulus (Fig. 1; Kennedy et al 1988) and after chronic stimulation of rabbit TA, clusters of ribosomes are interspersed between the myofibrils (Fig. 2B; Eisenberg & Salmons 1981). The myosin mRNA accumulates at specific cell locations before translation; then diffusion of isomyosin and rapid exchange into myofibrils follows. Regulation of growth could be at the transcriptional, translational and assembly stages.

Acknowledgements

We thank Dr M.P. Wenderoth for participation in many discussions and for the critical evaluation of this manuscript. This research was supported by grants from the National Institutes of Health and the American Heart Association.

References

Dix DJ, Eisenberg BR 1988a Spatial distribution of myosin mRNA in cardiac tissue by *in situ* hybridization techniques. In: Clark WA et al (eds) Biology of the isolated adult cardiac myocyte. Elsevier Science Publishers, Amsterdam, p 147–160

Dix DJ, Eisenberg BR 1988b *In situ* hybridization and immunocytochemistry in serial sections of rabbit skeletal muscle to detect myosin expression. J Histochem Cytochem, in press

Dlugosz AA, Antin PB, Nachmias VT, Holtzer H 1984 The relationship between stress fibre-like structures and nascent myofibrils in cultured cardiac myocytes. J Cell Biol 99:2268–2278

Eisenberg BR 1983 Quantitative ultrastructure of mammalian skeletal muscle. In: Peachey LD, Adrian RH (eds) Handbook of physiology. American Physiological Society, Bethesda. Section 10, 73–112

Eisenberg BR, Salmons S 1981 The reorganization of subcellular structure in muscle undergoing fast-to-slow type transformation: a stereological study. Cell Tissue Res 220:449–471

Eisenberg BR, Edwards JA, Zak R 1985 Transmural distribution of isomyosin in rabbit ventricle during maturation examined by immunofluorescence and staining for calcium-activated adenosine triphosphate. Circ Res 56:548–555

Endo, T, Nadal-Ginard B 1987 Three types of muscle specific gene expression in fusion-blocked rat skeletal muscle cells: translational control in EGTA-treated cells. Cell 49:515–526

Fitts RH, Metzger JM, Riley DA, Unsworth BR 1986 Models of disuse: a comparison of hindlimb suspension and immobilization. J Appl Physiol 60:1946–1953

Fulton AB, Wan KM, Penman S 1980 The spatial distribution of polyribosomes in 3T3 cells and the associated assembly of proteins into the skeletal framework. Cell 20:849–857

Gay DA, Yen TJ, Lau TY, Cleveland DW 1987 Sequences that confer β-tubulin autoregulation through modulated mRNA stability reside within exon of a β-tubulin mRNA. Cell 50:671–679

John HA, Patrinou-Georgoulas M, Jones KW 1977 Detection of myosin heavy chain mRNA during myogenesis in tissue culutre by *in vitro* and *in situ* hybridization. Cell 12:501–508

Kennedy JM, Kamel S, Tambone W, Vrbová G, Zak R 1986 Expression of myosin heavy chain isoforms in normal and hypertrophied chicken slow muscle. J Cell Biol 103:977–983

Kennedy JM, Eisenberg, BR, Reid SK, Sweeney LJ, Zak R 1988 Nascent muscle fiber appearance in overloaded chicken slow-tonic muscle. Am J Anat 181:203–215

Lawrence JB, Singer RH 1986 Intracellular localization of messenger RNAs for cytoskeletal proteins. Cell 45:407–415

Maeda K, Sczakiel G, Wittinghofer A 1987 Characterization of cDNA coding for the complete light meromyosin portion of a fast skeletal muscle myosin heavy chain. Eur J Biochem 167:97–102

Merlie JP, Sanes JR 1985 Concentration of acetylcholine receptor mRNA in synaptic regions of adult muscle fibres. Nature (Lond) 317:66–68

Pette D, Vrbová G 1985 Neural control of phenotypic expression in mammalian muscle fibers. Muscle & Nerve 8:676–689

Pramanik SK, Walsh RW, Bag J 1986 Association of messenger RNA with the cytoskeletal framework in rat L6 myogenic cells. Eur J Biochem 160:221–230

Ruzdijic SD, Bird R, Jacobs FA, Sells BH 1985 Specific mRNP complexes: characterization of the proteins bound to histone H4 mRNAs isolated from L6 myoblasts. Eur J Biochem 153:587–594

Salmons S, Henriksson J 1981 The adaptive response of skeletal muscle to increased use. Muscle & Nerve 4:94–105

Salviati G, Biasia E, Aloisi M 1986 Synthesis of fast myosin induced by fast ectopic innervation of rat soleus muscle is restricted to the ectopic endplate region. Nature (Lond) 322:637–639

Sinha AM, Umeda PK, Kavinsky CJ et al 1982 Molecular cloning of mRNA sequences for cardiac α- and β-form myosin heavy chains: expression in ventricles of normal, hypothyroid and thyrotoxic rabbits. Proc Natl Acad Sci USA 79:5847–5851

Staron RS, Gohlsch B, Pette D 1987 Myosin polymorphism in single fibers of chronically stimulated rabbit fast-twitch muscle. Pfluegers Arch 408:444–450

Wenderoth MP, Eisenberg BR 1987 Incorporation of nascent myosin heavy chains into thick filaments of cardiac myocytes in thyroid treated rabbits. J Cell Biol 108:2771–2780

Wydro RM, Nguyen HT, Gubits RM, Nadal-Ginard B 1983 Characterization of sarcomeric myosin heavy chain genes. J Biol Chem 258:670–678

Zak R 1984 Growth in the heart in health and disease. Raven Press, New York

DISCUSSION

Zak: One aspect of your paper is particularly fascinating, namely the need for the control of gene expression. This control could occur at many stages—transcription, translation, product degradation, and assembly. There are apparently no pools of myosin molecules that have not been assembled into thick filaments. How is this regulation achieved? Is it the product, myosin, which regulates the process, by 'siphoning' off the newly made myosin molecules and assembling them immediately, or is regulation at the level of transcription, with a feedback mechanism to determine how much protein to make?

Eisenberg: Your point about lack of waste is not necessarily true. There are examples, both pathological ones and in experimental situations, where the coordination between the different proteins is not complete. Thus in dog latissimus dorsi muscle, subjected to 12 months of intermittent high frequency stimulation (Acker et al 1987), I have recently observed large masses of actin filaments. It seems as if the muscle can make enough actin but cannot make enough myosin.

Zak: I agree with that; I mean coordination within the pool of myosin molecules. One seldom sees a pool of unassembled myosin molecules. There are data that indicate the presence of a pool of myosin light chains, but one never finds this for myosin heavy chains. Perhaps it's a technical problem.

Eisenberg: By 'pool', do you mean detected by your biochemical methods, and not by anatomical ones? I agree that the free pool of unassembled protein is different for different proteins, and that the size of the free pool could well be involved in regulation. The best example of this form of translational control is with tubulin, where an excess of monomeric tubulin regulates translation of further tubulin (Gay et al 1987).

Zak: All the evidence suggests that the assembly and disassembly of myosin molecules is a random process, yet we see a non-random distribution of myosin mRNA. There must be some diffusion of the messenger RNA, or of the synthesized protein, therefore.

Eisenberg: I prefer to think of a 'regional' feedback mechanism. I don't think the nucleus is able to say 'the periphery needs more protein' or 'the neuromuscular junction needs a different protein', and then to change its addressing system to deliver the mRNA to the particular location. It more likely works from 'local' feedback within the cytoplasm, so that when the microenvironment in one part of the cell is changed or damaged, not everything in the cell is repaired, but just that damaged part.

Kernell: In relation to this point, and your picture of 'old' myosin molecules being replaced by new ones, this replacement has to be coordinated. When a new molecule becomes available, might that stimulate the removal or disintegration of an old fragment of myosin?

Eisenberg: It's difficult to answer your question of how the new pool of myosin molecules influences the exchange process. In our first model the overloaded chick muscle is growing and in the second model the stimulated rabbit muscle is atrophying. When a muscle grows, one can say that the cell down-regulates the unwanted gene and up-regulates the new gene, to give an excess of new protein which could control the exchange process. But as a muscle atrophies, the ratio of the two mRNAs changes, but could still result in a decrease in the size of the myosin pool. At present I cannot solve the problem that you pose.

Rubinstein: Probably myosins are randomly coming in and out of the thick filament all the time, even in normal muscle. When you increase, for example, the synthesis of slow myosin, that is the one that is inserted, by a random process.

What interests me about this randomness is how a myosin molecule reaches the middle of a myosin filament, and also how it works: in a myosin filament that contains fast and slow myosin molecules, with different ATPase activities, how does this filament contract without damaging one or the other myosin molecule?

Eisenberg: The answer to the latter question can be given. Rick Moss's group take isolated segments of muscle and first measure the physiological parameters; they then assay the ratio of isomyosins. They find a good correlation between the maximum velocity of unloaded shortening and the isoform ratio in the fibre (Reiser et al 1985). So even though the classical sliding filament model of contraction is hard to envisage with cross-bridges of mixed isoforms along a thick filament, the muscle evidently knows how it's done. On your first point, myosin molecules in the shaft of the thick filament are only ionically bonded, and it is quite easy for one to be pulled out and another put back in.

Van Essen: So the exchange of myosin molecules is not an active process?

Eisenberg: Exchange happens in isolated thick filaments in the test tube. It's a physicochemical reaction and does not require energy. However, the role of the other thick filament proteins, apart from the heavy and light chains of myosin, seems to be important—for example, C protein. In our exchange studies the myosin in the ends of the thick filaments turned over more rapidly than in the middle of the A band where the C proteins are found. Experiments need to be done in which those auxilliary protein elements are controlled, in order to demonstrate their role in governing myosin exchange rates.

Van Essen: If the different isoforms can exchange, are the rates equivalent, or might there be some specificity, such that the probability of replacement of a slow molecule by another slow one is greater than that of a slow by a fast isoform?

Eisenberg: The 'packing' part of the protein (i.e., the tail) seems to be able to 'mix and match', even between myosins from smooth, cardiac and skeletal muscles, or between chicken and mouse. Thus, the tail region is highly conserved with three kinds of repeating features found in all myosins (Warrick & Spudich 1987).

Zak: I have the impression, from the experiments being discussed, that muscle is in a constant state of flux. However, myofibrils are very rigid structures. When Hugh Huxley tried to disassemble myofibrils some years ago, he had to use very drastic homogenization conditions. Only about 10% of the myofibril mass is dissociated to filaments by this procedure. So there is both fluidity and rigidity. In studies by Fischman and his associates (Saad et al 1986), the half-life of the myosin molecule was measured in seconds, but if this were true, the myofibrils would dissolve when isolated and put in saline; they are actually very difficult to dissolve.

Nadal-Ginard: In terms of replacement and the 'plasticity' of muscle proteins, you looked at isoform switches in a model in which there is no new *net* sarcomere formation. What happens when new sarcomeres are being formed, when the muscle fibre is increasing in length or in width? Is it true that the newly formed sarcomeres contain mainly newly synthesized protein?

Eisenberg: The new sarcomeres don't contain new myosin protein; as a new thick filament or a sarcomere is built, it immediately starts exchanging with the old. Because of this, there is never a completely new part.

Nadal-Ginard: In a muscle that has been both stretched and chronically stimulated, where there are new sarcomeres at the end of the fibre, near the tendon, these sarcomeres have been shown to have a different isoform composition to the old ones.

Eisenberg: Goldspink's group showed this at short times after the stretch (Williams et al 1986). We are repeating some of their experiments. It's also now known that in stretch you are recruiting satellite cell nuclei into the growing fibres (K.C. Darr, personal communication). Skeletal muscles are long, so it takes time for the exchange process to travel to the centre of the fibre.

Buller: How short is 'short'?

Eisenberg: Four days after the initial stretch is applied to a stimulated muscle, slow myosin is seen a few millimetres from the end of the muscle, as detected by ATPase histochemistry, whereas it is not seen in the middle. By a week or two, the slow myosin is seen throughout the fibre.

Nadal-Ginard: Your data on new myosin synthesis, and Donald Fischman's fluorescent energy transfer data (Saad et al 1986) (if the numbers he reports for subunit interchange are at all close to reality), suggest that it should be practically impossible to detect a boundary between old and new filaments. The complete mixing in a very long fibre may take a long time, as you say, but the boundaries between old and new filaments should disappear almost instantly, because the interchange is a local process at the boundary. How do your data fit with these rapid exchange rates?

Eisenberg: Quite well. Cardiac cells in mammals are about 150 μm in diameter, and the diffusion coefficient for myosin molecules is about 1.2×10^{-7}

cm^2s^{-1}, so the diffusion rate for myosin can be calculated. Within 1–2 minutes a myosin molecule could diffuse anywhere within this relatively short cell. Times might be slower if something is impeding diffusion, such as the exchange process. In skeletal muscle the distances are longer; for example, from tendon to the belly of a rabbit tibialis anterior is about 3cm. It would take over 100 days for a myosin molecule to travel this distance by free diffusion. Something else may also be coming into play there. Not every nucleus in multinucleate skeletal muscle is expressing the same genes at the same rates. I can conceive that the central part of a muscle fibre could be minimally up-regulated with the new isoform, whereas the rapidly growing tip is making only the new isoform. That would change the baseline and make exchange at the ends much faster than in the middle.

Nadal-Ginard: You showed that most myosin mRNA seems to be located in the periphery, Dr Eisenberg, where the nuclei are found in a normal fibre. Do you see a higher concentration of mRNA around the nuclei in the fibre and, if so, is there a gradient from the nuclei which becomes shallower towards the periphery?

Eisenberg: Cardiac muscle has centrally located nuclei, whereas skeletal muscle has peripheral nuclei; the myosin mRNA is central in cardiac muscle and peripheral in skeletal. The perinuclear poles show the highest concentration of mRNA, but simple diffusion would not explain the concentration gradient found beyond that.

Pette: Do gradients exist around the nuclei of a given skeletal muscle fibre, or is there an overall mRNA gradient within the fibre, for example in the stimulated muscle?

Eisenberg: We haven't looked at stimulated muscle with the mRNA assay. You have done the histochemistry of chronically stimulated fibres and have shown that different segments of the fibre are in different histochemical states, and we hope to confirm your finding.

Pette: There were focal differences in myosin ATPase activity and myosin heavy chains along the fibre (Staron & Pette 1987), which could mean that the exchange of fast myosin heavy chains with slow myosin heavy chains is not so rapid at the protein level.

Eisenberg: Or that a nucleus isn't transcribing the mRNA for a given isoform.

Pette: Or that the nuclei are not transcribing in an identical mode.

Kernell: A very general question: are there any systematic differences between different kinds of skeletal muscle fibres in the density of their nuclei?

Eisenberg: Skeletal muscles differ in nuclear density. Male fibres have more nuclei than female fibres! Fast-glycolytic fibres apparently have fewest nuclei, and the biggest diameter (Eisenberg 1983).

Kernell: So the lengthwise density might be the same? Is anything known about the total number of nuclei?

Eisenberg: The nuclear density along the length of soleus fibres is about double that in EDL fibres of the rat (K.C. Darr & E. Schultz, personal communication). One needs to study the absolute number, however, because muscle mass is also changing, in the experimental models.

Lowrie: Kelly (1978) reported that fast muscle fibres have a lower total number of nuclei per fibre than slow muscle fibres at all stages of development (from birth). This difference cannot be accounted for by differential growth of the muscle fibres.

Eisenberg: Nuclear density looks to me to be greatest in the neuromuscular junction and in the myotendinous region in skeletal muscle fibres. So, to summarize, nuclear density is regulated, and there are fibre-type, age-dependent and use-dependent differences.

Kernell: Might nuclear density be a factor of importance for the ease with which you can change the protein synthesis of a muscle fibre?

Eisenberg: Nuclear density may have more to do with the rate of protein synthesis that a fibre requires to keep its environment turning over and healthy and to repair damage.

Pette: The use of chronic stimulation provides a good model for studying changes in nuclear density. We have observed that chronic stimulation induces a pronounced increase in the number of nuclei (Maier et al 1986). It appears likely that the increase in nuclear density is to some extent due to an augmented number of satellite cells.

Carlson: An interesting correlate to nuclear density is satellite cell number, because there are more satellite cells in slow muscle fibres than in fast. After cross-transplantation or cross-innervation between fast and slow muscle, the satellite cell number corresponds to that of the transformed muscle (Schultz 1984). It is nice to see that there is regulation of the potential precursor population as well.

Dr Eisenberg, have you any ideas about the potential adaptive value of central nucleation with respect to the localization of protein synthesis that you have shown? Central nucleation seems to appear in some circumstances where the muscle has to change its forms of myosin more often than in others. Could there be something about being in the centre of the muscle fibre that facilitates qualitative or quantitative changes in gene expression?

Eisenberg: In normal skeletal muscle, the nucleus is peripheral, whereas in abnormal dystrophic or regenerating muscle it is definitely central. We don't know what holds nuclei in position or what gets them there, so I cannot answer this question.

Vrbová: I wanted to ask a question which probably also cannot be answered, but maybe you can speculate! What is it that changes the microenvironment of the muscle cell with stretch? Is it due to the fact that the cross-bridges are not allowed to slide so much?

Eisenberg: In experiments done with John Kennedy in Radovan Zak's laboratory we looked ultrastructurally at the chicken ALD within the first few hours and few days after stretch (Kennedy et al 1988). Certain regions of the fibre are over-stretched, so a few sarcomeres will be 'extra-digitated', so to speak; so that is one microenvironmental change.

Vrbová: Is it at the place where you find the exchange of myosin first?

Eisenberg: We don't know, because we cannot follow the same fibres. But other changes might be more significant: in a few places, adjacent to these overstretched regions, the plasma membrane has very small holes; more often the muscle membrane is intact but the basement membrane is broken. Then the muscle has access to a whole new world of stimulating factors, namely the fibroblast and tissue growth factors that normally can't get close to the muscle fibre. I believe that microenvironmental stimuli for growth might enter by this route.

Nadal-Ginard: We can say something more in answer to the question of what changes the muscle cell microenvironment in stretch. We know that stretch activates sodium channels very quickly in muscle cells. In fact, most types of cell have stretch-activated sodium channels. One other response to stretch in skeletal (or cardiac) muscle is a rapid (5–10 minutes) and significant induction of all the proto-oncogenes that are known to be involved in cell cycle regulation, namely c-*fos*, c-*myc* and c-*mos*. In many ways this response is the same cascade of responses that most cells exhibit in response to growth factors. I suppose that these early events triggered by stretch are involved in the isoform switching.

Vrbová: The skeletal muscle with least sodium channels is the one most sensitive to stretch: the ALD has few sodium channels and is the most sensitive skeletal muscle to stretch; the fast muscles that have many sodium channels hardly respond to stretch at all.

Nadal-Ginard: These sodium channels are voltage-gated sodium channels, however. This does not appear to be the case for the stretch-activated ones.

Eisenberg: Stretch opens almost all channels, as been studied with patch-clamp techniques; so what you say is correct, but we don't have to limit it to the sodium channels. Other channels also respond to stretch.

Nadal-Ginard: The sodium channel I am referring to is not the voltage-gated sodium channel that is also activated by stretch. It is a different channel where the normal activation is believed to be by stretch.

Pette: H.H. Vandenburgh and S. Kaufman (1980) have shown that, in stretched myotubules, there is an increase in amino acid uptake and protein synthesis. Unfortunately, these authors did not study the effect of stretch on the expression of specific proteins, such as myosin. I also recall a study of Alan Kelly and Neal Rubinstein (1980), who showed that local mechanical damage caused a switch in myosin expression along the length of the muscle fibre. I wonder to what degree stretch can change the ionic milieu, which would then possibly induce changes in myosin expression.

Eisenberg: In the chicken ALD, the fibre membrane has only a few small holes but there may be leaky patches that can't be detected. Moreover, the same mechanism that starts muscle repair may also stimulate the satellite cells to divide and fuse. In stimulation experiments, another problem is how the isoform switching occurs. We could debate whether damage occurs, and how much. In the thyroid hormone-induced isoform switch in cardiac muscle there are no satellite cells and there does not appear to be any membrane damage. That appears to be a direct transcriptional switch.

Hoffman: Do changes in mRNA levels in these cells reflect altered rates of transcription or changes in mRNA turnover?

Nadal-Ginard: Our results suggest that most changes in myosin mRNA levels in normal skeletal and cardiac muscle are transcriptionally controlled, and this is also found by Radovan Zak's group. In skeletal muscle there is some translational control, especially of the light chains. In a normal cell, however, it is possible to produce a situation where gene expression is regulated by a strict translational control; for example, contractile proteins expression is highly sensitive to calcium levels and it is possible to regulate the translation of all their messages by regulating the calcium level in the extracellular (and, as a consequence, intracellular) environment. Nevertheless, in the normal muscle, especially *in situ*, most changes in contractile protein expression are likely to be transcriptional changes. We have not seen dramatic changes in the stability of the mRNAs for sarcomeric proteins. This doesn't mean that there are no small modulations.

Eisenberg: And it doesn't rule out the possibility of microenvironmental (regional) regulation, either?

Nadal-Ginard: No; it could be that translation and stability are regulated at the 'regional' level in the cell, where we would be looking at averages with the available techniques.

Pette: Translational control is likely to exist for myosin light chain LC3f. This is evident from *in vitro* translation studies in which LC3f is strongly translated, whereas its protein content found *in vivo* is very low (Heilig & Pette 1982).

References

Acker M, Anderson WA, Hammond RL et al 1987 Oxygen consumption of chronically stimulated skeletal muscle. J Thorac Cardiovasc Surg 94:702–709

Eisenberg BR 1983 Quantitative ultrastructure of mammalian skeletal muscle. In: Peachey LD, Adrian RH (eds) Handbook of Physiology. Williams & Wilkins, Baltimore (Am Physiol Soc, Bethesda) Section 10, p 73-112

Gay DA, Yen TJ, Lau TY, Cleveland DW 1987 Sequences that confer β-tubulin autoregulation through modulated mRNA stability reside within exon of a β-tubulin mRNA. Cell 50:671–679

Heilig A, Pette D 1982 Changes in transcriptional activity of chronically stimulated fast twitch muscle. FEBS (Fed Eur Biochem Soc) Lett 151:211–214

Kelly AM 1978 Satellite cells and myofiber growth in the rat soleus and extensor digitorum longus muscles. Dev Biol 65:1–10

Kelly AM, Rubinstein NA 1980 Patterns of myosin synthesis in regenerating normal and denervated muscles of the rat. In: Pette D (ed) Plasticity of muscle. Walter de Gruyter, Berlin/New York, p 161–175

Kennedy JM, Eisenberg BR, Reid SK, Sweeney LJ, Zak R 1988 Nascent muscle fiber appearance in overloaded chicken slow-tonic muscle. Am J Anat 181:203–215

Maier A, Gambke B, Pette D 1986 Degeneration-regeneration as a mechanism contributing to the fast to slow conversion of chronically stimulated fast-twitch rabbit muscle. Cell Tissue Res 244:635–643

Reiser PJ, Moss RL, Guilian GG, Greaser ML 1985 Shortening velocity in single fibers from adult rabbit soleus muscles is correlated with myosin heavy chain composition. J Biol Chem 260:9077–9080

Saad AD, Pardee JD, Fischman DA 1986 Dynamic exchange of myosin molecules between thick filaments. Proc Natl Acad Sci USA 83:9483–9487

Schultz E 1984 A quantitative study of satellite cells in regenerated soleus and extensor digitorum longus muscles. Anat Rec 208:501–506

Staron RS, Pette D 1987 Nonuniform myosin expression along single fibers of chronically stimulated and contralateral rabbit tibialis anterior muscles. Pfluegers Arch Eur J Physiol 409:67–73

Vandenburgh HH, Kaufman S 1980 In vitro skeletal muscle hypertrophy and Na pump activity. In: Pette D (ed) Plasticity of muscle. Walter de Gruyter, Berlin/New York, p 493–506

Warrick HM, Spudich JA 1987 Myosin: structure and function in cell motility. Annu Rev Cell Biol 3:379–421

Williams P, Watt P, Bicik V, Goldspink G 1986 Effect of stretch combined with electrical stimulation on the type of sarcomeres produced at the ends of muscle fibers. Exp Neurol 93:500

Molecular basis of the phenotypic characteristics of mammalian muscle fibres

Dirk Pette and Robert S. Staron*

Fakultät für Biologie, Universität Konstanz Postfach 5560, D-7750 Konstanz, West Germany

Abstract. Adult mammalian skeletal muscle fibres can be separated into two distinct groups, fast and slow. Within each group there is a continuum of metabolic enzyme activity levels. In addition there are fast and slow isoforms of various myofibrillar proteins such as myosin, tropomyosin and troponin. These proteins are multimeric and multiple isoforms of their subunits assemble to create a continuum of subtypes within each major group. Fibres which coexpress both fast and slow subunit isoforms have an increased number of possible isoform combinations such that an entire spectrum of fibre 'types' is found between the two extremes, fast and slow. Numerous myosin heavy chain and fast troponin T isoforms further increase the diversity of muscle fibres. Such cellular diversity helps to explain the dynamic nature of skeletal muscle. Each individual fibre is able to respond to various functional demands by appropriate changes in its phenotypic expression of specific proteins.

1988 Plasticity of the neuromuscular system. Wiley, Chichester (Ciba Foundation Symposium 138) p 22–34

The heterogeneity of skeletal muscle is reflected by its variable fibre composition and by the diversity of its individual fibres. Contrary to the original assumption of only a few fibre types in mammalian skeletal muscle, we now know that a continuum exists. This continuum results from a pronounced polymorphism of thick and thin filament proteins and a gradation of metabolic properties. The following discussion will be confined to selected examples which illustrate the molecular basis for phenotypic expression and the dynamic nature of skeletal muscle fibres.

Myosin polymorphism

Myosin is an asymmetric, hexameric molecule consisting of a pair of heavy chains (HC) and two pairs of light chains (LC). One of the LC pairs consists

* *Permanent address:* Department of Zoological and Biomedical Sciences, College of Osteopathic Medicine, Ohio University, Athens, Ohio 45701–2979, USA.

of two phosphorylatable light chains, while the other pair is either a heterodimer or one of two possible homodimers of two different alkali light chains. With a fixed pair of heavy chains, these three LC combinations give rise to three isomyosins (d'Albis et al 1979, Hoh & Yeoh 1979). However, the number of possible isomyosins is greater because both light and heavy chains exist in fast and slow or additional isoforms.

Fast (LC1f, LC2f, LC3f) and slow (LC1s, LC2s) light chain isoforms predominate in fast- and slow-twitch muscles, respectively (Lowey & Risby 1971). An additional light chain isoform LC_{emb} (corresponding to the alkali LC found in cardiac atrium) is expressed in embryonic and newborn muscle tissue (Whalen et al 1978, 1982). A slow and two fast myosin HC isoforms exist in adult mammalian skeletal muscles. Slow (type I) fibres contain HCI and fast fibre types IIA and IIB contain HCIIa and HCIIb, respectively (Billeter et al 1981, Danieli Betto et al 1986, Staron & Pette 1987a, b).* Certain fibres express more than one HC isoform. Type IIAB fibres coexpress HCIIa and HCIIb (Danieli Betto et al 1986, Staron & Pette 1987b), and type IC and type IIC fibres coexpress HCI and HCIIa (Staron & Pette 1986, 1987a, Staron et al 1987). These C fibres may also coexpress embryonic (HC_{emb}) and neonatal (HC_{neo}) heavy chains (Maier et al 1988). In addition, distinct heavy chains have been found in developing muscles (Whalen et al 1981) and in specific muscles such as extraocular muscles (Wieczorek et al 1985, Sartore et al 1987) and some muscles of mastication (Rowlerson et al 1981).

Theoretically, these LC and HC isoforms could combine to form a large number of isomyosins. Considering only the heavy chains HCI, HCIIa and HCIIb, and the slow (LC1s and LC2s) and the fast (LC1f, LC2f, LC3f) light chains, a maximum of 60 isomyosins is possible in adult skeletal muscle fibres (Staron & Pette 1987a, b). Single fibre analyses have demonstrated the coexistence of fast and slow LC isoforms together with heavy chains HCI and HCIIa in C fibres of rabbit soleus and tibialis anterior muscles (Staron & Pette 1986, 1987a, b, Staron et al 1987). Therefore, C fibres may contain up to 54 different isomyosins (Table 1). Of these 54 possible isomyosins, one is expressed in type I fibres and three are expressed in type IIA fibres. Six additional fast isomyosins (Staron & Pette 1987b) result from the three possible fast LC combinations with the HCIIb homodimer (expressed in IIB fibres) and from the three fast LC combinations with the HCIIa/HCIIb heterodimer (expressed in IIAB fibres). The number of possible isomyosins is greater than 60 if the two slow alkali light chains LC1sa and LC1sb (Sréter et al 1972), as well as the embryonic LC (Whalen et al 1978) and the additional HC isoforms, are considered. On the other hand, the possible existence of

---

* *Note added in proof*: Recently, an additional HC, tentatively designated as HCIId, has been detected in rat muscles endowed for sustained contractile activity. Its highest level was found in diaphragm (A. Bär & D. Pette 1988. Three fast myosin heavy chains in adult rat skeletal muscle. FEBS Lett 235:153–155).

TABLE 1 Combinations between myosin light and heavy chains derived from single fibre studies in the rabbit (Staron & Pette 1978a,b)

	Heavy chain combinations		
Light chain combinations	*A*	*B*	*C*
1 $(LC1s)_2\ (LC2s)_2$	$(HCI)_2$	(HCI) (HCIIa)	$(HCIIa)_2$
2 (LC1s) (LC1f) $(LC2s)_2$	$(HCI)_2$	(HCI) (HCIIa)	$(HCIIa)_2$
3 $(LC1f)_2\ (LC2s)_2$	$(HCI)_2$	(HCI) (HCIIa)	$(HCIIa)_2$
4 $(LC1s)_2$ (LC2s) (LC2f)	$(HCI)_2$	(HCI) (HCIIa)	$(HCIIa)_2$
5 (LC1s) (LC1f) (LC2s) (LC2f)	$(HCI)_2$	(HCI) (HCIIa)	$(HCIIa)_2$
6 $(LC1f)_2$ (LC2s) (LC2f)	$(HCI)_2$	(HCI) (HCIIa)	$(HCIIa)_2$
7 $(LC1s)_2\ (LC2f)_2$	$(HCI)_2$	(HCI) (HCIIa)	$(HCIIa)_2$
8 (LC1s) (LC1f) $(LC2f)_2$	$(HCI)_2$	(HCI) (HCIIa)	$(HCIIa)_2$
9 $(LC3f)_2$ (LC2s) (LC2f)	$(HCI)_2$	(HCI) (HCIIa)	$(HCIIa)_2$
10 $(LC3f)_2\ (LC2s)_2$	$(HCI)_2$	(HCI) (HCIIa)	$(HCIIa)_2$
11 (LC1s) (LC3f) $(LC2s)_2$	$(HCI)_2$	(HCI) (HCIIa)	$(HCIIa)_2$
12 (LC1f) (LC3f) $(LC2s)_2$	$(HCI)_2$	(HCI) (HCIIa)	$(HCIIa)_2$
13 (LC1f) (LC3f) (LC2s) (LC2f)	$(HCI)_2$	(HCI) (HCIIa)	$(HCIIa)_2$
14 (LC1s) (LC3f) $(LC2f)_2$	$(HCI)_2$	(HCI) (HCIIa)	$(HCIIa)_2$
15 (LC1s) (LC3f) (LC2s) (LC2f)	$(HCI)_2$	(HCI) (HCIIa)	$(HCIIa)_2$
16 $(LC1f)_2\ (LC2f)_2$	$(HCI)_2$	(HCI) (HCIIa)	$(HCIIa)_2$
17 (LC1f) (LC3f) $(LC2f)_2$	$(HCI)_2$	(HCI) (HCIIa)	$(HCIIa)_2$
18 $(LC3f)_2\ (LC2f)_2$	$(HCI)_2$	(HCI) (HCIIa)	$(HCIIa)_2$
19 $(LC1f)_2\ (LC2f)_2$		(HCIIa) (HCIIb)	$(HCIIb)_2$
20 (LC1f) (LC3f) $(LC2f)_2$		(HCIIa) (HCIIb)	$(HCIIb)_2$
21 $(LC3f)_2\ (LC2f)_2$		(HCIIa) (HCIIb)	$(HCIIb)_2$

Assuming light and heavy chain heterodimers, and not considering the LC1sa/LC1sb heterogeneity, a total of 60 theoretical isomyosins are possible. Type I fibres contain one combination (1A). Types IC and IIC contain 54 combinations (1A,B,C–18A,B,C). Type IIA fibres contain three combinations (16C,17C,18C). Type IIB fibres also contain three combinations 19C, 20C, 21C). Type IIAB fibres contain nine combinations (16C–21C plus 19B, 20B, 21B).

preferential affinities between specific light and heavy chain isoforms could result in a predominance of certain isomyosins. Therefore, fewer isomyosins may exist than the number of theoretical LC/HC combinations. Regardless of the exact number of isomyosins, their coexpression in variable ratios (e.g. variable HC ratios; see Fig. 1) adds to the heterogeneity of skeletal muscle fibres.

Polymorphism of tropomyosin and troponin

Mammalian muscle fibres also display polymorphism of the regulatory proteins associated with the thin filament. Tropomyosin (TM), a dimeric protein consisting of α and β subunits, exists in homodimers and heterodimers (Dhoot & Perry 1979, Bronson & Schachat 1982). In addition, slow and fast forms of the α-subunit have been identified (Steinbach et al 1980, Bronson & Schachat 1982). The distribution of various TM patterns appears to be muscle-specific. Two TM patterns are prevalent in specific rabbit muscles.

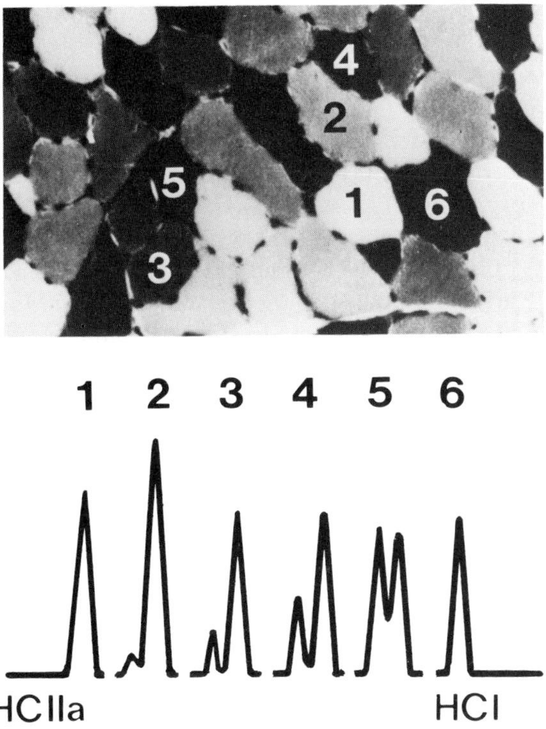

FIG. 1. Correlation between histochemically assessed myofibrillar actomyosin ATPase (mATPase) and myosin heavy chain content in single fibres. *Upper panel:* cross-section stained for mATPase (preincubation at pH 4.3) of a 30 d stimulated (24 hours daily at 10 Hz) tibialis anterior muscle of the rabbit. *Lower panel:* tracings of densitometric evaluations of electrophoretically separated myosin heavy chains from fragments of the same fibres numbered in the upper panel (for further details see Staron et al 1987). 1, type IIA fibre; 2–5, C fibres; 6, type I fibre.

The α-TM homodimer predominates in some fast muscles (m. longissimus dorsi, m. psoas), whereas other fast muscles, such as m. plantaris, assemble the α/β-TM heterodimer (Bronson & Schachat 1982).

Fast and slow isoforms also exist for the three troponin (Tn) subunits TnT, TnI and TnC (see Dhoot & Perry 1979). In addition, there is a pronounced heterogeneity of the TnT subunit which exists in at least five fast and two slow isoforms (Briggs et al 1987). Nevertheless, the number of fast TnT isoforms identified at the protein level so far is much smaller than the potential of the fast TnT gene to generate up to 64 isoforms at the transcriptional level (Breitbart & Nadal-Ginard 1986).

The expression of the various fast TnT isoforms (TnT_f) in rabbit muscles seems to follow a restricted pattern in association with the TM subunits. The TnT_{1f} isoform is found in combination with the α/β-TM heterodimer, the

TnT_{2f} is preferentially coexpressed with the α-TM homodimer, and the TnT_{3f} is found in combination with the α/β-TM heterodimer and the β-TM homodimer (Moore & Schachat 1985, Moore et al 1987). This distribution does not appear to correlate with individual fast fibre types where the TnT_f/TM combinations are expressed in a continuum (Schachat et al 1985). Therefore, fast-twitch glycolytic (FG) fibres of m. erector spinae and m. plantaris predominantly express TnT_{2f}, whereas TnT_{1f} is the major isoform in FG fibres of the diaphragm muscle. In contrast, the fast-twitch oxidative-glycolytic (FOG) fibres of different muscles express TnT_{1f}, TnT_{2f}, TnT_{3f} and TnT_{4f} in highly variable proportions (Schachat et al 1985, Moore et al 1987). However, these authors used a fibre classification scheme which is primarily based upon metabolic properties and, therefore, their attempt to correlate the distribution of the fast TnT subunits with fibre types may have led to ambiguous results.

It is evident from microbiochemical analyses that fibres classified by qualitative histochemistry span an entire metabolic spectrum (for reviews see Pette 1985, Pette & Spamer 1986). Furthermore, a correlation between metabolic properties and the expression of myofibrillar protein isoforms is not necessarily expected. However, the myofibrillar protein composition may correlate in some manner with the myosin ATPase (mATPase)-based histochemical fibre types. Indeed, the myosin HC composition of single fibres correlates with the mATPase histochemistry (Staron & Pette 1986) (see Fig. 1) and also with the contractile properties (Reiser et al 1985). These relationships do not appear to apply to the light chains (Reiser et al 1985, Moore et al 1987, Staron & Pette 1987a). It remains to be seen how the distribution of the regulatory proteins of the thin filament relates to the myosin composition in fast fibres. To answer this question, an mATPase-based fibre classification scheme should reveal more information than fibre typing according to metabolic properties.

Metabolic diversity

Highly variable activity levels of enzymes of anaerobic and aerobic metabolic pathways are found in adult skeletal muscle fibres (see Pette & Spamer 1986). Discriminative enzyme activity ratios do delineate two major fibre populations in rabbit skeletal muscles corresponding to fast and slow fibres (Fig. 2). However, each of these two populations is heterogeneous with regard to numerous enzymes and displays a continuum of metabolic activity. This metabolic heterogeneity appears to evolve during postnatal development (Dangain et al 1987). Measurements of metabolic enzyme activities in fibre types distinguished by the use of mATPase histochemistry indicate that fast IIB and IIA fibre subtypes cannot be clearly separated on the basis of their metabolic, especially aerobic-oxidative, potentials (see Pette 1985).

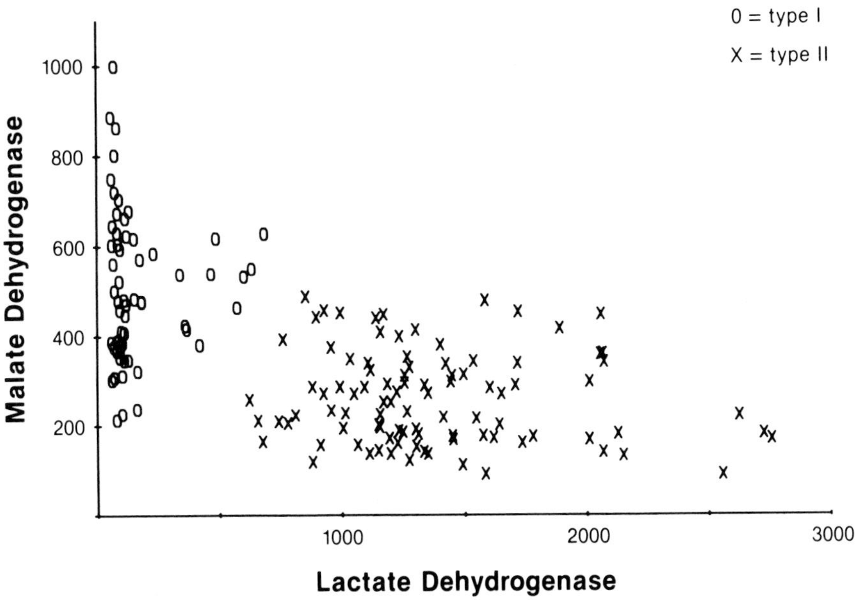

FIG. 2. Plot of malate dehydrogenase activity (aerobic-oxidative marker) against lactate dehydrogenase activity (glycolytic marker) determined in individual fast- (×) and slow-twitch (O) fibres of rabbit psoas and soleus muscles (for further details see Pette & Spamer 1986). Note the separation into two groups, each containing a continuum of activity levels. (From Pette & Spamer 1986 by permission of *Federation Proceedings*.)

Fibre 'type' continuum

The distinction of fibre types beyond the two major groups of fast and slow fibres, although scientifically useful, is an oversimplification. Qualitative and quantitative analyses have revealed an entire spectrum between these two extremes (Staron & Pette 1986, Schachat et al 1985, Pette & Spamer 1986, Moore et al 1987). The C fibres, which coexpress several isomyosins in various proportions (Fig. 1), best illustrate the concept of a fibre continuum. Nevertheless, the classification of fibre types has proved to be a useful tool in many studies on normal muscle. It must be kept in mind, however, that within each defined category is an unlimited number of possible combinations of protein isoforms. Moreover, histochemically defined fibre types differ between species (Carraro et al 1981), and between (Schachat et al 1980, Mizusawa et al 1982, Mabuchi et al 1984) and within (Moore & Schachat 1985, Pette & Spamer 1986, Staron & Pette 1987a) muscles of the same species. Therefore, in our view, attempts to create categories of static fibre types neglect the versatile nature of muscle fibres.

Conclusions

Muscle fibres are in a dynamic state which emphasizes their plasticity and explains their ability to respond specifically to altered functional demands by appropriate changes in phenotypic expression. Multigene families of major myofibrillar proteins, such as the myosin heavy chains (see Mahdavi et al 1986), permit the differential expression of various protein subunit isoforms. In addition, the mechanism of alternative splicing, e.g. of myosin alkali light chains (see Barton & Buckingham 1985), troponin T (Breitbart & Nadal-Ginard 1986) and possibly myosin heavy chains (Jandreski et al 1987), contributes to the multiplicity of protein subunit isoforms in muscle. Functionally different isoforms of the multimeric myofibrillar proteins result from multiple combinations of their subunits and combine to form a great number of specific myofibrillar structures. The total number of possible myofibrillar protein isoforms may be restricted because of preferential combinatorial patterns and this may explain why specific fibre groups can be delineated. Obviously, the predominant isoforms of these multimeric proteins depend upon the amounts of specific subunits maintained. This involves multiple regulatory steps at the levels of transcription, translation and protein degradation. The significance of neural control on some of these processes has recently been reviewed (see Pette & Vrbová 1985).

Acknowledgements

This study was supported by the Deutsche Forschungsgemeinschaft, Sonderforschungsbereich 156, and by the Alexander von Humboldt-Stiftung.

References

Barton PJR, Buckingham ME 1985 The myosin alkali light chain proteins and their genes. Biochem J 231:249–261

Billeter R, Heizmann CW, Howald H, Jenny E 1981 Analysis of myosin light and heavy chain types in single human skeletal muscle fibres. Eur J Biochem 116:389–395

Breitbart RE, Nadal-Ginard B 1986 Complete nucleotide sequence of the fast skeletal troponin T gene. Alternatively spliced exons exhibit unusual interspecies divergence. J Mol Biol 188:313–324

Briggs MM, Lin JJ-C, Schachat FH 1987 The extent of amino-terminal heterogeneity in rabbit fast skeletal muscle troponin T. J Muscle Res Cell Motil 8:1–12

Bronson DD, Schachat FH 1982 Heterogeneity of contractile proteins. Differences in tropomyosin in fast, mixed, and slow skeletal muscles of the rabbit. J Biol Chem 257:3937–3944

Carraro U, Dalla Libera L, Catani C 1981 Myosin light chains of avian and mammalian slow muscles: evidence of intraspecific polymorphism. J Muscle Res Cell Motil 2:335–342

d'Albis A, Pantaloni C, Bechet J-J 1979 An electrophoretic study of native myosin isozymes and of their subunit content. Eur J Biochem 99:261–272

Dangain J, Pette D, Vrbová G 1987 Developmental changes in succinate dehydrogenase activity in muscle fibres from normal and dystrophic mice. Exp Neurol 95:224–234

Danieli Betto D, Zerbato E, Betto R 1986 Type 1, 2A, and 2B myosin heavy chain electrophoretic analysis of rat muscle fibers. Biochem Biophys Res Commun 138:981–987

Dhoot GK, Perry SV 1979 Distribution of polymorphic forms of troponin components and tropomyosin in skeletal muscle. Nature (Lond) 278:714–718

Hoh JFY, Yeoh GPS 1979 Rabbit skeletal myosin isoenzymes from fetal, fast-twitch and slow-twitch muscles. Nature (Lond) 280:321–322

Jandreski MA, Sole MJ, Liew C-C 1987 Two different forms of beta myosin heavy chain are expressed in human striated muscle. Hum Genet 77:127–131

Lowey S, Risby D 1971 Light chains from fast and slow muscle myosins. Nature (Lond) 278:81–85

Mabuchi K, Pinter K, Mabuchi MS, Sréter F, Gergely J 1984 Characterization of rabbit masseter muscle fibers. Muscle & Nerve 7:431–438

Mahdavi V, Strehler EE, Periasamy M, Wieczorek DF, Izumo S, Nadal-Ginard B 1986 Sarcomeric myosin heavy chain gene family: organization and pattern of expression. Med Sci Sports Exercise 18:299–308

Maier A, Gorza L, Schiaffino S, Pette D 1988 A combined histochemical and immunohistochemical study on the dynamics of fast-to-slow fiber transformation in chronically stimulated rabbit muscle. Cell Tissue Res, in press

Mizusawa H, Takagi A, Sugita H, Toyokura Y 1982 Coexistence of fast and slow types of myosin light chains in a single fiber of rat soleus muscle. J Biochem (Tokyo) 91:423–425

Moore GE, Briggs MM, Schachat FH 1987 Patterns of troponin T expression in mammalian fast, slow and promiscuous muscle fibres. J Muscle Res Cell Motil 8:13–22

Moore GE, Schachat FH 1985 Molecular heterogeneity of histochemical fibre types: a comparison of fast fibres. J Muscle Res Cell Motil 6:513–524

Pette D 1985 Metabolic heterogeneity of muscle fibres. J Exp Biol 115:179–189

Pette D, Spamer C 1986 Metabolic properties of muscle fibers. Fed Proc 45:2910–2914

Pette D, Vrbová G 1985 Neural control of phenotypic expression in mammalian muscle fibers. Muscle & Nerve 8:676–689

Reiser PJ, Moss RL, Giulian GG, Greaser ML 1985 Shortening velocity in single fibers from adult rabbit soleus muscles is correlated with myosin heavy chain composition. J Biol Chem 260:9077–9080

Rowlerson A, Pope B, Murray J, Whalen RB, Weed AG 1981 A novel myosin present in cat jaw-closing muscles. J Muscle Res Cell Motil 2:415–438

Sartore S, Mascarello F, Rowlerson A, Gorza L, Ausoni S, Vianello M, Schiaffino S 1987 Fibre types in extraocular muscles: a new myosin isoform in the fast fibres. J Muscle Res Cell Motil 8:161–172

Schachat FA, Bronson DD, McDonald OB 1980 Two kinds of slow skeletal muscle fibers which differ in their myosin light chain complements. FEBS (Fed Eur Biochem Soc) Lett 122:80–82

Schachat FH, Bronson DD, McDonald OB 1985 Heterogeneity of contractile proteins. A continuum of troponin–tropomyosin expression in mammalian skeletal muscle. J Biol Chem 260:1108–1113

Sréter FA, Sarkar S, Gergely J 1972 Myosin light chains of slow twitch (red) muscle. Nat New Biol (Lond) 239:124–125

Staron RS, Pette D 1986 Correlation between myofibrillar ATPase activity and myosin heavy chain composition in rabbit muscle fibers. Histochemistry 86:19–23

Staron RS, Pette D 1987a The multiplicity of myosin light and heavy chain combinations in histochemically typed single fibres of rabbit soleus muscle. Biochem J 243:687–693

Staron RS, Pette D 1987b The multiplicity of myosin light and heavy chain combinations in histochemically typed single fibres of rabbit tibialis anterior muscle. Biochem J 243:695–699

Staron RS, Gohlsch B, Pette D 1987 Myosin polymorphism in single fibers of chronically stimulated rabbit fast-twitch muscle. Pfluegers Arch Eur J Physiol 408:444–450

Steinbach JH, Schubert D, Eldridge L 1980 Changes in cat muscle contractile proteins after prolonged muscle inactivity. Exp Neurol 67:655–669

Whalen RG, Butler-Browne GS, Gros F 1978 Identification of a novel form of myosin light chain present in embryonic muscle tissue and cultured muscle cells. J Mol Biol 126:415–431

Whalen RG, Sell SM, Butler-Browne GS, Schwartz K, Bouveret P, Pinset-Härström I 1981 Three myosin heavy-chain isozymes appear sequentially in rat muscle development. Nature (Lond) 292:805–809

Whalen RG, Sell SM, Eriksson A, Thornell L-E 1982 Myosin subunit types in skeletal and cardiac tissues and their developmental distribution. Dev Biol 91:478–484

Wieczorek DF, Periasamy M, Butler-Browne GS, Whalen RG, Nadal-Ginard B 1985 Co-expression of multiple myosin heavy chain genes, in addition to a tissue-specific one, in extraocular musculature. J Cell Biol 101:618–629

DISCUSSION

Nadal-Ginard: I sympathize with your quandary over the multiplicity of isoforms. Have you considered calculating the number of different sarcomeres that could be formed, theoretically, in a vertebrate muscle, say, from what we now know? I have myself done this calculation.

Pette: We know too little about the expression of different and probably still unidentified myosin heavy chains in different muscles to make that type of calculation. Unfortunately, only a few different muscles are traditionally examined. Perhaps one would see many more myosin and other myofibrillar protein isoforms if one looked at a greater variety of skeletal muscles.

Nadal-Ginard: That seems probable. For example, we found the myosin heavy chain isoform expressed exclusively in the extraocular muscles purely by chance. This may not be the only isoform with a very restricted pattern of expression. In skeletal muscles of most adult mammals there is trace expression of embryonic and neonatal isoforms in the normal state, and it is greatly increased when muscle function is perturbed by overload or stimulation. But the thin filament is even more complex. There are several different isoforms of tropomyosin, of both the alpha and the beta subunits, as you mentioned. So, from the information that we already have, there could, in theory, be about 1500 million different sarcomeres. Perhaps not all the possible combinations are used, but even if only a small fraction exist, it is a very large number.

Eisenberg: Even that estimate must give a lower limit on sarcomere diversity. If you include sarcoplasmic reticulum and metabolic proteins it will approach infinity!

Nadal-Ginard: My estimate includes all the contractile proteins, and all the genes known to exist for them, plus the known alternative splicing patterns.

Vrbová: We know that the myosin heavy chains are controlled by separate genes. Are the troponins and the thin filament proteins also separately controlled, or are they modified after translation?

Nadal-Ginard: There are two known forms of actin in skeletal muscle, the 'cardiac' and 'skeletal' actins (both in fact occur in both types of muscle); they are encoded by separate genes. The fast and slow isoforms of the three troponin subunits are encoded by separate genes. For troponin T (TnT), the fast isoforms are controlled by a different gene from the slow isoforms. All the TnT genes that have been examined so far are alternatively spliced; the fast TnT gene produces more isoforms at the mRNA level than any other gene that we know about, 64 in total (Breitbart & Nadal-Ginard 1986), as Dr Pette said.

The α-tropomyosin gene codes for the α isoforms in all striated muscles, both cardiac and skeletal, and for the α isoform in smooth muscle, and also for at least three non-muscle isoforms. The β gene seems to make at least two isoforms in skeletal muscle. For both α and β tropomyosins there are (at least) two genes, including a so-called 'fast' α gene and another called 'slow' for want of a better name.

Zak: Dr Pette, you suggested initially that co-expression of two isoforms within one fibre means the random assembly or combination of myosin subunits, but later you spoke of preferential combination. I would also like to suggest that co-expression does not necessarily mean random association between, say, the different classes of myosin heavy chain. We know two situations where this doesn't happen. One is in heart muscle, which has the two classes of heavy chain that combine to form two homodimers and one heterodimer (Dechesne et al 1987). If there were random combination, the heterodimer would be most abundant, but this is never the case.

The second example is even more striking. This concerns SM1 and SM2, the two variants of myosin that exist in avian slow muscle. Each variant is composed of its specific heavy chains. They are co-expressed within the same fibre but they never form a heterodimer. So there is regulation at this point. Consequently, although multiple forms of heavy chain are co-expressed, there are not necessarily so many different isoforms of myosin.

Pette: I agree, and this is why I spoke of preferential affinities. Nevertheless, the number of combinations is large enough to give a continuum of fibre types. The question arises as to the location of different isoforms. Are they separately distributed in different sarcomeres or myofibrils, or are they randomly distributed? Perhaps Brenda Eisenberg could answer this, from her work with immunoelectron microscopy?

Eisenberg: Specific monoclonal antibodies have been used to distinguish the isoforms. When they are co-expressed, in all mammalian species they seem to be randomly mixed throughout the muscle fibre. In the thick filaments of nematode worms that is not the case. The central portion from the M band outwards is the A isoform and the ends of the thick filament are the B isoform (Epstein et al 1982). There is an assembly core down the centre of the thick filament in the nematode and insect which perhaps controls this regional distribution of isoforms.

We should not forget the associated proteins, such as C-, H-, X- and M-proteins, titin and nebulin, because they also may limit the assembly of myosin into a thick filament.

Gordon: Dr Pette implied that the number of fibre types which are recognized really depends on how many one is looking for. The dichotomy of 'fast' and 'slow' fibres arose because of the recognition, by Dr Buller and others, of the functional differences between fast and slow muscles. In trying to understand function, we have emphasized the differences between fast and slow muscle. Henneman and his colleagues in the 1960s (see Henneman & Mendell 1981) recognized what Dr Pette is saying, that functionally any skeletal muscle is heterogeneous, with a continuum of fibre types. The emphasis on fast and slow muscles however, led one to look for just three types histochemically. When one considers the function of a muscle and the way motor units are recruited in order of size and contractile speed, one is not surprised by the heterogeneity, or by the fact that different muscles have different speeds. Analysis of contractile speed of the motor units in most skeletal muscles varies over a two- to three-fold range. Unfortunately the dichotomous view has resulted in confusion in functional terms too. So, to what extent should we continue to use the words 'fast' and 'slow' in our classifications? In physiological experiments, the slow and fast motor units are classified on the basis of their different susceptibility to fatigue and ability to maintain force during unfused tetanic contractions. These classifications essentially subdivide a continuous distribution with respect to contractile speed.

Eisenberg: Dr Pette reminds us that we should not equate histochemical and physiological nomenclatures: for example, IIA and 'FOG' (fast oxidative-glycolytic) should not be equated. Nevertheless, both in textbooks and in our teaching we have to simplify the picture, and so we classify into groups what we know to be a continuous distribution. The only way to deal with this problem is always to state the variables we are using and how the determination of the subtyping, or the clustering, is made. In my quantitative assessment of the morphological subcompartments in mammalian muscle I found that although any one variable shows overlap between fibres from different muscles, when I started grouping them in paired variables to give a two-dimensional scattergram, clusters were formed that were slow or fast. One then finds that some clusters happen only during a transitional adaptive state. When we understand the mechanisms which coordinate gene expression, we may find why these preferred clusters exist.

Kernell: With respect to the discussion relating to contractile speed, it should be remembered that this is a complex concept. Shortening speed and isometric aspects of speed do, for instance, largely depend on different mechanisms.

With respect to the various myosin isoforms and subfragments, it should be emphasized that many different contractile properties co-vary between muscle fibres. Might not some of the myosin subfragments have a functional role primarily related to something else than shortening speed (such as force production or, perhaps, endurance)?

Pette: This is a difficult question to answer. It appears that myosin ATPase activity, as well as contractile speed, is determined by the heavy chain composition (Staron & Pette 1986, Reiser et al 1985). The myosin light chain composition does not seem to be related to those properties. The myosin ATPase-based fibre typing reflects, in adult skeletal muscle fibres, the complement of heavy chains; that is, IIB fibres contain myosin heavy chain HCIIb, IIA fibres HCIIa, IIC fibres heavy chains HCIIa plus HCI, and typeI fibres contain HCI (Staron & Pette 1986). The fibre typing nomenclature should be restricted to the method used. Therefore, the frequently interchanged nomenclatures of myosin-based and metabolic enzyme-based fibre types must be separated. Thus, the IIA fibre is not necessarily identical with the FOG (fast-twitch oxidative-glycolytic) fibre type. Similarly, the IIB fibre is not necessarily identical with the FG (fast-twitch glycolytic) fibre type (Reichmann & Pette 1982, 1984). The only exception seems to be the correspondence between type I and the SO (slow-twitch oxidative) fibre. However, it must be kept in mind that the underlying histochemical methods have a very low analytical resolution. Thus, the myosin ATPase-based fibre typing does not distinguish differences in light chain composition (Staron & Pette 1987 a,b). In addition, species-specific differences have been described (Reichmann & Pette 1982, 1984).

Eisenberg: In dog, all the fibres are high oxidative. At the beginning of a paper one must therefore state the basis for the classification scheme to be used.

Van Essen: Have you any evidence on the uniformity of composition of a given muscle fibre along its length, Dr Pette?

Pette: In normal adult muscle fibres, the composition appears to be uniform, but fibres undergoing transformation display non-uniformity along their length, and different heavy chain and light chain combinations appear in different parts of the fibre (Staron & Pette 1987c).

Vrbová: Let me defend the stability of myosin here! If myosin is involved in the same function, then over a whole range of species it is very similar. Thus Joe Hoh (personal communication) finds that the myosin of the jaw muscle of the shark, lion and cat (all predators that have to seize their prey quickly) is very similar, whereas the dog has a completely different type of myosin in the jaw muscle. So myosin is a highly conserved molecule where it has the same function; if you find myosin diversity, then that type of muscle fibre has a different function, in relation to load or to some other property.

References

Breitbart RE, Nadal-Ginard B 1986 Complete nucleotide sequence of the fast skeletal troponin T gene. Alternatively spliced axons exhibit unusual interspecies divergence. J Mol Biol 188:313–324

Dechesne CA, Bouvagnet P, Walzthony D, Leger JJ 1987 Visualization of cardiac ventricular myosin heavy chain homodimers and heterodimers by monoclonal antibody epitope mapping. J Cell Biol 105:3031–3037

Epstein HF, Berman SA, Miller DM III 1982 Myosin synthesis and assembly in nematode body-wall muscle. In: Pearson ML, Epstein HF (eds) Muscle development: molecular and cellular control. Cold Spring Harbor Laboratory, Cold Spring Harbor, NY, p 419–427

Henneman E, Mendell L 1981 Functional organization of motoneuron pool and its inputs. In: Brooks V (ed) Motor control. Williams & Wilkins, Baltimore. Handbook of Physiology, section 1, vol 11:423–507

Reichmann H, Pette D 1982 A comparative microphotometric study of succinate dehydrogenase activity levels in type I, IIA and IIB fibres of mammalian and human muscles. Histochemistry 74:27–41

Reichmann H, Pette D 1984 Glycerolphosphate oxidase and succinate dehydrogenase activities in IIA and IIB fibres of mouse and rabbit tibialis anterior muscles. Histochemistry 80:429–433

Reiser PJ, Moss RL, Giulian GG, Greaser ML 1985 Shortening velocity in single fibers from adult rabbit soleus muscles is correlated with myosin heavy chain composition. J Biol Chem 260:9077–9080

Staron RS, Pette D 1986 Correlation between myofibrillar ATPase activity and myosin heavy chain composition in rabbit muscle fibers. Histochemistry 86:19–23

Staron RS, Pette D 1987a The multiplicity of myosin light and heavy chain combinations in histochemically typed single fibres of rabbit soleus muscle. Biochem J 243:687-693

Staron RS, Pette D 1987b The multiplicity of myosin light and heavy chain combinations in histochemically typed single fibres of rabbit tibialis anterior muscle. Biochem J 243:695–699

Staron RS, Pette D 1987c Nonuniform myosin expression along single fibers of chronically stimulated and contralateral rabbit tibialis anterior muscles. Pfluegers Arch Eur J Physiol 409:67–73

Hormonal control of myosin heavy chain genes during development of skeletal muscles

Neal A. Rubinstein, Gary E. Lyons and Alan M. Kelly*

Department of Anatomy, School of Medicine, and **Department of Pathobiology, School of Veterinary Medicine, University of Pennsylvania, Philadelphia, Pennsylvania 19104, USA*

Abstract. A pattern of myosin heavy chain (MHC) switching is a hallmark of developing muscles. Factors responsible for these changes in gene expression include endogenous signals, motoneurons and hormones, especially thyroid hormones. After perturbing the innervation and/or thyroid hormone levels we have examined the neonatal–IIb MHC transition during rat development. First, denervation does not qualitatively affect the transition at either the transcriptional or translational level. Second, hypothyroidism prevents the appearance of IIb MHC and its mRNA in the innervated limb; in the denervated hypothyroid limb IIb MHC is synthesized at moderately high levels. Third, hyperthyroidism causes a precocious increase in IIb MHC in both innervated and denervated muscles. These results suggest that the transition from neonatal to adult IIb myosin synthesis is endogenously programmed during development, but is closely orchestrated by the changing neuronal and hormonal status of the animal. Thyroid hormone may exert its influence by effects both on the muscle fibre and on the developing motoneuron.

In the guinea-pig the temporalis muscle is sexually dimorphic: it contains a fast-red MHC in the female but a fast-white MHC in the male. This dimorphism has been shown to be mediated by testosterone, since the castrated male synthesizes the fast-red MHC while the testosterone-supplemented female contains the fast-white MHC. During development male and female muscles initially synthesize the fast-red isoform. The male switches to the fast-white form at puberty.

1988 Plasticity of the neuromuscular system. Wiley, Chichester (Ciba Foundation Symposium 138) p 35–51

The distinct properties within individual adult fast and slow muscle fibres are partly the result of innervation by different types of motoneurons. This was first suggested by Eccles et al (1957), who demonstrated that the motoneurons innervating fast and slow-twitch skeletal muscles have different frequencies of impulse activity. In fact, Buller et al (1960) later showed that cross-innervation of fast and slow muscles causes a reciprocal transformation of the muscles' properties. The effects of cross-innervation of a fast muscle with a slow motoneuron can be mimicked by chronic stimulation of the fast muscle's own intact motoneurons at the frequency of activity of a slow motoneuron (Salmons & Vrbová 1969, Pette et al 1976). Since these changes occur within individual, pre-existing muscle fibres, an adult fibre's genetic programme cannot be rigidly determined (Rubinstein et al 1978). This plasticity that occurs in chronically stimulated or cross-innervated muscles can be demonstrated by changes in contractile protein isoforms, alterations in the levels of glycolytic versus oxidative metabolic enzymes, switches in proteins of the sarcoplasmic reticulum, and modulation of a wide variety of additional properties.

Thyroid hormones are also important in maintaining adult fibre types and regulating contractile protein isozymes and metabolic enzymes. Treatment of experimental animals with triiodothyronine (T3) and thyroxine (T4) increases skeletal muscle mitochondria and mitochondrial enzymes; in the predominantly slow-twitch soleus muscle, this treatment increases the myosin ATPase activity, shifts the light chains from the slow to fast types (Ianuzzo et al 1980), and causes an increase in fast (IIb) myosin heavy chain gene transcription (Izumo et al 1986). Hypothyroidism has the opposite effect in the soleus: it causes a marked slowing of contraction and half-relaxation times, a conversion of fast twitch to slow twitch fibres, the disappearance of fast myosin light chains (Johnson et al 1980), and an increase in slow myosin heavy chain gene transcription (Izumo et al 1986).

The anabolic steroid testosterone also has an effect on skeletal muscle fibres. The action of steroids is generally believed to involve binding of the steroid to a cytoplasmic receptor and then movement of the steroid–receptor complex into the nucleus to bind to the DNA. High affinity, androgen-binding proteins exist in the cytosol of a variety of rat limb muscles. Gutmann et al (1970) demonstrated an effect of testosterone on the metabolism of the guinea-pig temporalis muscle. Testosterone caused a shift from an oxidative to a glycolytic metabolic pattern.

During myogenesis, muscle fibres also undergo changes in myosin isozymes. When the myosins from bulk muscles are analysed, there appears to be a sequential synthesis of an embryonic, then a neonatal, myosin heavy chain, prior to the synthesis of an adult fast or slow myosin heavy chain (Whalen et al 1981). Because of the role hormones and motoneurons play in the postnatal determination of fibre types and myosin isoforms, it has been natural to regard the initiation of diversity among muscle fibres as a result of these extrinsic factors acting upon a homogeneous population of embryonic fibres. Yet, evidence is mounting that this may not be entirely true. For example, using immunohistochemistry or histochemical ATPase reactions, several groups have demonstrated that diversity among fibres begins only one or two days after myotube formation in the embryo (Butler & Cosmos 1981,

Rubinstein & Kelly 1981), well before the appearance of thyroid hormones (Gambke et al 1983), and at a time when electromyographic (EMG) activity is similar among muscles of different destinies (Navarrete & Vrbová 1983). Recently, diversity among fibres has been demonstrated in chick embryos whose neural tube has been removed before muscle innervation (Philips & Bennett 1984). Finally, Stockdale & Miller (1987) have demonstrated differences among fibres grown in tissue culture, in the absence of innervation and in a homogeneous hormonal milieu.

Despite the absence of evidence that either innervation or hormonal concentration plays a role in the *initiation* of diversity, we have put forward ample evidence that these cues assume an important function in regulating contractile protein isoforms during development; and that, in their absence, the normal expression of diversity is perturbed.

Thyroid control of myosin heavy chain transitions

During the development of many rat fast muscles, an embryonic myosin heavy chain is the first heavy chain synthesized; and this remains the dominant heavy chain until just before birth, when the neonatal myosin heavy chain appears. By 5–8 days *post partum*, this neonatal heavy chain is replaced — in most fibres — by the adult fast myosin heavy chain, which becomes the dominant heavy chain by 15 days (Whalen et al 1981, Periasamy et al 1984). We have recently studied the effects of innervation and thyroid hormones on the transcription of neonatal and adult fast (IIb) myosin heavy chain during the period between birth and 15 days *post partum*. For this study we have used two synthetic oligonucleotide probes specific to the 3′ untranslated regions of the neonatal and IIb myosin heavy chain mRNAs. The sequences of these 20 nt long probes are identical to those published by Gustafson et al (1985) using sequences derived from the work of Nadal-Ginard and his colleagues. mRNAs were isolated from the fast gastrocnemius muscle at various stages after birth and the presence of neonatal and IIb myosin heavy chain mRNAs was established by Northern blots using these oligonucleotide probes. Quantitative values were derived from dot blots, followed by scintillation counting.

The sequential synthesis of several myosin heavy chain isoforms during development is the product of the sequential transcription of distinct myosin heavy chain mRNAs (Weydert et al 1983). Fig. 1A shows a series of Northern blots of total gastrocnemius RNA probed with the oligonucleotide specific to neonatal mRNA. This mRNA can be easily resolved at one and 10 days (lanes a and b), but is difficult to detect at 20 and 30 days (lanes c and d). This is identical to the results of Periasamy et al (1984) using S1 nuclease digestion. IIb mRNA shows a complementary pattern. Northern blots (Fig. 1B) show that the fast (IIb) myosin heavy chain mRNA is not detectable at one day

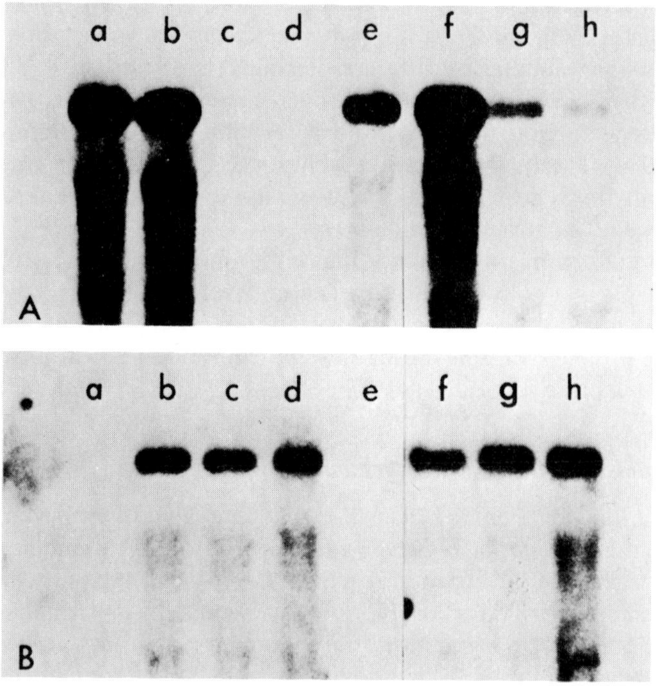

FIG. 1. Northern blots of neonatal and IIb myosin heavy chain mRNAs during development. Total RNAs were isolated from the rat gastrocnemius muscle at various stages of development. RNAs were separated on formaldehyde–1% agarose gels, transferred to Genescreen, and reacted with ^{32}P-labelled oligonucleotide probes specific to the 3′ untranslated region of the neonatal (A) or IIb (B) myosin heavy chain. RNAs were isolated from animals at (a) one day *post partum*, (b) 10 days *post partum*, (c) 20 days and (d) 30 days. Some animals were denervated at birth and their RNAs isolated at (e) one day *post partum*, (f) 10 days, (g) 20 days and (h) 30 days *post partum*.

post partum (lane a), but is present in large amounts at 10 days (lane b). Quantitative studies have shown no IIb myosin heavy chain mRNA at five days after birth. Hence, there is a large increase in mRNA between five and 10 days *post partum*, and these increased levels are stable at least through 30 days (lanes c and d).

The transition from neonatal to adult fast myosin heavy chains is not qualitatively disturbed by changes in innervation in the euthyroid animal. When the distal hindlimb of the rat is denervated at birth, the increase in IIb myosin heavy chain mRNA occurs normally, at least until 20 days *post partum* (Fig. 1B, lanes e, f, g). By 30 days (lane h), however, there is a large decrease in the level of this mRNA. Quantitative study has shown that IIb myosin heavy chain mRNA in the 30-day denervated muscle is less than 20% of the

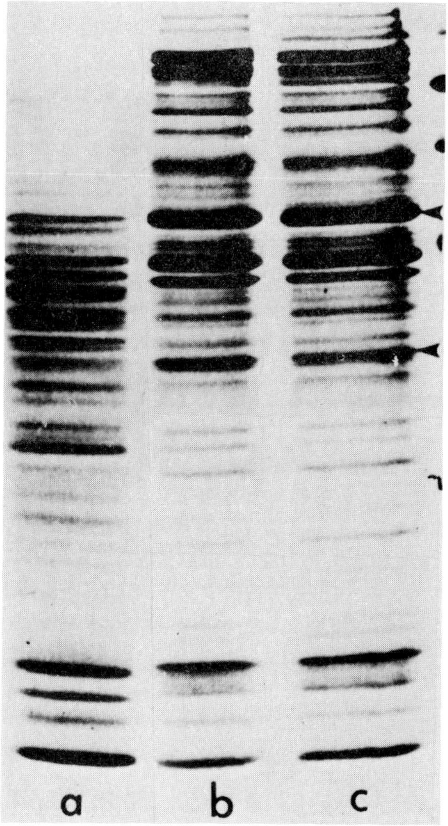

FIG. 2. Peptide maps of myosin heavy chains isolated from innervated and denervated gastrocnemius muscles. Myosins were isolated and subjected to partial proteolysis with chymotrypsin. Peptides were separated on 15% acrylamide–SDS gels. Myosins were isolated from muscles (a) five days *post partum*, (b) 25 days *post partum* and (c) 25 days *post partum*, denervated at birth.

level in innervated muscles of the same age. These lower levels persist at least until 60 days. Although lowered total levels of mRNA may reflect, in part, atrophy of the leg muscles, the quantitative studies measured IIb mRNA as a proportion of total RNA. Hence, innervation is at least partly responsible for maintaining increased levels of this mRNA during development, even though it is not likely to be responsible for initiating its synthesis. Denervation delays, but does not prevent, the disappearance of the neonatal myosin heavy chain (Fig. 1A, lanes e–h).

The neural independence of IIb mRNA synthesis is accompanied by the neural independence of translation of this protein. Fig. 2 shows a set of peptide maps of isolated myosin heavy chains from the gastrocnemius at

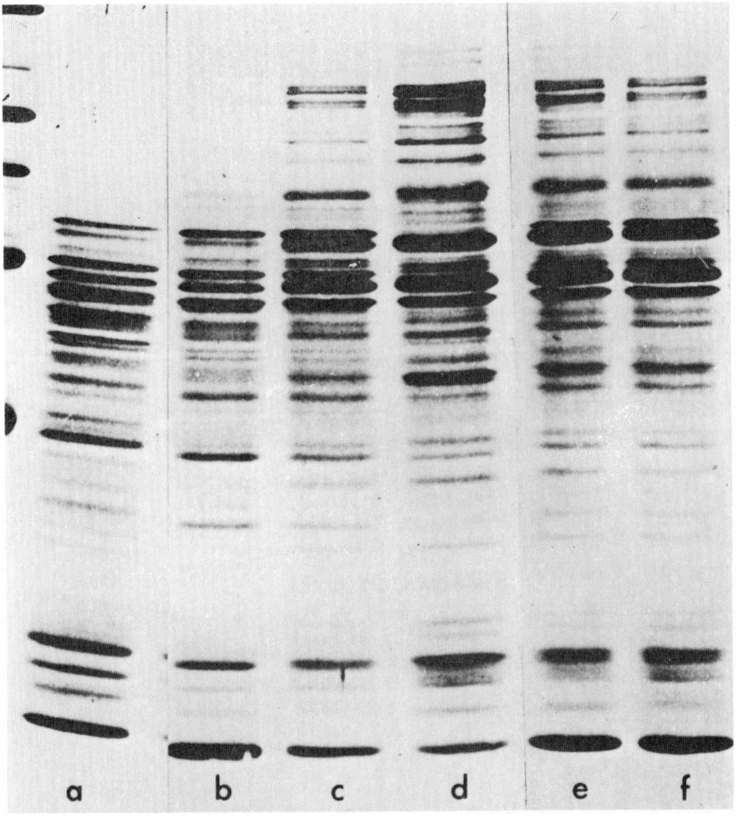

FIG. 3. Peptide maps of myosin heavy chains isolated from gastrocnemius muscles of thyroid hormone perturbed animals. Peptide maps were prepared as indicated in Fig. 2. Animals were made hypothyroid by the addition of propylthiouracil to the drinking water of their pregnant mothers. Propylthiouracil was continued in the drinking water of the pups after birth. Animals were made hyperthyroid by injections of T4 every other day, starting at one day *post partum*. Myosins were isolated from (a) euthyroid muscle, five days *post partum*; (b) hypothyroid muscle, 25 days *post partum*; (c) hypothyroid muscle, 25 days *post partum*, denervated at birth; (d) euthyroid muscle, 25 days *post partum*; (e) hyperthyroid muscle, 10 days *post partum* and (f) hyperthyroid muscle, 10 days *post partum*, denervated at birth.

several stages of development. At five days *post partum*, the muscle contains myosin heavy chains with a peptide pattern indicative of neonatal myosin heavy chains (lane a). By 25 days, however, there is a total switch to a pattern indicative of adult fast myosin heavy chains (lane b). Even when the animal is denervated at birth (lane c), the adult peptide map pattern appears.

The transition from neonatal to adult fast myosin heavy chains is however dependent on thyroid hormones. Fig. 3 shows an additional set of myosin peptide maps from gastrocnemius muscles of animals whose thyroid status

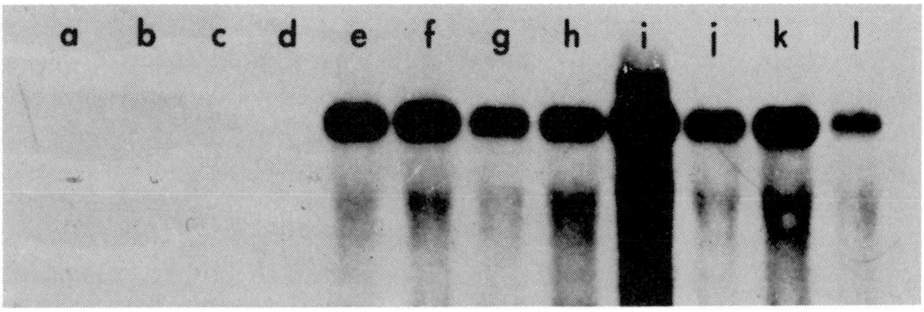

FIG. 4. Northern blots of IIb myosin heavy chain mRNA. RNAs were isolated, transferred, and analysed with an oligonucleotide specific to the IIb myosin heavy chain mRNA as described in Fig. 1. RNAs were isolated from innervated, hypothyroid gastrocnemius muscles at (a) 10, (b) 20 and (c) 30 days *post partum*; from hypothyroid muscles denervated at birth and analysed at (d) 10, (e) 20 and (f) 30 days *post partum*; from innervated, hyperthyroid muscles at (g) 10, (h) 20 and (i) 30 days *post partum*; and from hyperthyroid muscles denervated at birth and analysed at (j) 10, (k) 20 and (1) 30 days *post partum*.

had been compromised. Rat pups were made hypothyroid by adding propylthiouracil (PTU) to the drinking water of their pregnant mothers and to their own drinking water. Other pups were made hyperthyroid by intraperitoneal injections of T4 on alternate days, beginning on Day 1 *post partum*. Serum T4 levels were monitored by radioimmunoassay (Gambke et al 1983). Lane a (Fig. 3) shows the peptide map characteristic of neonatal myosin heavy chains. Although the 25-day euthyroid muscle showed a complete transition to adult fast myosin heavy chains (see Fig. 2), in the 25-day hypothyroid muscle the myosin peptide map is still identical to the neonatal pattern (Fig. 3, lane b), and not to the pattern indicative of adult fast myosin heavy chains (lane d). If, on the other hand, the animal is made hyperthyroid from birth by injections of T4, there is a precocious appearance of adult IIb myosin heavy chains (lane e). Curiously, while denervation does not affect the precocious appearance of fast myosin heavy chains in the hyperthyroid animal (lane f), it profoundly affects the response of the hypothyroid muscle. When an animal is made hypothyroid and the muscle is denervated at birth, the hypothyroid-induced inhibition of fast myosin heavy chains is released and peptides characteristic of adult fast myosin heavy chains can be seen (lane c).

This peculiar response of the hypothyroid muscle to denervation is corroborated by measurements of IIb myosin heavy chain mRNA. Fig. 4 demonstrates the Northern blots of mRNAs isolated from the gastrocnemius muscles of animals who had been made hypo- or hyperthyroid. Lanes a–c demonstrate that the hypothyroid innervated muscles do not produce adult fast IIb myosin heavy chain mRNA at any time between 10 and 30 days postpartum. If the animal is made hypothyroid and denervated at birth, fast IIb myosin heavy chain mRNA cannot be detected at 10 days (lane d), but it can be detected in large amounts at both 20 and 30 days *post partum* (lanes e, f). In the hyperthyroid animal, large amounts of IIb myosin heavy chain mRNA can be seen at 10, 20 and 30 days *post partum* (lanes g, h and i, respectively). In this case, however, denervation does not alter the precocious synthesis of the IIb mRNA (lanes j, k, l).

The period between birth and Day 15 of life is marked in the rat by a number of developmental milestones. First, there is a replacement of the neonatal myosin heavy chain with an adult fast or slow myosin heavy chain. Second, there is a change from a predominantly hypothyroid state to a physiologically hyperthyroid status. Third, there is maturation of motoneurons and the neuromuscular junction, including the disappearance of polyneuronal innervation, changes in extrajunctional sensitivity to acetylcholine, and changes in the frequency of miniature end-plate potentials and in the pattern of impulse activity, Kawa & Obata (1982) have demonstrated that the changing thyroid status is responsible for the changes in motoneuron maturation; and it is now clear that the changing thyroid status accounts for the switch in myosin heavy chain isoforms. In the hypothyroid animal the switch is blocked, while in the hyperthyroid animal it occurs precociously. The experiments in which thyroid status and innervation were altered simultaneously, however, suggest that the relationship between thyroid hormone levels and myosin isoforms is not simple. Thus, we might conclude that while thyroid hormones can alter the levels of IIb myosin heavy chains — and can even accelerate their appearance during development, as in the hyperthyroid animals — they are probably not the cue for the shift from neonatal to adult myosin heavy chains. This follows since the denervated rat can make the transition in the absence of thyroid hormones.

Our results suggest that the normal transitions from embryonic to neonatal to adult fast myosin heavy chains are endogenously programmed during development, but are closely orchestrated by the changing neuronal and hormonal status of the animal. Moreover, we suggest that the quantitative regulation of IIb myosin heavy chain and its mRNA in the normal animal is the result of an interaction among neural and hormonal factors affecting all elements of the motor unit. One possible explanation for our results is that the immature motoneuron actually inhibits the synthesis of IIb myosin heavy chain mRNA. This inhibition can be removed in one of two ways. Either the motoneuron can be forced to mature by the normal rise in T4 levels (this can occur precociously in hyperthyroid animals) or the motoneuron can itself be removed by denervation. These complex interactions may ensure that the properties of all elements of the developing motor unit can be closely coordinated.

Testosterone control of myosin heavy chain synthesis in guinea-pig temporalis muscles

As mentioned earlier, Gutmann et al (1970) demonstrated that testosterone caused a shift in metabolism of the guinea-pig temporalis muscle. It was shown that female guinea-pig temporalis muscles could be classified as fast twitch/oxidative (or fast-red), while the male muscle was fast twitch/glycolytic (or fast-white). Since it has been reported that fast-white and fast-red muscles contain distinct myosin heavy chain isoforms, we examined the isoforms in adult and developing temporalis muscles to determine the factor(s) responsible for the sexual dimorphism in the temporalis muscle.

Myosins were isolated and peptide maps prepared by limited proteolytic digestion (Fig. 5). These peptide maps demonstrate that the adult female and male temporalis muscles contain distinct myosin heavy chains which we call the fast-red (lane 6) and fast-white (lanes 1 and 8) heavy chains, respectively. Moreover, in the castrated male guinea-pig, peptides indicative of fast-red myosin heavy chains are evident (lane 7), while in the testosterone-supplemented female peptides characteristic of fast-white myosin heavy chains can be seen (lane 9). Testosterone, then, appears to modulate a

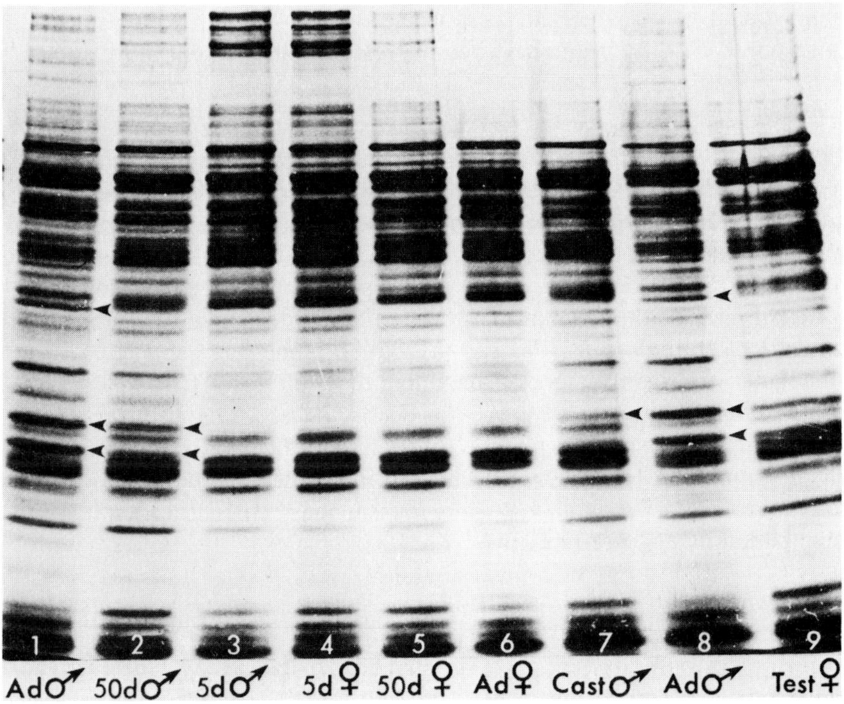

FIG. 5. Peptide maps of myosin heavy chains of male and female guinea-pig temporalis muscles. Myosins were isolated from temporalis muscles and analysed by partial proteolysis as described in Fig. 2. Myosins were isolated from (1) adult male temporalis, (2) 50-day male, (3) five-day male, (4) five-day female, (5) 50-day female, (6) adult female, (7) adult male castrated at 120 days, examined at 230 days *post partum*; (8) adult male and (9) adult female maintained on testosterone from 120 to 230 days *post partum*. Arrowheads indicate peptides characteristic of the myosin found in adult male temporalis muscle. (Reproduced by permission of the *Journal of Biological Chemistry*.)

fast-white/fast-red myosin transition in the temporalis muscle, possibly at the level of transcription.

During development, both male and female temporalis muscles initially contain the fast-red myosin heavy chains (Fig. 5, lanes 3 and 4). In the male, testosterone levels increase rapidly between 30 and 60 days *post partum*. At 50 days, while the female temporalis contains only the fast-red isoform (lane 5), the male already contains peptides indicative of the fast-white isoform (lane 2).

In the guinea-pig temporalis muscle, then, testosterone is responsible for both the initiation and maintenance of the fast-white myosin heavy chain isoform. In the absence of testosterone, only the fast-red isoform is synthesized in this muscle. Surprisingly, other fast-white and fast-red muscles — containing myosins identical to those found in the temporalis muscles — did not show sexual dimorphism in terms of myosin isoforms, and their myosins were unaffected by castration or testosterone supplementation. Thus, myosin heavy chain genes which are responsive to testosterone in one muscle are unresponsive to the hormone when expressed in a different muscle.

Summary

In developing fast muscles, changes in myosin heavy chains can be affected by a changing hormonal milieu. In the rat, increasing serum thyroid hormone concentrations are correlated with maturation of the motoneurons, neuromuscular junctions and muscle fibres. This coordinated development of all aspects of the motor unit probably ensures coordinated changes in the properties of each of the elements of the unit. In any case, muscle fibres in the absence of both motoneurons and thyroid hormones are capable of progressing from embryonic through neonatal to adult fast (IIb) myosin heavy chains. Hence, this progression is probably an endogenously programmed feature of fast muscle fibres and is independent of thyroidal and neural cues. In the guinea-pig, however, a transition in the developing male temporalis muscle from a fast-red to a fast-white myosin heavy chain is both initiated and maintained by testosterone.

Acknowlegements

The authors would like to thank Mrs Zs. Paltzmann for her excellent technical assistance. This work was supported by NIH grants NS14332 and HL 15835, and by grants from the Muscular Dystrophy Association of America. Part of this work was done while N.A.R. was an Established Investigator of the American Heart Association.

References

Buller AJ, Eccles JC, Eccles RM 1960 Interactions between motoneurones and muscle in respect of the characteristic speeds of their responses. J Physiol (Lond) 150:417–439

Butler J, Cosmos E 1981 Differentiation of the avian latissimus dorsi primordium: analysis of fiber type expression using the myosin ATPase histochemical reaction. J Exp Zool 218:219–232

Eccles JC, Eccles RM, Lundberg A 1957 The convergence of monosynaptic excitatory afferents on to many different species of alpha motoneurones. J Physiol (Lond) 137:22–50

Gambke B, Lyons G, Haselgrove J, Kelly AM, Rubinstein NA 1983 Thyroidal and neural control of myosin specialization in developing rat fast and slow muscles. FEBS (Fed Eur Biochem Soc) Lett 156:335–340

Gustafson TA, Markham EE, Morkin E 1985 Analysis of thyroid hormone effects on myosin heavy chain gene expression in cardiac and soleus muscles using a novel dot-blot mRNA assay. Biochem Biophys Res Commun 130:1161–1167

Gutmann E, Hanzlíková V, Lojda Z 1970 Effects of androgens on histochemical fibre type. Histochemie 24:287–291

Ianuzzo C, Patel P, Chen V, O'Brien P 1980 A possible thyroidal trophic influence on fast and slow skeletal muscle myosin. In: Pette D (ed) Plasticity of muscle. Walter de Gruyter, Berlin, p 593–605

Izumo S, Nadal-Ginard B, Mahdavi V 1986 All members of the MHC multigene family respond to thyroid hormone in a highly tissue-specific manner. Science (Wash DC) 231:597–600

Johnson M, Mastaglia F, Montgomery A, Pope B, Weeds A 1980 Changes in myosin light chains in the rat soleus after thyroidectomy. FEBS (Fed Eur Biochem Soc) Lett 110:230–235

Kawa K, Obata K 1982 Altered developmental changes of neuromuscular junction in hypo- and hyperthyroid rats. J Physiol (Lond) 329:143–161

Navarrete R, Vrbová G 1983 Changes of activity patterns in slow and fast muscles during postnatal development. Dev Brain Res 8:11–19

Periasamy M, Wieczorek D, Nadal-Ginard B 1984 Characterization of a developmentally regulated perinatal myosin heavy chain gene expressed in skeletal muscle. J Biol Chem 259:13573–13578

Pette D, Muller W, Leisner E, Vrbová G 1976 Time dependent effects on contractile properties, fibre population, myosin light chains and enzymes of energy metabolism in intermittently and continuously stimulated fast twitch muscles of the rabbit. Pfluegers Arch Eur J Physiol 364:103–112

Philips W, Bennett M 1984 Differentiation of fiber types in wing muscles during embryonic development: effect of neural tube removal. Dev Biol 106:457–468

Rubinstein NA, Kelly AM 1981 Development of muscle fiber specialization in the rat hindlimb. J Cell Biol 90:128–144

Rubinstein NA, Mabuchi K, Pepe FA, Salmons S, Gergely J, Streter F 1978 Use of type specific antimyosins to demonstrate the transformation of individual fibers in chronically stimulated rabbit fast muscles. J Cell Biol 79:252–261

Salmons S, Vrbová G 1969 The influence of activity on some contractile characteristics of mammalian fast and slow muscles. J Physiol (Lond) 201:535–549

Stockdale FE, Miller JB 1987 The cellular basis of myosin heavy chain isoform expression during development of avian skeletal muscles. Dev Biol 123:1–9

Weydert A, Daubas P, Caravatti M et al 1983 Sequential accumulation of mRNAs encoding different myosin heavy chain isoforms during skeletal muscle development *in vivo* detected with a recombinant plasmid identified as coding for an adult fast myosin heavy chain from mouse skeletal muscle. J Biol Chem 258:13867–13874

Whalen R, Sell S, Butler-Browne G, Schwartz K, Bouveret P, Pinset-Harstrom I 1981 Three myosin heavy chain isozymes appear sequentially in rat muscle development. Nature (Lond) 292:805–809

DISCUSSION

Vrbová: It is very pleasing to see that you are following up the work that Ernest Gutmann started in the 1960s on the guinea-pig temporalis muscle (Gutmann & Hanzlíková 1970). He observed guinea-pigs mating and saw that the male holds onto the female with his jaws. This is why he began to look at the temporalis muscle, because it seemed to be a 'sex-linked' muscle, and used by male guinea-pigs in mating. So again we see a combination of functional differentiation and myogenic specialization. Perhaps Bruce Carlson could remind us of his experiments with Dr Gutmann, where this myogenic specialization of the receptors for a particular hormone is even more obvious?

Carlson: We found, first of all, that the rat levator ani muscle, regenerating *in situ*, is sensitive to castration or added testosterone (Gutmann & Carlson 1978). Subsequent transplantation experiments (Carlson et al 1979) showed that the levator ani muscle, transplanted into the bed of the extensor digitorum longus muscle, retained its hormonal sensitivity in the presence or absence of innervation. Hanzlíková & Gutmann (1974) grafted the soleus muscle into the bed of the levator ani and found that the transplanted soleus muscle did not become hormonally sensitive. The sum of these experiments showed a strong myogenic bias in the hormonal sensitivity of the levator ani muscle.

The thumb muscle of most species of frog is another good example of a testosterone-sensitive muscle used in mating.

Guinea-pigs are said to be basically hypothyroid; is that true in your experience, Dr Rubinstein? If so, a reduced amount of thyroid hormone might relate to the effects that you see in the temporalis muscle?

Rubinstein: I am not sure; we have not examined thyroid hormone levels in the developing guinea-pig. In the male frog, the sternoradialis, a clasp muscle, has an increased number of slow tonic fibres, compared to the female muscle or to the same muscle in a castrated male (Rubinstein et al 1983). This is thought to be used for clasping the female for long periods during mating.

Thesleff: You implied that your experiments show that hormones influence the muscle directly. Is it not possible that hormones simply release something else that will produce the change—that is, they have an indirect effect? Experiments in tissue culture would clarify this.

Rubinstein: Yes; my graduate student, Becky Hoffman, is examining the effect of thyroid hormone on the expression of IIb myosin in tissue-cultured rat muscle. The preliminary results suggest that IIb myosin is increased in culture after the addition of T4.

Pette: Thyroid hormone has been reported to induce, in myotube cultures, an increase in transmembrane resting potential and also an increase in the frequency of spontaneously occurring action potentials generated by the myotubes (Bannet et al 1984).

Rubinstein: We can do that experiment, by paralysing the muscle and then assaying the effects of thyroid hormone on IIb myosin heavy chain synthesis.

Gordon: In relation to the functions of hormones and growth factors, one of the great mysteries is why there is so much nerve growth factor in the submandibular glands of male mice. Some recent experiments indicate that aggravated male mice during episodes of intraspecific fighting show elevated plasma levels of nerve growth factor, the only situation in which these levels become detectable (Levi-Montalcini & Calissano 1986). Perhaps, as Dr Thesleff suggests, there may be similarities between growth factors and hormones in the sense that some effects may be attributed to a direct action of the hormone or factor, whereas others may be secondary to more general developmental effects of the hormone in promoting maturation.

Rubinstein: There certainly may also be effects of hormones on some of the cellular 'oncogenes', such as c-*myc* and c-*fos*. It would be interesting to look at those in these cases of precocious changes in myosin isoforms.

Nadal-Ginard: Why is the myosin heavy chain gene induced by testosterone in one muscle, such as the temporalis, but not in another muscle? How is this thought to work, and can you reproduce it *in vitro*?

Rubinstein: We would like to reproduce the effects of testosterone on muscles *in vitro*. The testosterone/temporalis muscle system in the guinea-pig is so clear-cut (Lyons et al 1986). If the temporalis fast-red and fast-white MHCs are products of the same genes as in other muscles, and we know that all limb muscles will respond to testosterone by hypertrophying, all these muscles must take up the testosterone. Yet the myosins of the limb muscles don't change in response to the hormone. How are their myosin genes protected, when the same genes in the temporalis are not protected from the effects of testosterone? We have not been very successful in growing the temporalis muscle in tissue culture yet, but we hope to continue with that.

Nadal-Ginard: But you have no evidence that the testosterone receptor is acting directly on the heavy chain genes? It could be an indirect effect, therefore.

Vrbová: The levator ani muscle of the rat is dependent on testosterone in a much more dramatic way than is the temporalis. Without testosterone the levator ani muscle degenerates in the male, as it does in the female; but it can be maintained by injecting the female or castrated male rat with testosterone (Čihák 1970). There must be powerful receptors, or mechanisms of responding to this hormone, in that muscle. This doesn't occur in any other rat muscle.

Nadal-Ginard: What myosin does the rat elevator ani muscle make?

Vrbová: It is a fast muscle; it probably makes the IIa myosin.

Rubinstein: We have looked at this muscle and have not seen differences in myosins between male rats on the one hand and castrated male and testosterone-treated female rats on the other.

Buller: Have you done contractile studies alongside your studies of the effects of thyroid hormone on the transition from embryonic to neonatal to adult rat MHC?

Rubinstein: We haven't looked at the physiological properties of developing rat muscles.

Hoffman: The time at which you find the transition between rat neonatal and fast myosins is very similar to the time when we find a transition in cytoskeletal gene expression in neurons (N.A. Muma, E.H. Koo & P.N. Hoffman, unpublished work). If you make a rat hypothyroid and delay the appearance of the fast myosin, does that correlate with the prolongation of the time at which polyinnervation is found in these muscles?

Rubinstein: Polyinnervation persists for longer than usual in hypothyroid rats, but it still disappears, maybe a week later than in normal rats. You still do not see the transition from neonatal to adult IIb myosin heavy chains when polyinnervation disappears. I don't think other properties of the immature motoneuron have been examined, to see which might be playing a role in this transition.

Ribchester: Denervation of at least some muscles (for example, the rat lumbricals; Betz et al 1980) can arrest the emergence of secondary myotube populations. Is there any correlation between the biochemical changes that you see and the ratio of primary to secondary myotubes, or the emergence of the latter? Or are you looking at this at a time when the full adult complement of fibres has been produced, in the muscles that you have been studying?

Rubinstein: Pretty much the full complement of fibres is present in the fast muscles that we examine, the gastrocnemius and the EDL of the rat. The soleus muscle—a slow muscle in the rat—forms fibres for several weeks after birth, however, so it is more difficult to exclude the contributions of newly formed fibres to this process in the soleus. In the fast muscles, though, there is clearly no correlation between the biochemical changes and the fibre type ratios.

Vrbová: Marilyn Duxson, Jenny Ross and John Harris (1986) have shown that in the early stages of mammalian development, when all the axons of motoneurons have reached the muscle, the secondary and primary myotubes share axons. This makes it difficult to believe in a process whereby motoneurons seek out the primary myotubes of future slow muscle fibres. The important difference between primary and secondary myotubes is that the primary myotube arises by the fusion of many mononucleated cells and almost immediately is long enough to extend the whole length of the muscle. The secondary myotubes arise at the endplate region and remain very short until soon after birth in the rat lumbrical muscle, and they do not contribute to the tension developed by that muscle (A.J. Harris, personal communication). Moreover, these myotubes are not being stretched by the growth of the bone. So the major difference is not the innervation or the pattern of activity, because at this stage the myotubes are electrically coupled, but the length of the fibre and the load under which it operates. Have you considered the effect of this different mechanical situation on the myosin expression of these two very different types of muscle fibre?

Rubinstein: We haven't considered that. However, Dr Gordon and Dr Van Essen (1985) found that there is still specific innervation of fibre types at a time when the muscle is polyinnervated. They showed that individual motor units in the neonatal rabbit soleus were largely homogeneous, despite the presence of extensive polyinnervation. This suggests, first, that there is a very specific sorting out of motoneurons to match predetermined fibre types and, second, that the nerve endings on the primary and secondary generation cells that do come from the same motor neurons may not be equivalent. This would be an interesting area to investigate.

Van Essen: We are discussing two different time stages here; the specificity of innervation that has been demonstrated by Thompson et al (1984) and by Gordon & Van Essen (1985) is postnatal, whereas Dr Vrbová is describing a situation which is prenatal. Is it clear what the incidence is of overlap of innervation between primary and secondary myotubes?

Vrbová: According to John Harris, almost every cluster has shared axons between primary and secondary myotubes. Then, as the secondary myotube separates itself out, there is transfer of some of the axon terminals to the secondary myotube.

Rubinstein: The motoneurons have to sort themselves out rather rigorously, especially if we consider that muscle fibres are intrinsically determined. We think that motoneurons might be innately independent, so there must still be the correct sorting out, since the primary and secondary generation fibres do share motoneurons initially.

Vrbová: It depends how rigid you think your wiring diagram is.

Nadal-Ginard: How is this postulated predetermination of slow and fast muscle fibres compatible with the fact that, at least *in vitro*, myoblasts seem to form myotubes more or less at random, in that they fuse with their neighbours? This 'predetermination' implies a sorting-out of the predetermined slow and fast fibres. This model is very complicated developmentally, although it is possible that it is true. How do you distinguish it from a recruiting mechanism, where there may be a founder cell that determines what is going to be fast or slow, and everything that fuses to it will be slow or fast? From the work of Frank Stockdale and others, I don't think that that choice of mechanism has been resolved.

Rubinstein: Those two options are still open. But if there are 'founder' cells, then there still are cells which are already predetermined to be fast or slow. It doesn't seem any less complicated to generate specific founder cells than to generate specific (whole) populations.

Nadal-Ginard: This model could work with a very small population of determined cells and a vast majority of cells that can be indoctrinated.

Van Essen: In muscles that end up with very different ratios of fast and slow fibres, is anything known about the rate of formation of primary and secondary myotubes that would correlate with that eventual arrangement?

Rubinstein: We know that early on, at 14–15 days gestation, in the future fast EDL of the rat, a couple of hundred fibres are formed; those are the primary generation fibres. That is roughly the number of slow fibres in that muscle in the adult animal. Whereas when the rat soleus, a future slow muscle, is first distinguishable, almost all the fibres are primary generation and there are many more of them. The number of secondary generation fibres is much smaller than in the EDL.

Gordon: There is an anatomical compartmentalization of most skeletal muscles in the limbs, with more slow muscle fibres than fast fibres around the bone (Armstrong & Laughlin 1985). If primary and secondary generations of muscle fibres account for the emergence of the adult distribution of fibres, is it possible that there is an inductive effect of bone?

Rubinstein: This can't be excluded. We see, in the external parts of the tibialis anterior muscle and the EDL, that there are also primary generation fibres, and they stain lightly with our antibody to slow myosin. That staining disappears around the time of birth. Dr Dhoot (1985) has shown a similar result. There is almost a 'gradient', in terms of the amount of slow myosin, with higher amounts deep in the muscle and lower amounts in the external part; whether there is a correlation with the innervation and blood supply coming from the deeper parts of the muscle, or with a higher oxygen tension, is not known.

References

Armstrong RB, Laughlin MH 1985 Metabolic indicators of fibre recruitment in mammalian muscles during locomotion. J Exp Biol 115:201–213

Bannet RR, Sampson SR, Shainberg A 1984 Influence of thyroid hormone on some electrophysiological properties of developing rat skeletal muscle cells in culture. Brain Res 294:75–82

Betz WJ, Caldwell JH, Ribchester RR 1980 The effects of partial denervation at birth on the development of muscle fibres and motor units in rat lumbrical muscle. J Physiol (Lond) 303:265–279

Carlson BM, Herbrychova A, Gutmann E 1979 Retention of hormonal sensitivity in free grafts of the levator ani muscle. Exp Neurol 63:94–107

Čihák R, Gutmann E, Hanzlíková V 1970 Involution and hormone-induced persistence of the M. sphincter (levator)ani in female rats. J Anat 106:93–110

Dhoot GK 1985 Initiation of differentiation into skeletal muscle fiber types. Muscle & Nerve 8:307–316

Duxson MJ, Ross JJ, Harris AJ 1986 Transfer of differentiated synaptic terminals from primary myotubes to newly formed cells during embryonic development in the rat. Neurosci Lett 71:147–152

Gordon H, Van Essen DC 1985 Specific innervation of muscle fiber types in a developmentally polyinnervated muscle. Dev Biol 111:42–50

Gutmann E, Carlson BM 1978 The regeneration of hormone-sensitive muscle (levator ani) in the rat. Exp Neurol 58:535–548

Gutmann E, Hanzlíková V 1970 Involution and hormone induced persistence of the m. sphincter (levator) ani in female rats. J Anat 106:93–110

Hanzlíková V, Gutmann E 1974 The absence of androgen sensitivity in the grafted soleus muscle innervated by the pudendal nerve. Cell Tissue Res 154:121–129

Levi-Montalcini R, Calissano P 1986 Nerve growth factor as a paradigm for other polypeptide growth factors. Trends Neurosci 9:473–477

Lyons GE, Kelly AM, Rubinstein NA 1986 Testosterone-induced changes in contractile protein isoforms in the sexually dimorphic temporalis muscle of the guinea pig. J Biol Chem 261:13278–13284

Rubinstein NA, Erulkar S, Schneider G 1983 Sexual dimorphism in the fibers of a clasp muscle of *Xenopus laevis*. Exp Neurol 82:424–431

Thompson WJ, Sutton LA, Riley DA 1984 Fiber type composition of single motor units during synapse elimination in neonatal rat soleus muscle. Nature (Lond) 309:709–711

Studies of acetylcholine receptor subunit gene expression: chromatin structural changes during myogenesis

C. Michael Crowder and John P. Merlie

Department of Pharmacology, Washington University School of Medicine, 660 South Euclid, St. Louis, Missouri 63110, USA

Abstract. Myogenesis proceeds stepwise from pluripotential stem cell to differentiated myotube. The precise number of transitions that occur along the developmental pathway remains to be determined. We examined the myogenic pathway as modelled by mouse mesodermal stem cell and muscle cell lines for stage-specific alterations in the chromatin structure of the acetylcholine receptor δ and γ subunit genes. We reasoned that such an analysis would allow us to observe either the primary events in the activation of these muscle-specific genes or processes secondary to the binding of muscle-specific regulatory proteins. We probed chromatin structure with DNase I (deoxyribonuclease I) and precisely mapped to the 5′ ends of the δ and γ genes DNase I hypersensitive (DH) sites whose induction is unique to each myogenic stage. Putative mesodermal stem cells have the simplest pattern of DH sites with no sites near the 5′ ends of the δ and γ genes, whereas differentiated myotubes express the most complex pattern; the myoblast pattern is intermediate and of two types. In muscle cell lines where differentiation must be induced the myoblasts have a simple pattern (one more site than stem cells); in muscle lines where differentiation is spontaneous the myoblasts express a complex pattern of DH sites (one less site than myotubes). Inducible myoblasts seem to be arrested in an earlier step in the myogenic pathway than spontaneously differentiating myoblasts. Thus, myogenic activation of acetylcholine receptor subunit genes appears to be a stepwise process that can be detected by chromatin structural changes specific to four distinct stages of muscle development: stem cell, early myoblast, late myoblast, and differentiated myotube.

1988 Plasticity of the neuromuscular system. Wiley, Chichester (Ciba Foundation Symposium 138) p 52–70

Muscle differentiation is biochemically characterized by the expression of a set of muscle-specific proteins and isoforms. The similar kinetics of accumulation of those proteins suggest coordinate regulatory mechanisms (Devlin & Emerson 1978). Studies with the genes that code for several of these muscle-specific proteins demonstrate that the induction of these proteins during differentiation is under transcriptional control (Schwartz & Rothblum 1981,

Hastings & Emerson 1982, Medford et al 1983, Buonanno & Merlie 1986). Further, activation of muscle-specific genes can be achieved by diffusible *trans*-acting factors (Blau et al 1983) that are likely to be coded for by a small number of regulatory genes (Konieczny & Emerson 1984, Lassar et al 1986). Thus, myogenesis can be viewed as a process of selective and coordinate activation by a small number of diffusible regulatory factors of a group of genes whose products create the muscle phenotype.

In order to observe regulatory events for acetylcholine receptor subunit genes, we examined the myogenic pathway for changes in the chromatin structure of the δ and γ subunit genes. Chromatin structural changes such as cytosine demethylation and deoxyribonuclease I (DNase I) hypersensitivity have been correlated with genetic regulatory regions and are thought either to represent primary regulatory events or to result from the action or binding of regulatory molecules (Yisraeli & Szyf 1984, Thomas et al 1985). We have searched for DNase I hypersensitive (DH) sites around receptor subunit genes, since DH sites correlate highly with enhancer and promoter sequences and can be created by the binding of regulatory molecules (Thomas et al 1985). Further, hypersensitivity studies allow the simultaneous examination *in situ* of large stretches of DNA for chromatin perturbations far removed from the structural gene.

Using this assay we have previously reported that muscle-specific DH sites surround the δ subunit gene in mouse (Crowder & Merlie 1986). We also showed that in the BC3H-1 muscle cell line a subset of the sites are present before the onset of δ subunit gene transcription; other sites appear concomitantly with transcriptional activation. On the basis of δ/γ linkage in other species (Shibahara et al 1985, Nef et al 1984) and on δ/γ recombinational frequencies in mouse (Heidmann et al 1986), we predicted at that time that tissue-specific DH sites 3′ to the δ gene would lie near the 5′ end of the γ subunit gene. This work builds on our initial findings by showing definitively the δ–γ linkage in mouse, by more precisely mapping δ/γ muscle-specific DH sites relative to the 5′ ends of the δ and γ genes, and by examining the differentiation specificity of these sites in two muscle cell lines — one in which differentiation must be induced and the other where differentiation is spontaneous. Analysis of DH sites in the two cell lines allowed for dissection of three different steps in the activation of acetylcholine receptor subunit genes during myogenesis.

Results and discussion

The methods used to observe DH sites in the various cell lines have been described in detail previously (Crowder & Merlie 1986); but briefly, the cells are gently disrupted with a Dounce homogenizer and the free nuclei are purified by differential centrifugation, then treated with varying concentrations of DNase I. DNA is subsequently purified from these nuclei, digested with restriction endonucleases, size-fractionated by agarose gel electrophoresis, and transferred to a DNA-binding membrane (Southern blot). The blot is then hybridized with δ or γ subunit cDNA probes.

Fig. 1 shows the results of such an experiment with differentiated BC3H-1 cells; the blot was probed with a full-length δ subunit cDNA. In the zero DNase I lane, only the δ gene restriction fragments are seen except for a low intensity *Hin*dIII band migrating at 4.3 kilobases (kb); this faint band melts at higher stringency (0.1× SSC, 70 °C) and co-migrates with a γ subunit gene restriction fragment. With low concentrations of DNase I, subfragments are produced that represent cutting at one boundary by the restriction enzyme and by digestion within a 100–300 bp region at the other end by DNase I. This region of preferential cutting by DNase I is a hypersensitive site. With higher concentrations of DNase I, cutting throughout the δ gene becomes significant and restriction fragments and DH subfragments are randomly degraded. The DH sites in Fig. 1 are labelled according to their location and muscle specificity with numbers increasing from 5′ to 3′. The X sites are not restricted to muscle; the D site is muscle specific and located near the δ gene (Crowder & Merlie 1986); the G sites are also muscle specific and are located near the γ gene (see below). DH site X_3 was previously thought to be restricted to muscle since we did not detect it in C3H/10T$_{1/2}$ fibroblasts (Crowder & Merlie 1986); however, we have subsequently seen this subfragment as well as DH site X_2 in mouse liver nuclei. DH sites X_1, D_1, G_1 and G_2 are not detectable in liver.

Characterization of DH site D_1 at the 5′ end of the δ gene and definitive demonstration of δ–γ linkage in mouse required cloning the mouse δ gene and as much downstream sequence as possible. Fig. 2 shows the results of screening λ L47.1 (kindly provided by the laboratory of Stephen Weaver) and λ Charon 28 (kindly provided by the laboratory of Phil Leder) mouse genomic libraries. Charon 28 clones $\delta\gamma$ 2, 4 and 6 prove that the mouse δ and γ genes are tightly linked; approximately 6 kb separates the two structural genes. Southern blots and restriction mapping of the clones along with whole genomic blots with δ (Crowder & Merlie 1986) and γ (data not shown) subunit gene end-specific probes place G_1 and G_2 5′ to the γ structural gene.

Precise localization of D_1, G_1 and G_2 necessitated subcloning and sequencing the requisite restriction fragments. For the D_1 site, we subcloned a 975 bp *Bam*HI-*Acc*I fragment from the 5′ end of the λL47.1 δA-1 clone. The *Bam*HI site was formed during construction of the library; we calculate that the recreated *Bam*HI site lies approximately 100 bp from the upstream *Acc*I site shown in Fig. 2. We showed previously that this *Acc*I fragment contains the D_1 DH site (Crowder & Merlie 1986). Accurate restriction mapping of the A-1 clone and reproducible sizing of the D_1 fragment allows us with a precision of ± 100 bp to place D_1 on the sequence; the tight D_1 subband

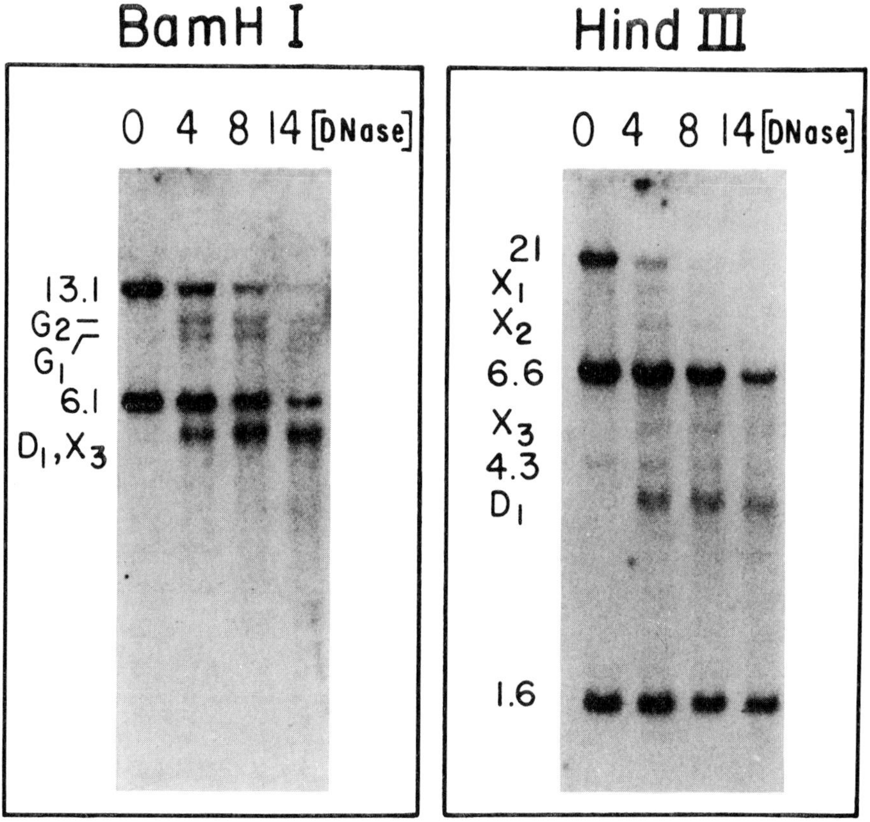

FIG. 1. DNase I hypersensitive sites found in the δ/γ locus after muscle differentiation. We detect six hypersensitive sites around the δ subunit gene in differentiated BC3H-1, C2, C2i and F-3 muscle cells; a BC3H-1 blot probed with a full-length δ subunit cDNA is shown here. DH sites not specific to muscle cells are designated X sites. Muscle-specific DH sites 5′ to the δ gene are designated D sites. Muscle-specific sites 5′ to the γ gene are designated G sites. DH sites are numbered from 5′ to 3′ relative to the δ and γ genes. The sizes of restriction fragments are given in kilobases; the 4.3 kb *Hin*dIII band disappears at higher stringency and co-migrates with a γ subunit gene restriction fragment. Concentrations of DNase I are given as units/mg DNA.

suggests that the site is small, on the order of 100 bp or less. Thus, the D_1 site is shown as a 300 bp region; however, we do not know the exact boundaries of the hypersensitivity region. The G_1 and G_2 sites are similarly shown as 300 bp regions. Fig. 3A illustrates the location of D_1 within the *Bam*HI-*Acc*I fragment. In both Fig. 3A and Fig. 3B, defined exons and introns are depicted as rectangular bars (exons, larger bars; introns, smaller bars) and undefined sequence (introns, non-coding exons or promoters) are drawn as cylinders.

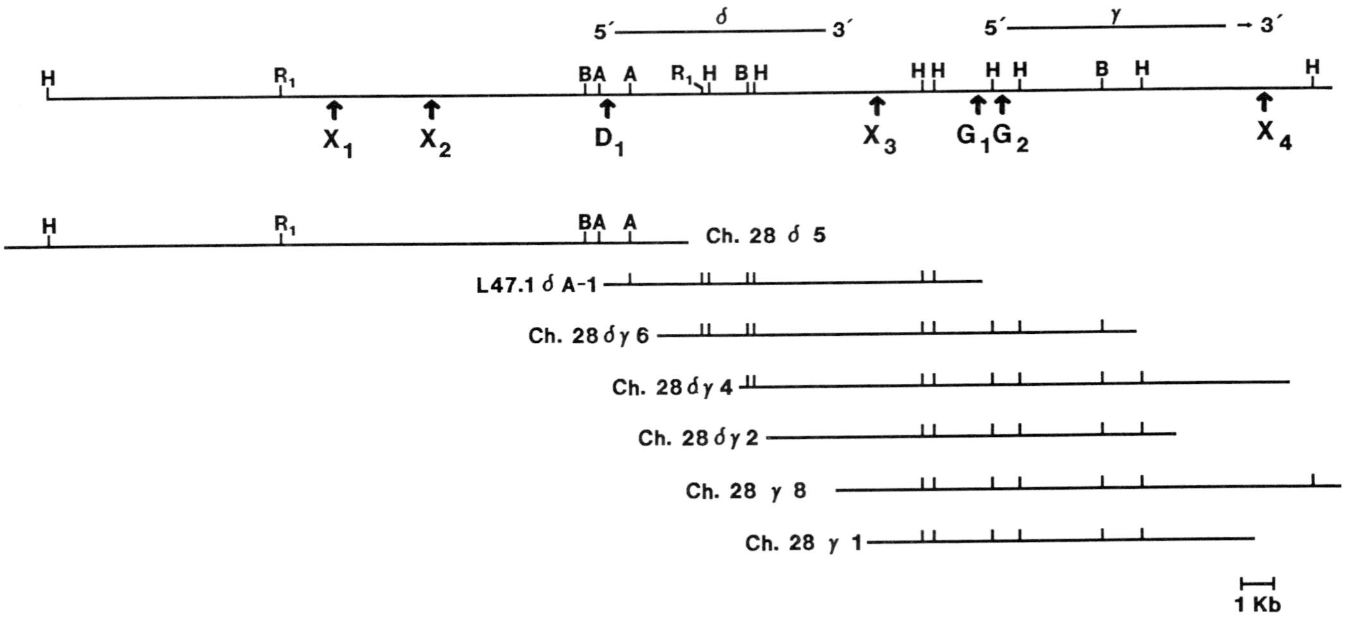

FIG. 2. The δ and γ subunit genes are tightly linked in the mouse genome. This linkage was demonstrated by isolating several individual clones that hybridize to both δ and γ subunit probes. The extent and orientation of the δ and γ genes are shown overlying the restriction map of the entire genomic region. The location of the hypersensitive sites in the δ/γ locus is drawn below. DH site X_4 is seen in blots hybridized with γ subunit probes; this site is present in all cell types so far examined, including liver. The bottom of the figure shows the restriction map of the clones obtained from screening a λ Charon 28 genomic library with a δ subunit cDNA probe.

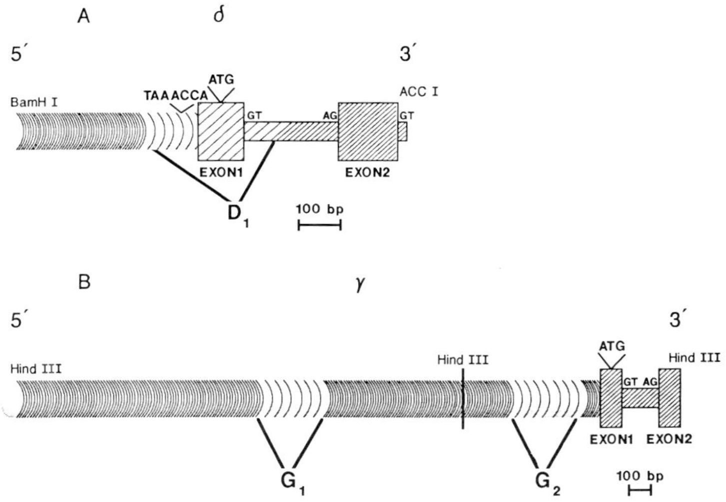

FIG. 3. Muscle-specific hypersensitive sites are located at the 5′ ends of the δ and γ genes. The DH sites were mapped to the indicated locations by precise sizing of DH subbands, parent restriction fragments, and distances between restriction sites. The precision of our placement of the centre of the DH sites is ± 100 bp. The narrow DH subbands indicate typically sized hypersensitivity regions of 100 bp or less. Thus, we show the DH sites as extending 300 bp. Exon–intron boundaries for these regions were mapped by comparison of published mouse cDNA sequences to the genomic sequences obtained from M13 subclones of λ genomic clones L47.1δA-1 and Ch.28δγ4. The *Bam*H I site at the 5′ end of the δ subclone was created during construction of the library. Rectangular bars indicate exons (wide bars) or introns (narrow bars), and cylinders indicate undefined sequence (exon, intron, or promoter region). The 5′ boundary of δ exon 1 was defined by RNase protection and primer extension analyses. A TATA-like sequence (TAAACCA) lies 27–33 bp upstream of the δ transcription start site. RNase protection experiments with γ constructs have not conclusively shown where the 5′ end of γ exon 1 lies, so we show the minimal size of γ exon 1 as determined by sequence comparison of genomic and cDNA clones.

The chromatin structure of the region is overlaid onto the exon–intron organization with hypersensitivity to DNase I indicated by lighter shading. The centre of the D_1 site lies approximately at the boundary of exon 1, 56 bp upstream of the initiator ATG. The 5′ boundary was defined by RNase protection (Melton et al 1984) and primer extension (Ghosh et al 1980) analyses. The agreement of these analyses defines the 5′ boundary of exon 1 as the transcription start site of the δ subunit gene. The sequence TAAACCA lies 27 bp upstream of the δ transcription start site and probably serves as the TATA box for the δ subunit gene.

To localize the G_1 and G_2 DH sites, we subcloned and sequenced two *Hin*dIII fragments, the 2.2 kb fragment containing G_1 and the 1.1 kb fragment for G_2. Both G_1 and G_2 lies 5′ to the initiator ATG within as yet undefined sequence. We do not show either the 5′ boundary of exon 1 or a putative TATA box since we have neither convincing RNase protection nor primer extension data for the γ gene. However, on the basis of their identical cell type specificities (see below), we speculate that G_2 like D_1 is centred near the transcription start site and that G_2 and D_1 represent homologous regulatory elements for the γ and δ genes. G_1, however, has no detectable spatial homologue 5′ to the δ gene and may either function in the regulation of both genes or serve a function unique to one. Alternatively, the δ homologue of G_1 may lie so close to D_1 as to be obscured by it.

The presence of DNase I hypersensitive sites at the 5′ ends of the δ and γ subunit genes affords the opportunity to directly observe myogenic changes in chromatin structure that are likely to participate in the regulation of these muscle-specific genes. As a model for the myogenic pathway, we chose to study three different cell lines: C3H10T½ (Reznikoff et al 1973), C2 (Yaffe & Saxel 1977), and C2i (Pinset et al 1988). C3H10T½ is a fibroblast-like line that has the interesting property of converting at high frequency into myocytes, adipocytes and chondrocytes when treated with 5-azacytidine, a cytosine methylation inhibitor (Taylor & Jones 1979). Thus. C3H10T½ are pluripotent mesodermal stem cells that require some degree of DNA hypomethylation to follow any one of the mesodermal differentiation pathways. The other two lines that we used are variants of the C2 muscle cell line. C2 is a skeletal muscle cell line that fuses and differentiates when confluent even in the presence of serum. The C2 inducible (C2i) cell line was subcloned from C2 cells. C2i, unlike the parent cell line, will not differentiate in the presence of serum, even when confluent and non-dividing (Pinset et al 1988). C2i myoblasts must be induced to differentiate by removing all serum and incubating in media with insulin and transferrin.

Fig. 4 summarizes our results with each cell line. C3H10T½ cells express few hypersensitive sites in the δ/γ locus. Only DH sites X_1 and X_2 are detectable. Site X_2 is also present in liver (data not shown) while X_1 is not (data not shown) and may represent a mesoderm-specific site.

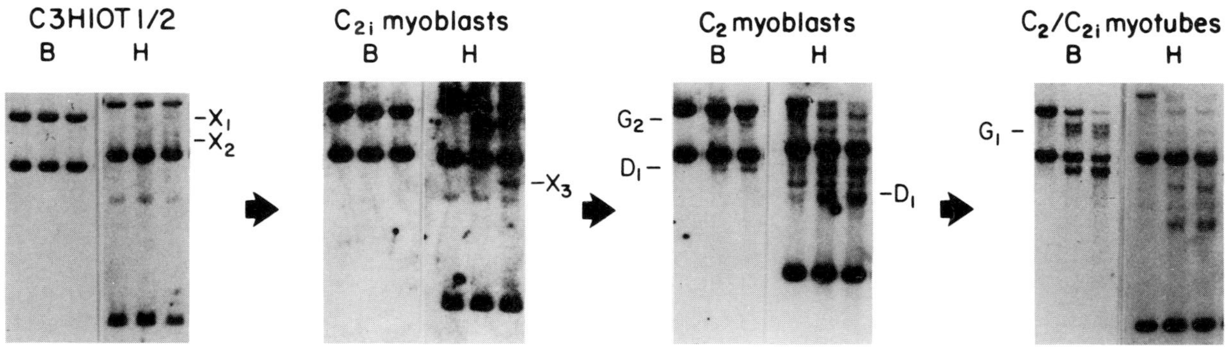

FIG. 4. The degree of chromatin structural activation of the δ/γ locus correlates with the potential to express the δ/γ genes. All blots were performed with a full-length δ subunit cDNA probe. The concentrations of DNase I are from left to right 0.4 and 8 units/mg DNA. Hypersensitive sites are indicated only where they first appear in the myogenic pathway. C3H10T½ ĉells are pluripotential mesodermal stem cells capable of converting into myocytes, adipocytes or chondrocytes when treated with cytosine methylation inhibitors. C2i myoblasts will not differentiate in the presence of serum even when confluent and non-dividing and must be induced to differentiate by removing serum. C2 myoblasts spontaneously differentiate into myotubes and seem to be at some more advanced stage of myogenesis. Both C2 and C2i myotubes have an identical DH site pattern. B, *Bam*HI;H, *Hin*dIII.

The next level of chromatin structural complexity is seen in C2i myoblasts, where DH site X_3 appears. This site is also present in liver (data not shown) and may be unrelated to myogenic activation of the δ/γ locus or may result from the binding of some multifunctional regulatory protein expressed in both liver and muscle cells. However, the clear absence of X_3 in undifferentiated BC3H-1 muscle cells (Crowder & Merlie 1986) indicates that high level expression of the δ and γ genes does not require this perturbation in chromatin structure prior to differentiation. One other site, X_4, is also present in C2i myoblasts; this site is present in every cell type thus far examined, including liver. We have not examined C3H/10$T_{1/2}$ cells for this site, which lies 3′ to the γ gene.

A further degree of chromatin activation of the δ/γ locus is seen in C2 myoblasts. The spatially homologous DH sites D_1 and G_2 appear at this stage of myogenesis. These cells are undoubtedly myoblasts, not just prematurely differentiating myotubes, since we see no myotubes at the time of harvesting. Further, the population of cells undergoes approximately three more divisions before reaching confluency and fusing. Finally, no transcription complexes can be measured in C2 myoblasts for the α and δ (Buonanno & Merlie 1986) and β and γ subunit genes (A. Buonanno, personal communication 1987). Thus, D_1 and G_2 appear in dividing myoblasts before the onset of transcription within the δ/γ locus.

Differentiation is accompanied by the appearance of only one more DH site, G_1. This site is specific to the transcriptionally active state of the δ/γ locus; it cannot be detected in liver, C3H10$T_{1/2}$ cells, undifferentiated BC3H-1 cells, C2i myoblasts or C2 myoblasts. G_1, then, is a good candidate for a *cis*-acting element required for or dependent upon δ/γ gene transcription.

The notion that chromatin structural changes are involved in myogenesis received its first experimental basis with the work of Yaffe and coworkers (Carmon et al 1982). They showed that the generalized DNase I sensitivity of the rat α-actin and myosin light chain 2 genes increases with differentiation of L8 myoblasts. Their work suggested that myoblasts, in a chromatin structural sense, are quiescent and that all of the activation of muscle-specific genes occurs during differentiation. Recently, the generality of this hypothesis has been challenged. Minty et al found that C2 myoblasts transfected with a human cardiac actin gene inappropriately expressed the construct (Minty et al 1986). On the other hand, transfection of the same construct into L8 myoblasts resulted in proper differentiation-specific expression. The authors suggest that the loose regulation of their transfected construct in C2 myoblasts is due to the presence of transcription factors prior to differentiation and that the endogenous gene is somehow unresponsive to or does not interact with these factors. Perhaps by opening up the chromatin structure, the authors suggest, differentiation converts the endogenous cardiac actin gene to a responsive state (Minty et al 1986). The developers of the C2i cell line, Pinset and coworkers, have also suggested that C2 myoblasts express regulatory factors before differentiation whereas C2i do not (Pinset et al 1988). Their proposal is based on measurements of the expression of several muscle-specific genes in C2i and C2 myoblasts. The absence of pretranscriptional DH sites in front of the δ and γ genes in C2i and their presence in C2 provides strong support for this hypothesis.

Pretranscriptional hypersensitive sites have been studied in detail in other systems (reviewed in Thomas et al 1985), most notably heat shock genes (Wu 1980) and globin genes (Weintraub et al 1981). *Drosophila* heat shock gene DH sites are probably the best characterized sites to date. Here, pretranscriptional DH sites are associated with the presence of TATA binding protein(s) TAB (Wu 1984). The binding of TAB and perhaps other factors may prime the promoter region for the arrival of heat shock transcription factor and RNA polymerase II, and the initiation of transcription. Although less well characterized than the heat shock genes, the developmentally regulated globin gene DH sites are highly analogous to δ/γ DH sites. Weintraub et al describe two types of avian erythroblasts, one where DH sites 5′ to the globin genes are not detectable before transcription is induced and the other where the sites are present pretranscriptionally (Weintraub et al 1981). The authors suggest that the former type of erythroblast is frozen at an immature developmental stage.

We propose a similar explanation for our results. C2i myoblasts are arrested at an early stage of myogenesis where little chromatin structural change has taken place within the δ/γ locus. Then, in a step we call priming, DH sites appear 5′ to the δ and γ genes, before the onset of transcription. The primed myoblasts would then be competent to bind other differentiation-specific transcription factors and begin transcription at the δ and γ start sites. Fig. 5 illustrates our model of the steps in myogenic activation of the δ and γ subunit genes. This model emphasizes three important conclusions from our results. First the creation of DH sites at the 5′ ends of the δ and γ genes is not an essential part of the determination of these genes; perhaps other chromatin modifications of receptor subunit genes and/or myogenic regulatory genes, such as cytosine hypomethylation, are responsible for determination, since 5-azacytidine treatment of C3H10T$_{1/2}$ is sufficient for its commitment to the muscle lineage. Second, these genes are not sequestered in an unresponsive state in C2 myoblasts, as was suggested for the cardiac actin gene; rather, in these myoblasts the non-transcribing δ and γ genes are already primed by 5′ DH sites that by analogy to other systems are probably associated with bound regulatory proteins. Finally, differentiation results in the appearance of only one more DH site within the δ/γ locus, at the resolution of our current experiments. Finer analyses, such as *in vitro* footprinting within the D_1 and G_2 regions, are required to detect any other DNA–protein interactions that might activate the transcription of δ and γ subunit genes. In addition, the G_1

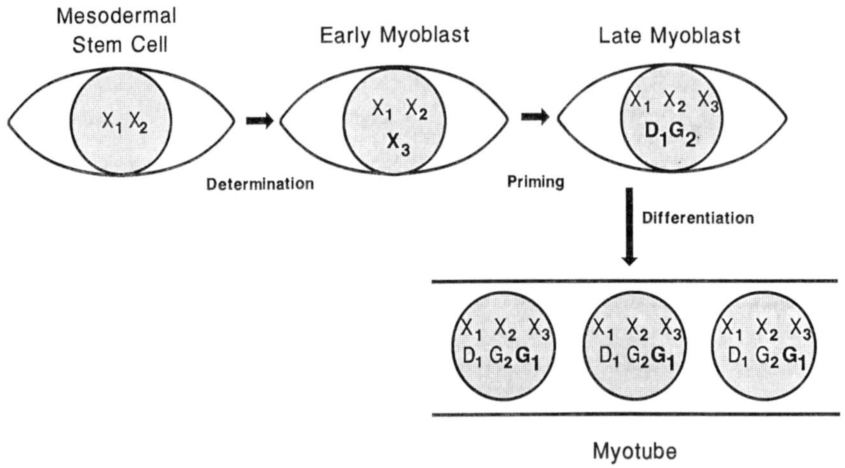

FIG. 5. A model for the activation of the δ/γ gene locus during myogenesis. Activation of the δ/γ locus is shown as a three-step process of determination, priming and differentiation. Each myogenic stage has a unique combination of hypersensitive sites that are shown in the nucleus of each cell type. The X_4 site, present at the 3′ end of the γ subunit gene, is not included in this model, since C3H/10T½, cells have not been examined for this site; the X_4 site is present in all cell types so far examined, including liver.

site might serve as a long-range enhancer for both the δ and γ genes and thus coordinate the expression of these two genes. 3′ regulatory elements have been demonstrated for a number of genes, including β-globin, which has a DH site associated with the 3′ enhancer (Emerson et al 1987).

We do not wish to imply that we have definitively demonstrated a three-step mechanism for the myogenic activation of the δ and γ subunit genes as it occurs *in vivo*, only that we can dissect four distinct stages of chromatin structural activation that fit into a stepwise scheme. However, in the future we would like to examine whole muscle DH sites to determine the relevance of our findings to the regulation of the δ and γ subunit genes *in vivo* and to assess the influence of innervation on the DH sites at the δ/γ locus.

Acknowledgements

We are grateful to Christian Pinset for the gift of the C2i cell line and Norman Davidson for the δ subunit cDNA, p6H. We thank John Kornhauser for cell culture work and Vandana Shah and Jacqueline Mudd for their help with sequencing. We acknowledge Tom Dietz and Liu Lin Thio for their help in cloning and mapping the δ/γ locus. We also want to thank Anne Dillon and the Washington University School of Medicine computer graphics facility for our illustrations. C.M.C. was supported by National Research Service Awards 5 T32 GM07200 and 5 T32 HL07275. Research was supported by funds from the Senator Jacob Javits Center of Excellence in the Neurosciences and by grants to J.P.M. from the National Institutes of Health, the Muscular Dystrophy Associations of America, and the Monsanto Company.

References

Blau HM, Chiu C-P, Webster C 1983 Cytoplasmic activation of human nuclear genes in stable heterokaryons, Cell 32:1171–1180

Buonanno A, Merlie JP 1986 Transcriptional regulation of nicotinic acetylcholine receptor genes during muscle development. J Biol Chem 261:11452–11455

Carmon Y, Czosnek H, Nudel U, Shani M, Yaffe D 1982 DNAase I sensitivity of genes expressed during myogenesis. Nucleic Acids Res 10:3085–3098

Crowder CM, Merlie JP 1986 DNase I-hypersensitive sites surround the mouse acetylcholine receptor δ-subunit gene. Proc Natl Acad Sci USA 83:8405–8409

Devlin RB, Emerson CP 1978 Coordinate regulation of contractile protein synthesis during myoblast differentiation. Cell 13:599–611

Emerson BM, Nickol JM, Jackson PD, Felsenfeld G 1987 Analysis of the tissue-specific enhancer at the 3′ end of the chicken adult β-globin gene. Proc Natl Acad Sci USA 84:4786–4790

Ghosh PK, Reddy VB, Piatak M, Lebowitz P, Weisman SM 1980 Determination of RNA sequences by primer directed synthesis and sequencing of their cDNA transcripts. Methods Enzymol 65:580–595

Hastings KEM, Emerson CP 1982 cDNA Clone analysis of six co-regulated mRNAs encoding skeletal muscle contractile proteins. Proc Natl Acad Sci USA 79:1553–1557

Heidmann O, Buonanno A, Geoffroy B, Robert B, Guenet J-L, Merlie JP, Changeux J-P 1986 Chromosomal localization of muscle nicotinic acetylcholine receptor genes in mouse. Science (Wash DC) 234: 866–868

Konieczny SF, Emerson CP 1984 5-Azacytidine induction of stable mesodermal stem cell lineages from 10T½ cells: evidence for regulatory genes controlling determination. Cell 38:791–800

Lassar AB, Paterson BM, Weintraub H 1986 Transfection of a DNA locus that mediates the conversion of 10T½ fibroblasts to myoblasts. Cell 47:649–656

Medford RM, Nguyen HT, Nadal-Ginard B 1983 Transcriptional and cell cycle-mediated regulation of myosin heavy chain gene expression during muscle cell differentiation. J Biol Chem 258:11063–11073

Melton DA, Krieg PA, Rebagliati MR, Maniatis T, Zinn K, Green MR 1984 Efficient *in vitro* synthesis of biologically active RNA and RNA hybridization probes from plasmids containing a bacteriophage SP6 promoter. Nucleic Acids Res 12:7035–7056

Minty A, Blau H, Kedes L 1986 Two level regulation of cardiac actin gene transcription: muscle-specific modulating factors can accumulate before gene activation. Mol Cell Biol 6:2137–2148

Nef P, Mauron A, Stalder R, Alliod C, Ballivet M 1984 Structure, linkage, and sequence of the two genes encoding the δ and γ subunits of the nicotinic acetylcholine receptor. Proc Natl Acad Sci USA 81: 7975–7979

Pinset C et al 1988 Differentiation, submitted

Reznikoff CA, Brankow DW, Heidelberger C 1973 Establishment and characterization of a cloned line of C3H mouse embryo cells sensitive to postconfluence inhibition of division. Cancer Res 33:3231–3238

Schwartz RJ, Rothblum KN 1981 Gene switching in myogenesis: differential expression of the chicken actin multigene family. Biochemistry 20:4122–4129

Shibahara S, Kubo T, Perski HJ, Takahashi H, Noda M, Numa S 1985 Cloning and sequence analysis of human genomic DNA encoding γ subunit precursor of muscle acetylcholine receptor. Eur J Biochem 146:15–22

Taylor SM, Jones PA 1979 Multiple new phenotypes induced in 10T½ and 3T3 cells treated with 5-azacytidine. Cell 17:771–779

Thomas GH, Siegfried E, Elgin SCR 1985 DNase I hypersensitive sites: a structural feature of chromatin associated with gene expression. In: Reeck G et al (eds) Chromosomal proteins and gene expression. Plenum Press, New York, p 77–101
Weintraub H, Beug H, Groudine M, Graf T 1981 Temperature-sensitive changes in the structure of globin chromatin in lines of red cell precursors transformed by ts-AEV. Cell 28:931–940
Wu C 1980 The 5′ ends of *Drosophila* heat shock genes in chromatin are hypersensitive to DNase I. Nature (Lond) 286:854–860
Wu C 1984 Two protein-binding sites in chromatin implicated in the activation of heat-shock genes. Nature (Lond) 309:229–234
Yaffe D, Saxel O 1977 Serial passaging and differentiation of myogenic cells isolated from dystrophic mouse muscle. Nature (Lond) 270:725–727
Yisraeli J, Szyf M 1984 Gene methylation patterns and expression. In: Razin A et al (eds) DNA methylation. Springer-Verlag, New York, p 352–378

DISCUSSION

Rubinstein: Why is this stepwise uncovering of the ACh receptor subunit gene necessary at a time when one is not even activating it?

Crowder: I don't know why this is needed. It may be that the chromatin structural changes are slow processes and for genes to be rapidly activated the chromatin changes need to occur first, over a longer time scale; then when the cell is ready to turn on the genes, it can do it rapidly. We compared the speeds at which genes are activated; the cell line C2i takes about two days longer than C2 to reach the same level of mRNA expression for δ and γ subunit genes. This kind of scenario is well-established for the *Drosophila* heat-shock genes, which need to be rapidly activated. There are hypersensitive sites around the promoter region of these genes before transcription as well. As I mentioned, these hypersensitive sites are associated with TATA binding protein (TAB) which is bound to the TATA box before transcription (Wu 1984). Others, such as John Lis and his coworkers, have shown that prior to induction of high level transcription by heat shock, RNA polymerase II has bound and transcribed a few bases, then stops for some reason (Gilmour & Lis 1986). Other gene systems (e.g. MMTV, globins) have documented chromatin structural or protein-binding events occurring before activation. So a stepwise mechanism may be utilized generally for rapid gene activation.

Henderson: Is it fair to say that the appearance of the DNaseI hypersensitive sites at certain stages of differentiation is thus far only a correlation?

Crowder: In this case, for receptor subunit genes, it remains a correlation, but for many well-studied genes a causative relationship of hypersensitive sites to gene regulation has been established. In only a few cases do you find DH sites that are not regulatory. In those instances they are created by a similar mechanism, such as a protein binding to the DNA, but the sites are doing something else, such as affecting replication or transposition.

Henderson: So the problem remains of knowing whether the DNAse I hypersensitive sites are concerned with transcription or not. There is a further problem: the presence of G_1 would seem in your system to be associated with ACh receptor expression, but it is possible that in other cell lines, that are primed but not differentiated, G_1 might also be present. Even if one does not question the idea that the hypersensitivity represents ongoing transcription, it is still necessary to show that a particular site is involved in regulating receptor expression.

Crowder: Yes, although in all four muscle cell lines studied, the G_1 site is not present until the gene is activated. We have looked at C2 and C2i, related cell types that differ in their regulation, and the cell line BC3H-1, which expresses ACh receptor, and the muscle cell line F3, derived from C3H/10½ by 5-azacytidine treatment. In all four lines, G_1 is restricted to the active state of the δ and γ genes. The power of the method is that it allows us to look over 40kb and find small regions that are likely to be regulatory. We can then characterize those regions by more classical techniques, such as transfection. But there are problems with transfection as a technique as well. The predictive value of hypersensitive sites is high and has recently proved to be crucial in understanding chicken β-globin locus regulation, which had been studied *ad nauseam* by transfection methods.

Holder: Do any of the DNase I-sensitive sites disappear in the area of the myotube that is not innervated?

Crowder: No. If a site disappeared with gene activation we might postulate some kind of gene suppression and then removal of that suppression. We don't see that. The resolution of DNaseI hypersensitivity experiments is low, however. We would like to do so-called foot-printing experiments, to look in detail at these protein-binding events that we think are responsible for creating the hypersensitive sites.

Nadal-Ginard: Has anyone shown that the putatively shared nucleotide sequences among the different subunits do bind to a nuclear protein, and, if they do, whether the different sequences compete for these binding factors?

Crowder: At this point it is merely a homology. We want to look at protein binding to the homologous region.

Nadal-Ginard: One possible pitfall is that unless one knows the protein-binding sequence (the minimum consensus required for the protein to bind), computer searches for common binding sites are difficult, because in a binding site of, say, eight nucleotides, where only four are constant, there can be a lot of scatter.

Crowder: This is why we don't know whether the homology is functional. However, homologies are useful for directing mutational studies.

Nadal-Ginard: Generally when deletion experiments are done, it is shown that the sequence does not affect the expression of the gene.

Crowder: Yes. Emerson and colleagues have pointed to a sequence in the quail troponin I gene whose homology with other muscle promoters was not as good as the receptor subunit homology; they have recently shown that deletion of the sequence doesn't affect regulation (Konieczny & Emerson 1987). But another sequence discovered by homology, the CAARG sequence, of about the same length as the receptor subunit homology, does seem to have some function in actin gene regulation (Minty & Kedes 1986).

Nadal-Ginard: You mentioned that in denervated but immature skeletal muscle the ACh receptor subunits are spread over the myofibre, whereas in the innervated muscle they cluster at the neuromuscular plate; but there is another change that happens at the same time, when the cell switches from expression of the γ subunit to the ε subunit. Does this switch in subunit composition have anything to do with the localization of the receptor in the neuromuscular plate?

Crowder: There is certainly an isoform switch in going from the embryo to the adult in the channel and biochemical properties of the ACh receptor. These two receptor types can be produced in an oocyte expression system. Numa, Sakmann and their coworkers (Mishina et al 1986) have shown that replacement of the γ subunit with ε converts the receptor from the embryonic to the adult type. Perhaps this subunit change occurs *in vivo*. The mRNA levels for the γ subunits decrease and ϵ mRNA levels increase during the time that the isoform switch takes place in cow. So mechanistically the switch is probably effected at transcriptional level. This emphasizes another advantage of studying the γ and δ subunits at the same time; we are studying two genes which *in vivo* may be regulated differently, but in cell lines are probably regulated similarly.

Buller: The extrajunctional ACh receptors were first shown to be different from those at the endplate after denervation by Ricardo Miledi.

Chowder: That is right. And we don't know yet if the ϵ subunit is localized to the synaptic receptors, with the γ subunit in extrasynaptic receptors.

Zak: In your search for expression of the mRNA by nuclear run-on experiments, did you try primary cultures? Apparently, satellite cells, which will be present in primary cultures, express one ACh receptor subunit, whereas the myocytes express another kind. Would you be able to differentiate, in your run-on experiments, between these two classes?

Crowder: I doubt it. In fact, nuclear run-on is not so very sensitive a method because of background problems. The best way to look at localized expression is by *in situ* hybridization. Also, to do your nuclear run-on experiment, you would have to isolate satellite cells from myotubes in large enough quantity.

We would like to look at hypersensitive sites in whole animals, to study the effects of innervation and denervation on the DNase I hypersensitivity pattern, and also to see if there are DH sites in muscle *in vivo*. We have tried purifying muscle nuclei, but the contractile apparatus seems to inhibit purification of the muscle nuclei in large quantities.

Eisenberg: Acetylcholine sensitivity is also high in the tendon regions of muscle. Is anybody looking at ACh receptor expression in the non-neuromuscular junction but high-expressing regions of the muscle fibre? Neurobiologists sometimes forget that things other than nerve can turn on a gene. This might help to narrow down the factors that are actually used.

Vrbová: Hans Brenner (personal communication) has looked at the channel properties of the ACh receptors of the tendon ends of muscles; their properties are different from those of the receptor at the neuromuscular junction.

Crowder: Presumably they are similar to those of extrasynaptic ACh receptors.

Vrbová: Yes.

Nadal-Ginard: In terms of the neuromuscular junction, the mRNAs encoding the ACh receptor subunits accumulate under the neuromuscular plate. These mRNAs are thought to be synthesized mainly by the nuclei lying under the plate. How is this regulated? If the muscle is denervated, do all the nuclei of the myofibre transcribe the AChR subunits? When you innervate the muscle, does it shut off all the nuclei that are farther away and only the ones that are under the endplate continue to transcribe these genes?

Crowder: My favourite theory is that in general, electrical or contractile activity in muscle fibres decreases all ACh receptor subunit expression at the level of the gene, but the nerve itself secretes some factor that releases that inhibition locally at the synapse. So perhaps these two things are both going on. We know that electrical activity down-regulates AChR expression.

Nadal-Ginard: If you treat the muscle with curare, does the innervated pattern of expression remain, or disappear?

Crowder: Curare treatment results in up-regulation of ACh receptors, as does any treatment that reduces electrical activity in the muscle.

Buller: That approach treats ACh receptors as a homogeneous group, but you say that they are not? You have shown the differential control of two different receptor subunits, namely ε and δ.

Crowder: Yes, but the receptor isoforms probably share subunits. So there perhaps is a general effect whereby all acetylcholine receptor subunits, except perhaps ε, are down-regulated by muscle electrical activity, and then a factor specific to the neuronal region might release that inhibition for all subunits except possibly the γ subunit. But this is speculative.

Henderson: On the putative localization of ACh receptor synthesis to the nuclei under the motor endplate, Fontaine and colleagues at the Institut Pasteur (Fontaine et al 1988), and other groups, have direct evidence by *in situ* hybridization that the messenger for the α subunit is in that position. This is not the same as showing that those nuclei are transcribing that mRNA, of course.

Crowder: The experiment that John Merlie and Josh Sanes published a few years ago (Merlie & Sanes 1985) agrees with your recent results but the Changeux group's experiment is more precise.

Henderson: Coordinated gene regulation is obviously important in ACh receptor expression, and also in a wider sense for those working on synaptic proteins. Given that the four subunit genes probably result from gene duplications, would one not expect to find the important regulatory sites in the same position in relation to each gene? I have two specific questions: (a) do you know anything about DNase I hypersensitive sites in the genes for the α- and β-subunits of the ACh receptor, which are on different chromosomes; and (b) is it not surprising that the G_1 site has a different topographical relationship to the δ and γ genes?

Crowder: It would certainly make a neater story if there was an analogue to the G_1 site, 5′ to the δ gene. One can postulate that G_1 is specific for γ gene expression, and that this chromatin structural change is the vanguard of something that regulates γ differently from δ in later stages of myogenesis. Alternatively, perhaps G_1 has an analogue, 5′ to δ, but it is so close to D_1 that it is basically the same hypersensitive site but is really due to two different protein–DNA interactions. Multiple proteins binding within one hypersensitive site has been documented in several cases.

Ribchester: Pursuing these points about coordinated regulation further, denervated muscles also express other membrane proteins, more or less synchronously with the expression of the extrajunctional ACh receptor. Is anything known about the promoter sequences of, say, the tetrodotoxin-resistant sodium channel, or N-CAM (the neural cell adhesion molecule)?

Crowder: Nobody has gone to that level yet with N-CAM, at least by looking at gene regulation in muscle. We are beginning to study examples of other synapse-specific proteins. One that has been cloned in our laboratory by Don Frail is the 43K protein, which is a cytoskeletal element that is localized fairly well to the subsynaptic membrane (Frail et al 1987). Our laboratory is now studying its regulation. This 43K protein doesn't seem to be regulated like ACh receptor, at least during myogenesis. It is perhaps not regulated at all during terminal differentiation, because we can detect 43K in C2 myoblasts at the levels of protein, message, and transcription.

Ribchester: Does it spread over the surface of denervated muscle fibres?

Crowder: We don't know.

Gordon: The incorporation of sodium channels into embryonic and neonatal muscle and the disappearance of extrajunctional ACh receptors follow different time courses, so one might not expect the time courses of the different proteins to be the same.

Ribchester: After denervation of an adult skeletal muscle, extrajunctional sensitivity is seen about three days later, and it is at that stage that you can record TTX-resistant action potentials (Harris & Maltin 1981, Lømo & Slater 1978).

Gordon: You see these ACh receptors at three days, but the time course of the incorporation of receptors reaches a plateau only three weeks later. And although the TTX-sensitive sodium channel reappears, it does not do so to the extent it does in embryonic development.

Crowder: The Na^+ channel never goes down as much as the ACh receptor does, on innervation, and it is not as localized to the subsynaptic membrane.

Ribchester: I am thinking about the initiation of these two quite separate events. Granted that the numbers of binding sites for α-bungarotoxin, TTX-binding sites and so on differ in their rates of synthesis and degradation, is the *initiation* of synthesis synchronized?

Crowder: According to the model that I mentioned, each gene might share some activity down-regulatory sequence but have unique regulatory elements responsive to gene-specific factors.

Sanes: When you removed serum from C2i myoblasts they began to express ACh receptor subunits. Is that an all-or-none event, and once the α-bungarotoxin sites are exposed, if you put serum back, are they down regulated?

Crowder: If serum is removed, the myoblasts fuse, so the process isn't reversible. But the activation is an all-or-none process; we fail to detect any expression of ACh receptor subunits, either mRNA or protein, in totally confluent C2i cultures until serum is removed. This tight control has also been shown for other genes that are expressed a higher levels, such as myosin light chain and heavy chains. Perhaps you are referring to serum 'switchback' experiments with the non-fusing line BC3H-1. It would be interesting to look at hypersensitive sites in these cells after adding back serum.

Vrbová: I believe that myoblasts were reported to have ACh receptors and bind α-bungarotoxin. Is that right, or is α-bungarotoxin bound to something else?

Crowder: That is true in chick; in that species, myoblast ACh receptors don't seem to be as tightly regulated as are the mammalian receptors. Mouse and rat myoblasts have few if any ACh receptors.

References

Fontaine B, Sassoon D, Buckingham M, Changeux JP 1988 Detection of the nicotinic acetylcholine receptor α-subunit mRNA by *in situ* hybridization of the neuromuscular junctions of fifteen-day old chick striated muscles. EMBO (Eur Mol Biol Organ) J 7:603–609

Frail D, Mudd J, Shah V, Carr C, Cohen JB, Merlie JP 1987 cDNAs for the 43000-dalton v_1 protein of *Torpedo* electric organ encode two proteins with different carboxy termini. Proc Natl Acad Sci USA 84:6302–6306

Gilmour D, Lis JT 1986 RNA Polymerase II interacts with the promoter region of the noninduced hsp 70 gene in *Drosophila melanogaster* cells. Mol Cell Biol 6:3984–3989

Harris JB, Maltin C 1981 The effect of the subcutaneous injection of the crude venom of the australian common brown snake, *Pseudonaja textilis* on the skeletal neuromuscular system. Br J Pharmacol 73:157–163

Konieczny SF, Emerson CP 1987 Complex regulation of the muscle-specific contractile protein (troponin I) gene. Mol Cell Biol 7:3065–3075
Lømo T, Slater CR 1978 Control of acetylcholine sensitivity and synapse formation by muscle activity. J Physiol (Lond) 275:391–402
Merlie JP, Sanes JR 1985 Concentration of acetylcholine receptor mRNA in synaptic regions of adult muscle fibres. Nature (Lond) 317:66–68
Minty A, Kedes L 1986 Upstream regions of the human cardiac actin gene that modulate its transcription in muscle cells: presence of an evolutionary conserved repeated motif. Mol Cell Biol 6:2125–2136
Mishina M, Takai T, Imoto K et al 1986 Molecular distinction between fetal and adult forms of muscle acetylcholine receptor. Nature (Lond) 321:406–411
Wu C 1984 Two protein-binding sites in chromatin implicated in the activation of heat shock genes. Nature (Lond) 309:229–234

General discussion I

Microenvironmental effects on myosin expression

Buller: In our earlier discussions of the myosin isoforms we concluded that any change in the microenvironment may produce changes in isoform. What are the microenvironmental changes that can do this? We spoke of hormones, including testosterone and thyroid hormone; we spoke of ionic environmental changes caused by cell surface damage. The microenvironment can presumably also be changed by contractile activity of the muscle fibre, and probably nerve activity. Are there other examples of changes that lead to changes in myosin isoform?

Pette: I wonder whether, with our current knowledge, we can always distinguish between effects brought about by specific stimulus patterns and effects brought about by changes in the overall amount of contractile activity. In my view, contractile activity by itself could be important in controlling the exposure of receptors, such as hormone receptors or receptors for growth factors. Therefore, some of the observations relating to contractile activity may reflect indirect effects of receptor exposition which alter the sensitivity of the muscle to specific agents.

Thesleff: We are seeing many influences on the muscle cell, of various kinds, but there surely must be a common second messenger by which these influences exert their effect on the genome. One suggestion has been the adenylate cyclase–cyclic AMP system, proposed by Greengard (1978); calcium has also been suggested as a common second messenger (Birnbaum et al 1980).

Mudge: On the nature of the second messengers involved in regulating the levels of acetylcholine receptor (AChR), Lee Rubin at the Rockefeller University has shown that intracellular calcium levels may mediate the effects of electrical activity. He found that the calcium ionophore, A23187, can reverse the effects of tetrodotoxin (TTX) on muscle; when cultured muscle cells are grown in the presence of A23187 and TTX, the levels of AChR are decreased and the levels of acetylcholine esterase are increased, when compared to cultures grown with TTX alone (Rubin 1985). On the other hand, both J.-P. Changeux and my own group have shown that calcitonin gene-related peptide (CGRP), which is found in motoneuron terminals, can increase the levels of AChR on cultured myotubes. CGRP seems to act by increasing cyclic AMP levels. It could be that nuclei at the neuromuscular junction escape the regulation by calcium that decreases the production of AChR extrajunctionally and, instead, are stimulated by cyclic AMP to increase AChR junctionally (see New & Mudge 1986, Fontaine et al 1986, Laufer & Changeux 1987).

Pette: The calcium ionophore A23187 has been applied to myotube cultures and it has been reported that, under its influence, an increase in slow myosin light chains occurs (Silver & Etlinger 1981).

Nadal-Ginard: On another level of regulation, one theory is that we shall find a gene that is induced during myogenesis and turns on all the many coordinately regulated proteins, from the receptors to the contractile proteins to the cell adhesion molecules (CAMs); in other words, a master switch. Unfortunately, the problem will not be resolved that way. Just to take the contractile proteins, it is already known that the different genes are regulated by different mechanisms. The induction of tissue specificity and the developmental induction of the myosin heavy chain is negatively regulated; there is a positive factor that regulates the level of expression, but the decision whether or not to express these genes in a cell is under negative regulation. There is a DNA sequence on each of these genes that binds a nuclear factor that represses their expression. When this piece of DNA is removed, the gene is expressed everywhere, not just in muscle. In contrast, the light chain gene is positively regulated, not at the promoter level (because the promoter is not muscle specific and is a weak and inefficient promoter) but regulated by an enhancer on the 3′ end of the gene. This is a strong, muscle-specific enhancer that does not affect the regulation of the myosin heavy chain or any other contractile protein, so far as we know.

There is a troponin I, studied by Charles Emerson, which has a muscle-specific enhancer that is different from the light chain enhancer, and is located in the middle of the gene, in an intervening sequence. Finally, the tropomyosin gene has a promoter that is not muscle specific; it is quite a good promoter in every cell of the animal. Nevertheless, muscle cells contain much more tropomyosin than non-muscle cells, so how does this happen? In fact, the tropomyosin gene has an enhancer different from those of the light chain and troponin I, located at the 5′ end of the tropomyosin gene. It is muscle specific, in that it enhances the expression of genes in muscle, but it acts as a negative regulator in non-muscle cells. So this inducer has a dual function.

These are just four examples, and each is apparently regulated by a different mechanism. So the search for the Holy Grail of muscle regulation does not look very promising!

Crowder: Weintraub & Lassar looked for your Holy Grail, a clone that might convert 10T½ cells into a muscle lineage cell type (Davis et al 1987). They found a clone which when transfected into these cells does convert them into muscle, at a frequency of about 50% conversion; they see contractile proteins expressed. They are careful in their conclusions, but it might be that this cDNA that they have cloned codes for the highest hierarchical regulatory protein, which regulates other regulatory proteins that then turn on the various groups of proteins which are regulated similarly. So their clone may be something like the Holy Grail of muscle protein regulation!

Kernell: There are reasons why one should talk about 'grails' in the plural. We know from experiments on the chronic electrical stimulation of hindlimb muscles in the cat that one can affect different contractile properties in different ways by varying certain aspects of the activity. For instance, different patterns of stimulation may cause the same effects on speed but still have different effects on maximum tetanic force or endurance (Kernell et al 1987a,b). Such a relative independence between different types of effects of chronic stimulation implies that there should exist multiple links between the contractile (and electrical?) activity of a muscle fibre and the factors regulating its protein synthesis and other biochemical functions. Long-term usage-related changes in muscle properties could not all be mediated by, for instance, changes in intracellular calcium concentration.

Vrbová: Bernardo Nadal-Ginard is speaking about regulation at the gene level, whereas you, Professor Buller, were asking how the microenvironment reaches that level and affects the gene. I think that here the regulation is also multifactorial. Andrew Huxley originally suggested to you that the reason for the conversion of fast into slow muscle is vibratory stress. I think there was a lot in this idea, even though it may turn out not to be vibratory stress; the interaction of the cross-bridges, which is very different from what it was before, could change the microenvironment of the muscle cells. This is an example of how the microenvironment can be changed at a localized site in the cell. A similar situation occurs when loading a muscle; then the interaction between the cross-bridges again becomes different. This must produce a localized change in the microenvironment of the muscle cell which could act on these different regulatory 'Holy Grails'.

Eisenberg: Many experiments can be cited in support of one or other 'trophic' factor. For instance, the denervated and then doubly reinnervated model was recently used by Salviati et al (1986). In that situation, ectopically, under just one of the neuromuscular junctions, there is preferential expression of one myosin heavy chain, and under the other junction both myosin heavy chains are expressed. That could be explained by nerve trophic factors, or local concentrations of ionic activity, or some microenvironmental changes. Also, in the tendon–muscle junction, and near the neuromuscular junction, there is local expression of special proteins. All these studies suggest that individual muscle nuclei can be re-programmed differently from their neighbours. This is hard to explain in terms of vibrations or sarcomeric forces if uniformity of mechanical and electrical properties is assumed along the fibre. A recent paper questions these assumptions of mechanical uniformity (Edman et al 1988).

Kernell: With respect to the linkage between muscle usage and the regulation of protein synthesis (e.g., myosin), it might be of interest to consider the possible role of various activity-related metabolites. Depending on the local balance between energy-producing and energy-consuming processes, metabolite concentrations might conceivably even vary along the length of a single active muscle fibre. Such factors might cause different behaviours to appear in different nuclei along a fibre.

Holder: As a more general point, I have been struck by the pattern of different myotube types that differentiates, independent of innervation, during the early stage of development (see, for example, Laing & Lamb 1983, Butler et al 1982). Has Dr Rubinstein, or anyone else, studied the effect of position-dependent molecules, such as retinoic acid, on the myosin isoforms seen?

Rubinstein: We haven't done that ourselves.

Lance-Jones: We have been interested in the question of whether a muscle's somitic level of origin along the embryonic axis determines its fibre type composition. We have made assessments of fibre type distribution in embryonic chick hindlimbs after exchanging somites known to contribute to predominantly slow limb muscles and somites that contribute to predominantly fast limb muscles. We moved somites prior to muscle cell migration and used antibodies to myosin heavy chain subunits obtained from Dr Michael Crowe to demonstrate fibre type patterns. Preliminary results suggest that one does not change the pattern of fibre type distribution by somite exchange. Butler & Cosmos (1987) have reported similar results for the brachial region. These observations suggest that fibre type patterns are likely to be determined in some manner by cell position in the limb after migration (see also Crowe & Stockdale 1986). We do not know what characteristics of the limb environment are responsible for setting up these patterns.

Carlson: Many of the different types of unusual muscle protein molecules mentioned earlier by Dirk Pette seem to appear in the craniofacial muscles. Is there any biochemical evidence for a fundamental difference between craniofacial and other skeletal muscles, or is the difference just due to sampling?

Pette: The craniofacial muscles may represent a distinct group. However, a systematic evaluation of other than the few commonly studied skeletal muscles needs to be made.

Eisenberg: In that regard, the muscle spindles are usually the exception, and contain unusual combinations of muscle proteins.

Carlson: This again is a confounding effect, but it does look like a function of sampling variation rather than a biological variation.

Pette: To return to the question of microenvironments, I would like to focus on metabolites. It is generally agreed that mitochondria and glycogen are unevenly distributed along the length of skeletal muscle fibres. This distribution could create local differences in ions and metabolites. For example, where there is more glycogen, there will be more lactate formation, and hence shifts in pH due to an increase in proton concentration.

Eisenberg: It is difficult to build up high gradients of ions, because they diffuse so readily. For example, the chloride ion has a diffusion coefficient of 18.9×10^{-6} cm^2 s^{-1} in dilute aqueous solutions, so the gradient will have dispersed over a millimetre within a few minutes.

Pette: This is true; however, lactate can accumulate 10–20-fold in a muscle fibre (e.g. Meyer & Terjung 1979). Such an acid load could have direct effects on enzymes, such as phosphofructosekinase, as well as inhibiting calcium uptake by the sarcoplasmic reticulum.

Buller: We need to know more about the time constants. I agree with Brenda Eisenberg that one thinks of diffusion as having a time constant which, while it may stretch into seconds or more, is not necessarily going to act as an effective switch to change the production of a myosin.

Pette: A fast-twitch glycolytic fibre, when continuously stimulated at 10Hz, may be rapidly depleted of glycogen (Maier & Pette 1987). Thus, the energy charge may decrease, leading to a collapse of ATP-driven ion pumps. This could lead to fibre deterioration or, alternatively, could create signals leading to fibre transformation—for example, an increase in mitochondria, changes in myosin expression, and so on.

Eisenberg: I agree with that scenario, except that I don't see how you get different parts of a muscle fibre going ahead of the other parts. One sees only a fibre segment 2mm long on a grid in the electron microscope, admittedly, but the structure looks the same over this length in normal fibres. In fibres stimulated for a few hours, the SR swelled along the entire 2mm segment, but we did not see regional changes.

Zak: When you attempt to transform a muscle from slow to fast or vice versa, I believe that it is not the total amount of contractions forced on the muscle but rather the pattern of the stimulation. If this is true, I assume that the change in metabolites differs according to whether the sequence of contractions occurs rapidly or whether the stimuli are distributed more evenly, giving the opportunity for the diffusion of metabolites and the elimination of any kind of gradient.

Vrbová: On the question of why the effects of high frequency stimulation are similar to those of low frequency stimulation, we received a comment from Dr K. Kaplove asking whether we had looked at the sort of activity that the muscle sees. If you stimulate at 40Hz for a long period, does the muscle actually 'see' that 40Hz?

Eisenberg: The muscle cells have protective mechanisms, and there are many stages of fatigue—the neuromuscular junction, the T-tubule system of conduction, and the T-to-SR junction mechanism. One protective mechanism is that calcium is sequestered back into the terminal cisternae but the release mechanism fails, so when you try to drive the fibre, calcium isn't released and you get protective fatigue. Thus a biological defence mechanism is operating against us when we drive the muscle with external stimulators!

Pette: There are rapidly occurring changes in the SR after the onset of stimulation. After one day, there is a 50% decrease in Ca^{2+} uptake by the sarcoplasmic reticulum, and this reduction in Ca^{2+} sequestration persists with prolonged stimulation (Leberer et al 1987).

Zak: I think you are avoiding my question! I was asking whether the pattern of stimulation determines the type of muscle response.

Vrbová: Stimulation will always increase oxidative capacity, but I wasn't really avoiding the question. We are saying that it is possible that the contractile machinery of the muscle sees the same pattern of activity because the neuromuscular junction, membrane, and so on do not permit activation of the muscle at high frequency for long periods.

Eisenberg: The load that is actually moved or not moved may be the most relevant factor in switching isoforms.

Vrbová: This is what I am saying; the contractile machinery is performing the same function in response to both types of activity.

Kernell: In our chronic stimulation experiments, which were performed on fast hindlimb muscle of the cat, high and low pulse rates had the same slowing effect (Eerbeek et al 1984, Kernell et al 1987a, b). Dr Kaplove has suggested that such results might have been produced because many of the high-rate spikes were not transmitted across the neuromuscular junction. Hence, in such a case, the experimental muscle would actually have been activated by slow pulse rates, whatever the stimulation frequency given to the nerve (Kaplove 1987). As we have recently pointed out (Kernell & Eerbeek 1988), this is not a likely explanation for our own findings because, with respect to other properties than isometric speed, the high and low pulse rates did produce different effects on the muscles. Maximum force and fibre diameter were better preserved after treatment with high than with low rates. Furthermore, after considerable daily amounts of chronic activation, the twitch versus tetanus ratio was found to have been decreased by high rates and increased by low rates (Eerbeek et al 1984). It was also evident from direct observations during the chronic treatment that high pulse rates of motor nerve stimulation indeed caused stronger contractions to occur than those seen with slower frequencies of activation.

References

Birnbaum M, Reis MA, Shainberg A 1980 Role of calcium in the regulation of acetylcholine receptor synthesis in cultured muscle cells. Pfluegers Arch Eur J Physiol 385:37–43

Butler J, Cosmos E 1987 Positional information influences embryonic muscle fibertype. Soc Neurosci Abstr 13:893

Butler J, Cosmos E, Brierley J 1982 Differentiation of muscle fibre types in aneurogenic brachial muscles of the chick embryo. J Exp Zool 224:65–80

Crowe MT, Stockdale FE 1986 Myosin expression and specialization among the earliest muscle fibers of the developing avian limb. Dev Biol 113:238–254

Davis RL, Weintraub H, Lasser AB 1987 Expression of a single transfected cDNA converts fibroblasts to myoblasts. Cell 51:987–1000

Edman KAP, Reggiani C, Schiaffino S, Te Kronnie G 1988 Maximum velocity of shortening related to myosin isoform composition in frog skeletal muscle fibres. J Physiol (Lond) 395:679–694

Eerbeek O, Kernell D, Verhey BA 1984 Effects of fast and slow patterns of tonic long-term stimulation on contractile properties of fast muscle in the cat. J Physiol (Lond) 352:73–90

Fontaine B, Klarsfeld A, Hökfelt T, Changeux J-P 1986 Calcitonin gene-related peptide, a peptide present in spinal cord motoneurons, increases the number of acetylcholine receptors in primary cultures of chick embryo myotubes. Neurosci Lett 71:59–65

Greengard P 1978 Phosphorylated proteins as physiological effectors. Science (Wash DC) 199:146–152

Kaplove KA 1987 Reanalysis: impulse activity and fiber type transformation. Muscle & Nerve 10:375–376

Kernell D, Eerbeek O 1988 Different rates of long-term motor-nerve stimulation produce different effects on muscle properties. Muscle & Nerve 11:89-90

Kernell D, Donselaar Y, Eerbeek O 1987a Effects of physiological amounts of high- and low-rate chronic stimulation on fast twitch muscle of the cat hindlimb. II. Endurance-related properties. J Neurophysiol (Bethesda) 58:614–627

Kernell D, Eerbeek O, Verhey BA, Donselaar Y 1987b Effects of physiological amounts of high and low-rate chronic stimulation on fast-twitch muscle of the cat hindlimb. I. Speed- and force-related properties. J Neurophysiol (Bethesda) 58:598–613

Laing NG, Lamb AH 1983 The distribution of muscle fibre types in chick embryo wings transplanted to the pelvic region is normal. J Embryol Exp Morphol 78:67–82

Laufer R, Changeux J-P 1987 Calcitonin gene-related peptide elevates cyclic AMP levels in chick skeletal muscle: possible neurotrophic role for a coexisting neuronal messenger. EMBO (Eur Mol Biol Organ) J 6:901–906

Leberer E, Härtner KT, Pette D 1987 Reversible inhibition of sarcoplasmic reticulum Ca-ATPase by altered neuromuscular activity in rabbit fast-twitch muscle. Eur J Biochem 162:555–561

Maier A, Pette D 1987 The time course of glycogen depletion in single fibers of chronically stimulated rabbit fast-twitch muscle. Pfluegers Arch Eur J Physiol 408:444–450

Meyer RA, Terjung RL 1979 Differences in ammonia and adenylate metabolism in contracting fast and slow muscle. Am J Physiol 237:C111–C118

New HV, Mudge AW 1986 CGRP regulates muscle acetylcholine receptor synthesis. Nature (Lond) 323:809–811

Salviati G, Biasia E, Aloisi M 1986 Synthesis of fast myosin induced by fast ectopic innervation of rat soleus muscle is restricted to the ectopic endplate region. Nature (Lond) 322:637–639

Silver G, Etlinger JD 1981 Regulation of myosin light chain synthesis and accumulation in skeletal muscle cultures by the calcium ionophore A23187. J Cell Biol 91:351a

Mechanism of interaction between motoneurons and muscles

M.C. Brown and E.R. Lunn

University Laboratory of Physiology, Parks Road, Oxford OX1 3PT, UK

Abstract. Possible ways in which muscle fibres can affect motoneuron properties and motoneurons can affect muscle properties are discussed. A more detailed analysis of the local control by muscle of motor nerve terminal growth is then given. Experimental results from the gluteus maximus muscle of the mouse show that nerve terminal growth occurs very quickly if the muscle under the endplate is killed and that the growth can still be elicited after all communication with the cell body of the motoneuron has been eliminated. These observations show that all the materials needed for axon growth are normally present in axon terminals and that the control of terminal growth by the muscle is exercised at a local level. It is argued that this inhibitory control works by preventing the incorporation of these growth materials rather than by continual degradation of a persistently growing terminal. One reason for taking this view is that recent evidence suggests that the axonal cytoskeleton consists of a largely stationary array of discontinuous neurofilaments and microtubules.

1988 Plasticity of the neuromuscular system. Wiley, Chichester (Ciba Foundation Symposium 138) p 78–96

A summary of some of the results of long-term interactions between motor nerves and skeletal muscle fibres is given in Tables 1 and 2. The positive effects of motor nerves on muscle and of muscle on motor nerves are given, as well as the negative effects of separating the two tissues. In the embryo, contact at critical stages in development is necessary for the very survival of both partners. In the adult, the interactions between muscle fibres and neurites have less overwhelming consequences but are still of considerable importance. They seem to be aimed at stopping the elongation of the individual nerve branches, promoting muscle fibre growth, and precisely adapting the contractile properties to the discharge pattern received from the motoneuron. The effects of denervation, in contrast, promote reunion of the two partners by encouraging motor nerve sprouting and synapse formation.

The tables also indicate something that has become increasingly clear over the years, that both chemical factors and electrical activity play a part in mediating the trophic interactions between nerve and muscle.

Space clearly does not allow for adequate discussion of all these mechan-

TABLE 1 Some 'trophic' effects of motor nerves on skeletal muscle fibres

			Likely mediator of positive effect	
Part of muscle	*Effects of denervation*	*Positive effects of nerve*	*Chemical factors*	*Electrical activity*
Extrajunctional region	TTX-resistant action potentials appear Appearance of extrajunctional AChR Appearance of N-CAM, Thy-1 Acceptance of innervation Small increase in lysosomal activity Small increase in endocytotic activity	Regulation of extrajunctional membrane properties	–	+
Contractile machinery and excitation/contraction coupling	Atrophy	Fibre growth	–	+
	Slowing of contractile speed	Regulation of contractile speed	–	+
	Failure of secondary myotube development in embryo	In embryo, generation of secondary myotubes	?	?
Endplate region	Endplate breakdown	AChR aggregation	+	–
	Increased endocytosis Increased lysosomal activity	AChR production – junctional type	+	–
	Loss of AChE	AChE production	+	+
		In embryo, development of junctional basal lamina	?	?

TTX, tetrodotoxin; N-CAM, neural cell adhesion molecule; Thy-1, an antigen originally isolated from thymus-derived lymphocytes; AChE, acetylcholinesterase; AChR, acetylcholine receptor.

isms, many of which in any case will be dealt with by other contributors. We shall confine ourselves to making a few general points before concentrating

TABLE 2 Some 'trophic' effects of muscle fibres on motor nerves

Part of motoneuron	*Effect of disconnection from muscle*	*Positive effects of muscle*	*Likely mediators*	
			Chemical factors	*Electrical activity*
Axon terminal	Sprouting	Cessation of growth-cone elongation	?	?
Cell body, axon and dendrites	Dendritic shrinkage	Dendritic growth	+	−
	Receptor loss			
	Axonal shrinkage	Axonal growth	+	−
	Chromatolysis including GAP production	Control of protein synthesis	+	−
	Motoneuron death in embryo	In embryo, promotes survival	+	−

GAP, growth-associated proteins.

on just one aspect of long-term communication between nerve and muscle, namely, the way in which contact with the muscle controls motor nerve growth. The restriction on the number of references allowed has meant that we have not always been able to do justice to our colleagues in the field.

(1) In the adult innervated muscle fibre useful nerve–muscle interactions occur only at the synapse

This is obviously true where effects are mediated by the transmission of electrical activity from nerve to muscle. Less obviously, it also seems to be true for retrograde effects exerted by muscle on nerve, where presumably the mediator must be a chemical. For example, a foreign motor nerve transplanted on an innervated muscle outside the endplate region clearly obtains none of the retrograde chemical signals available to synaptically attached motoneurons. This special feature of the synapse might arise because muscles manufacture and/or release chemicals only at synaptic sites or because uptake by non-synaptically placed axons not involved in transmitter release is defective. In contrast, the whole of a *denervated* muscle's surface seems to turn into one large, if unspecialized, postsynaptic structure, because fruitful preliminary interactions can occur anywhere on its surface. Interestingly, Libelius & Tågerud (1984) have observed a large increase in endocytotic, and presumably therefore also exocytotic, activity in denervated muscle fibres. This occurs particularly around the endplate region but to some extent also along the whole length of the fibres and could be a means for delivering and receiving chemical signals.

(2) Do intrinsic differences resistant to neural influence exist between classes of muscle fibre?

(a) Intrinsic differences in contractile properties

The exchange of motor nerves between mammalian muscles of very different contractile speeds can, even in the adult, cause an apparent total change in a muscle's phenotype (especially from fast to slow) (e.g. Pette et al 1985). For many years it was, therefore, assumed that muscle fibre differentiation during development depended entirely on the motor nerves. Nevertheless, recent evidence shows that the initial differentiation of muscle fibre types occurs in the absence of motor nerves (e.g. Miller & Stockdale 1986). Thus interaction between motoneurons and muscle fibres is not necessary for the latter's initial differentiation, although innervation is needed if slow-twitch fibres are to continue to produce adult slow myosin light chains (Rubinstein et al 1985).

The paradox has been resolved by recent experiments which show that in spite of major changes in virtually every aspect of a muscle's biochemistry and physiology brought about by various regimes of imposed activity, initially different fibres continue to express some myosin components of their original type (Hennig & Lømo 1987, Pette et al 1985). This could explain how it is that transplanted foreign motor nerves can themselves undergo some changes in properties, presumably under the influence of some residual non-convertible trait in the muscle (Lewis et al 1977).

(b) Intrinsic differences in surface recognition molecules

Motor nerves probably do not make synapses at random on muscles during development. Synaptic preferences are shown for muscle fibres of particular type and for fibres in particular places in the muscle.

(i) Type matching. It has now been demonstrated that motoneurons either place their initial synapses on particular types of muscle fibre (Gordon & Van Essen 1985) or preferentially lose synapses from muscle fibres of inappropriate type (Jones et al 1987) during the period of synapse elimination. In lower vertebrates the preference shown by different motoneurons for particular muscle fibre types is even stronger and is maintained in the adult (Sayers & Tonge 1982). In mammals the capacity to discriminate fibre types in the adult is, however, very limited (Ip & Vrbová 1983).

(ii) Place matching. In spite of the fairly widespread distribution of an individual motoneuron's branches within a muscle, recent work suggests that

this distribution is not made randomly. There are weak but definite correlations between the position of a motoneuron's cell soma in the motor pool and the position in the muscle where its branches are distributed (e.g. Brown & Booth 1983). These correlations probably arise because of some form of molecular labelling, for in neonatal animals regenerating motor axons can more or less re-establish their correct terminal distribution when the only clues available to them would seem to be a recognition mechanism with a spatial component in it (Hardman & Brown 1987). These spatial clues may persist in the adult (Wigston & Sanes 1985) but motor nerves for one reason or another seem to lose their ability to detect them (Hardman & Brown 1987).

(3) Are the actions of nerve-derived chemical mediators confined to the endplate region of muscle?

Both chemical factors released by or present on motor nerves, and electrical activity evoked by synaptic transmission, mediate the trophic actions which motor nerves have on muscle. In Table 1 a tentative assignment of roles is given.

In trying to rationalize how motor nerves control different muscle properties an obvious simplification appears to be that structures which may be situated at a considerable distance from the endplate (contractile and control proteins, metabolic proteins, surface membrane) are manipulated only by electrical activity, possibly mediated by the level of Ca^{2+} ions (Forrest et al 1981). Local events at the junction can, however, be directly influenced by nerve-borne chemicals, although here too activity plays an additional and sometimes an alternative part (Brenner et al 1987).

Caution is called for, however. Hennig (1987) has found that motor nerves can have an influence at a distance along muscle fibres that is not mediated by action potentials; a foreign nerve which has established synapses at ectopic sites is a far better inhibitor of reinnervation by the original nerve than is direct muscle stimulation by implanted electrodes!

Effects of skeletal muscle on motor nerves

In the embryo, contact with muscle is necessary for motoneuron survival and it is assumed, partly by analogy with nerve growth factor and its role in the survival and maintenance of certain neural crest derivatives, that muscle manufactures an equivalent 'retrophin' for motoneurons. Other articles in this volume will enlarge on this. We shall concentrate on the way synaptic contact in the adult controls the behaviour of the terminal branches of motoneurons.

Contrary to many previous reports (Barker & Ip 1966) it seems that mammalian motor terminals are probably very constant in appearance (Licht-

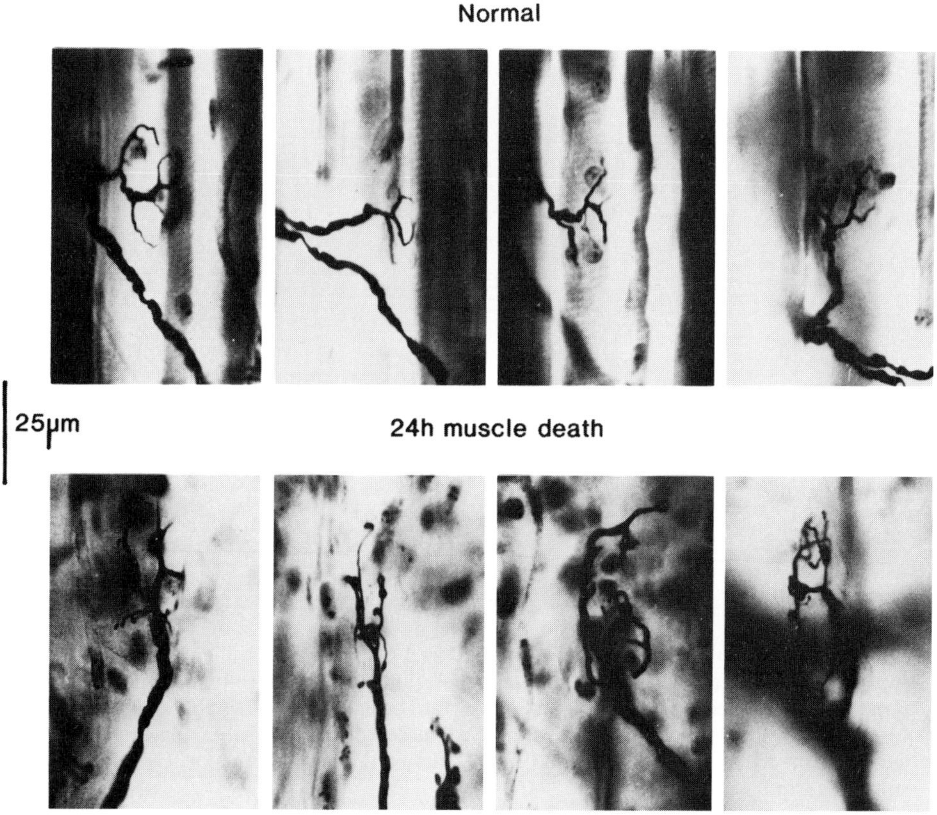

FIG. 1. Silver-stained motor nerve terminals in a whole-mount preparation of the mouse gluteus muscle. *Above*: from a normal muscle. *Below*: 24 hours after crushing the muscle fibres on either side of the endplate band.

man et al 1987) in a normal fully adult muscle. Growth however can be readily elicited. This can be done directly by axotomy (when the terminal end of the proximal stump sprouts) or indirectly by axotomy of neighbouring axons (collateral sprouting). It is generally held that such growth only starts after fast retrograde axonal transport of a message from the periphery to the cell body. This message triggers the cell's biochemical machinery and enables it to send new material rapidly to the periphery (Graftstein & McQuarrie 1978). We shall describe a simple experiment which demonstrates that communication with the cell body is not needed either for deciding that growth should occur or for the supply of new material required to initiate growth.

If muscle fibres under an endplate are killed by crushing them on either side of the endplate, the motor terminals, which are not directly damaged, sprout (Huang & Keynes 1983). This phenomenon can be used to chart the

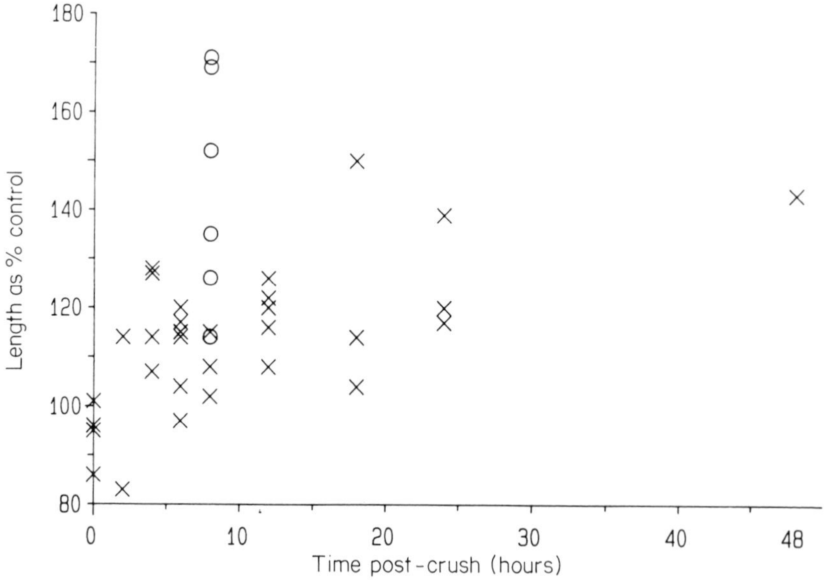

FIG. 2. Change in length with time of silver-stained motor nerve terminal branches, as a percentage of control, after muscle crush at time 0. Each point plots the mean total branch length of 17 terminals expressed as a percentage of the equivalent figure measured in 17 terminals in the contralateral gluteus. Each terminal's total branch length was obtained by projecting its image onto a digitizing tablet; the lengths of all individual branches occurring after the last node of Ranvier were measured and summed. ×, data after muscle crush only. ○, data after crushing the muscle nerve 3–4 hours after muscle crush.

early onset of nerve growth and to see if it still occurs after all connections with the cell body have been severed. The muscle under study was the gluteus maximus of the mouse whose sheet-like structure allows easy visualization of growing terminals with simple conventional histological methods. All operations were done under chloral hydrate anaesthesia with instruments sterilized in 70% alcohol.

Fig. 1 shows the appearance of normal silver-stained endplates and the obvious changes in morphology 24 hours after muscle crush. To quantify the changes the numbers and lengths of the terminal branches of endplates situated on dead muscle fibres were measured and compared with the equivalent figures from normal endplates from undamaged fibres in contralateral muscles or muscles from siblings. When this is done, growth can be seen to start some four hours after the operation (Fig. 2). Growth is seen not only as an increase in mean total terminal branch length but also as a rise in the number of branches (Fig. 3). Terminal branch length increases only slowly up to three days, when there is a considerable acceleration (Fig. 4).

Most significantly, the growth normally observed eight hours after crushing

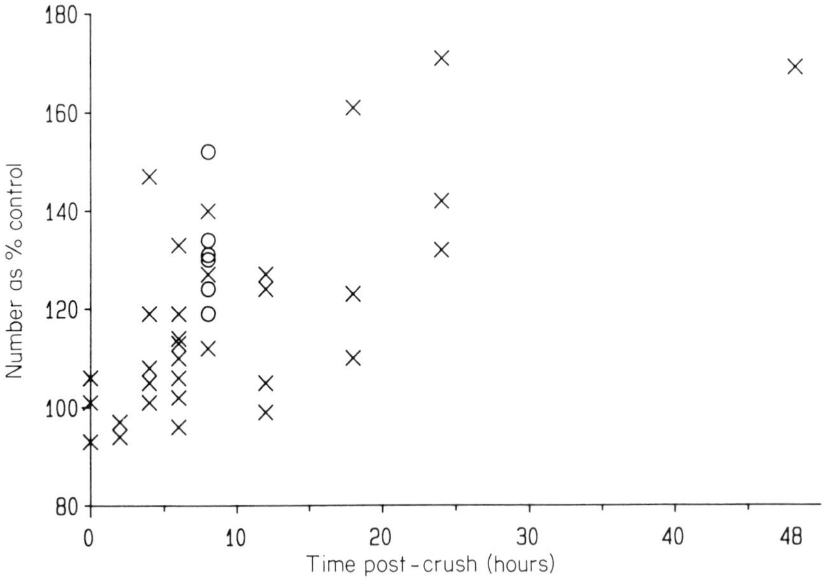

FIG. 3. Change in the number of nerve terminal branches, as a percentage of control, after muscle crush at time 0. Data from same junctions as in Fig. 2 and normalized in the same way. Symbols as in Fig. 2.

the muscle is still seen if the nerve to the muscle is crushed 3–4 hours after the initial muscle operation (Figs 2 and 3). This operation severs the link between the motoneuron somata and their terminals. The five or four hours that then had to elapse before the animal was killed and the muscle fixed for histology was short enough for the axons not to have degenerated. The 3–4 hour gap before the nerve was crushed was itself too short for communication between the terminals and the cell body. The reasoning goes like this: the distance from the gluteus muscle to the lumbar cord in our mice is between 25 and 30 mm. Fast anterograde axonal transport travels at 400 mm/day and retrograde transport at 300 mm/day. So the minimum transit time from muscle to cord and back again is 3.5 hours. Therefore in total only 0.5 hours could be available for the following three further processes needed to complete the chain of information coming from muscle to cord, namely degeneration of the muscle, uptake by the terminals of a message, and a response from the cell soma. In the case of the three hour nerve crushes, no time at all was available.

This experiment thus makes it highly unlikely that the initial growth requires a response from the motoneuron soma and new material to be transported by the fast axoplasmic route to the terminal. As Hoffman & Lasek (1980) have pointed out, the first elongation of microtubules and neurofilaments in growing nerves must occur on the basis of the normally transmitted supplies of precursor material, for these are transported so slowly that no

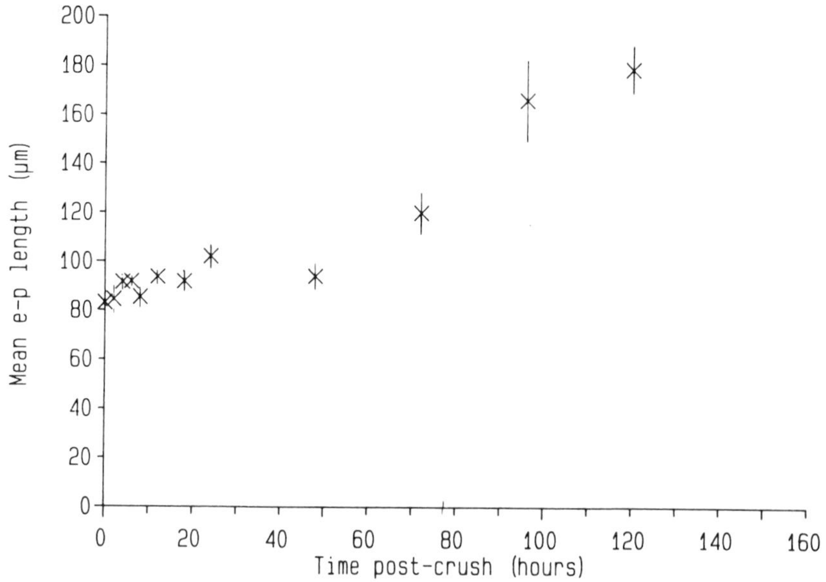

FIG. 4. Mean total branch lengths of motor nerve terminals following muscle crush at time 0. Terminal branch lengths increase with age. There is a highly significant linear relationship between total branch length and weight (E.R. Lunn & M.C. Brown, unpublished). We therefore reduced scatter in this graph by standardizing the data from each mouse to that of a 25 g mouse. Error bars are ± one standard error of the mean.

additional microtubule or neurofilament precursors would reach our muscle for eight days. The apparent acceleration of growth at about three days might be caused by the arrival of specially synthesized material carried by faster routes. It might alternatively be the result of a more favourable local environment developing in the muscle.

The nature of local control of motor nerve growth

At normal neuromuscular junctions there must be inhibitory mechanisms at work which prevent the usually transported filament and membrane precursors from being utilized for growth. One can be confident about this now because the experiment just described demonstrates that the material normally transported along axons *is* capable of supporting growth if the muscle at the endplate is destroyed. It must therefore be prevented from doing so at intact junctions.

Is there continual disassembly of a centrally manufactured cytoskeleton at nerve terminals?

Lasek & Black (1977) proposed that growth was inhibited because of activation of intracellular proteases by entry of Ca^{2+} ions into the presynaptic terminals. This caused a nibbling away at the advancing cytoskeleton which at that time was regarded as a continuous lattice of neurofilaments and microtubules assembled in the cell body and pushed forward into the axon from there. Vrbová and her colleagues have also favoured this view and have demonstrated that both calcium-chelating agents and protease inhibitors can, when applied to developing neuromuscular junctions, cause failure of redundant branches to retract (O'Brien et al 1984). Calcium ions normally enter nerve terminals when they are depolarized by the action potential. Additional entry might occur as a result of local accumulation in the synaptic cleft of potassium ions liberated by active muscle. These could further depolarize the axon and allow additional Ca^{2+} entry (Vrbová 1987).

Some observations already in the literature argue against this idea. Jansen & Van Essen (1975) found that neuromuscular junctions re-forming after phrenic nerve section in the rat had the same appearance whether they were made on normal muscle fibres which could contract or on fibres in which all transmission was prevented by the postsynaptic blocker α-bungarotoxin. Another point to consider is that afferent axons as well as motor axons stop growing when they reach their appropriate targets and they must be halted by other means. Likewise in culture, Kapfhammer & Raper (1987) have demonstrated that the meeting of certain pairs of neurites leads to rapid cessation of growth without synapse formation whereas other pairs of neurites will happily elongate in contact with one another. Furthermore, in partly denervated muscles terminal sprouts arise from intact endplates which are still activating their muscle fibre and so should still be subject to autolytic digestive processes triggered by potassium-induced calcium entry.

Finally, a continuously active autolytic mechanism such as that proposed would have to be of extraordinary precision to keep the architectures of motor nerve terminals as constant as they apparently are (Lichtman et al 1987).

Control of nerve growth by rate of assembly of components in the terminal

Evidence now exists for an alternative view of the cytoskeleton. On this view it does not consist of a continually advancing monolithic lattice which has to be eaten away at the terminals if axons are to maintain their shapes. Firstly, recent serial electron microscopy (Tsukita & Ishikawa 1981) has shown that neither microtubules nor neurofilaments are continuous polymers stretching from the cell body all the way down the axon to the terminal. Secondly, other studies (Nixon 1987) have shown that much of the cytoskeleton may be stationary in the axon with its component molecules having a very long half-life. Neurofilament precursors are probably transported along this stable framework, repairing it where necessary and, on reaching the terminal, are recirculated back to the cell body. Thirdly, it is also known that new tubulin monomers are added to the distal ends (which are the + ends) of microtubules during neurite growth in culture, rather than in the cell body (Bamburg et al 1986).

Taking the above observations into account it seems possible that growth control is mediated by highly specific recognition mechanisms in motor terminals rather than simple changes in ionic environment. These recognition mechanisms, in some way as yet unknown, control not degradation but synthesis. If, as a result of recognition, all cytoskeletal components ending in the nerve terminal were to become capped, no growth could occur and the observed stability of the presynaptic ending would be easily explained. Lack of appropriate recognition could lead to uncapping. Instant incorporation of normally transported growth materials could then occur. Failure of recognition would be very likely under the following conditions: (i) when a nerve is cut; (ii) when a muscle is killed; (iii) when a muscle is paralysed, for this leads to rapid endplate breakdown, at least in mice (Brown et al 1982); and (iv) in a partly denervated muscle, because in these muscles even innervated muscle fibres undergo the same changes in properties as do the denervated ones (Cangiano & Lutzemburger 1977) and might therefore temporarily lose their recognition molecules. All these are circumstances in which terminal sprouting occurs.

Conclusion

It is probably impossible to explain all the retrograde effects of muscle fibres on motoneurons by one mechanism. Positive trophic feedback seems to be needed as well as some sort of locally acting inhibition of growth. A case can be made, as has been argued here, that the inhibitory mechanism works by preventing the incorporation into the axon tip of normally transported growth components rather than by the enzymic degradation of a continually growing axon.

Acknowledgements

E.R. Lunn is a Medical Research Council Scholar. The Wellcome Trust generously supported the experiments described here.

References

Bamburg JR, Bray D, Chapman K 1986 Assembly of microtubules at the tip of growing axons. Nature (Lond) 321:788–790

Barker D, Ip MC 1966 Sprouting and degeneration of mammalian motor axons in normal and de-afferented skeletal muscle. Proc R Soc Lond B Biol Sci 163:538–554

Brenner HR, Lømo T, Williamson R 1987 Control of end-plate properties by neuro-

trophic effects and by muscle activity in rat. J Physiol (Lond) 388:367–381
Brown MC, Booth CM 1983 Postnatal development of the adult pattern of motor axon distribution in rat muscle. Nature (Lond) 304:741–742
Brown MC, Hopkins WG, Keynes RJ, White I 1982 A comparison of early morphological changes at denervated and paralysed endplates in fast and slow muscles of the mouse. Brain Res 248:382–386
Cangiano A, Lutzemburger L 1977 Partial denervation affects both denervated and innervated fibres in the mammalian skeletal muscles. Science (Wash DC) 196:542–545
Forrest JW, Mills RG, Bray JJ, Hubbard JI 1981 Calcium-dependent regulation of the membrane potential and extrajunctional acetylcholine receptors of rat skeletal muscle. Neuroscience 6:741–749
Gordon H, Van Essen DC 1985 Specific innervation of muscle fibre types in a developmentally polyinnervated muscle. Dev Biol 111:42–50
Grafstein B, McQuarrie IG 1978 Role of the cell body in axonal regeneration. In: Cotman CW (ed) Neuronal plasticity. Raven Press, New York, p 155–195
Hardman VJ, Brown MC 1987 Accuracy of re-innervation of rat intercostal muscles by their own segmental nerves. J Neurosci 7:1031–1036
Hennig R 1987 Late reinnervation of the rat soleus muscle is differentially suppressed by chronic stimulation and by ectopic innervation. Acta Physiol Scand 130:153–160
Hennig R, Lømo T 1987 Effects of chronic stimulation on the size and speed of long term denervated and innervated rat fast and slow skeletal muscles. Acta Physiol Scand 130:115–131
Hoffman PN, Lasek RJ 1980 Axonal transport of the cytoskeleton in regenerating motor neurons: constancy and change. Brain Res 202:317–333
Huang CLH, Keynes RJ 1983 Terminal sprouting of mouse motor nerves when the post-synaptic membrane degenerates. Brain Res 274:225–229
Ip MC, Vrbová G 1983 Re-innervation of the soleus muscle by its own or by an alien nerve. Neuroscience 10:1463–1469
Jansen JKS, Van Essen DC 1975 Reinnervation of rat skeletal muscle in the presence of α-bungarotoxin. J Physiol (Lond) 250:651–667
Jones SP, Ridge RMAP, Rowlerson A 1987 The non-selective innervation of muscle fibres and mixed composition of motor units in a muscle of neonatal rat. J Physiol (Lond) 386:377–394
Kapfhammer JP, Raper JA 1987 Collapse of growth cone structure on contact with specific neurites in culture. J Neurosci 7:201–212
Lasek RJ, Black MM 1977 How do axons stop growing? Some clues from the metabolism of the proteins in the slow component of axonal transport. In: Roberts S et al (eds) Mechanisms, regulation and special functions of protein synthesis. Elsevier Science Publishers, Amsterdam, p 161–169
Lewis DM, Bagust J, Webb SN, Westerman RA, Finol HJ 1977 Axon conduction velocity modified by re-innervation of mammalian muscle. Nature (Lond) 270:745–746
Libelius R, Tågerud S 1984 Uptake of horseradish peroxidase in denervated skeletal muscle occurs primarily at the endplate region. J Neurol Sci 66:273–281
Lichtman JW, Magrassi L, Purves D 1987 Visualization of neuromuscular junctions over periods of several months in living mice. J Neurosci 7:1215–1222
Miller JB, Stockdale FE 1986 Developmental regulation of the multiple myogenic cell lineages of the avian embryo. J Cell Biol 103:2197–2208
Nixon RA 1987 The axonal transport of cytoskeletal proteins: a reappraisal. In: Smith RS, Bisby MA (eds) Axonal transport. Alan R Liss, New York p 175–200

O'Brien RAD, Ostberg AJC, Vrbová G 1984 Protease inhibitors reduce the loss of nerve terminals induced by activity and calcium in developing rat soleus muscles *in vitro*. Neuroscience 12:637–646

Pette D, Heilig A, Klug G, Reichmann H, Seedorf U, Wiehrer W 1985 Alterations in phenotype expression of muscle by chronic nerve stimulation. Adv Exp Med Biol 182:169–178

Rubinstein NA, Lyons GE, Gamble B, Kelly A 1985 Control of myosin isozymes during myogenesis in the rat. Adv Exp Med Biol 182:141–153

Sayers H, Tonge DA 1982 Differences between foreign and original innervation of skeletal muscle in the frog. J Physiol (Lond) 330:57–68

Tsukita S, Ishikawa H 1981 The cytoskeleton in myelinated axons: serial section study. Biomed Res 2:424–437

Vrbová G 1987 Reorganisation of nerve-muscle synapses during development. In: Winlow W, McCrohan CR (eds) Growth and plasticity of neural connections. Manchester University Press, Manchester, p 22–35

Wigston DJ, Sanes JR 1985 Selective reinnervation of intercostal muscle transplanted from different segmental levels to a common site. J Neurosci 5:1208–1221

DISCUSSION

Thesleff: When you say, in Table 1, that there are no chemical influences from the motor nerve on contractile activity, this implies that blocking muscle activity by curare or denervating the muscle (with either a long or a short stump) should have the same effect; yet that doesn't appear always to be the case.

Brown: It is known that the rate of onset of appearance of denervation changes is delayed if the motor nerve is cut some distance from rather than near to the muscle (Luco & Eyzaguirre 1955). Terje Lømo's view (Cangiano et al 1978) of this 'short and long stump' experiment is that the difference is not the result of chemical trophic agents within the nerve that are continuing to be liberated for longer if a longer stump is left, but has something to do with the fact that when nerves degenerate there is 'inflammation': for example, the cut sciatic nerve becomes loaded with inflammatory cells, including monocytes and polymorphs. These cells accelerate the changes that inactivity will on its own eventually bring about. The influx of inflammatory cells into a denervated muscle is delayed if the nerve is cut some distance from (long stump) rather than near to (short stump) the muscle.

Vrbová: Jones and I (1974) did the experiment of putting a piece of degenerating nerve on an intact muscle, and showed that there are changes in ACh sensitivity that resemble those after denervation, in that the ACh receptors appeared in the part of the muscle under the degenerating nerve. This is the basis of Dr Lømo's interpretation of the 'short and long stump' experiment. But I still think one has to explain it in chemical terms. Peng et al (1981) induced ACh receptors by latex beads, and these 'contact' phenomena are probably important in the regulation of membrane proteins in an activity-independent manner.

Brown: I said that they were—for here we are talking about the neuromuscular junction region of the muscle, and not about parts of the muscle situated some distance from it!

Eisenberg: The charge on the beads creates a leakiness for current through channels underneath them, and that is the link between excitation and contraction. The bead causes current to flow through the membrane.

Vrbová: The different currents across the membrane do not induce impulse activity.

Eisenberg: A protein cannot know what made a current flow through it, so it is really the same thing!

Vrbová: I agree that the mechanism by which the change is induced is the same; that is, the current flow caused by contact phenomena could activate or inhibit the same intracellular messenger as impulse activity, or lack of it.

Crowder: Beads affect receptors in exactly the opposite way to activity. If ion flow into the neuron is being induced, in the normal state that would down-regulate ACh receptors, but here the beads are up-regulating them.

Brown: I was thinking of Hennig's (1987) experiments, which suggested that a motor nerve has an influence along a muscle fibre that is distinct from any contractile activity elicited. In this experiment a foreign nerve is implanted on a denervated muscle and it makes endplates. You then allow the original nerve to grow back and see how many successful endplates it forms. You compare this degree of reinnervation with that achieved when there are no ectopic synapses but the denervated muscle is kept active by direct electrical stimulation throughout the period of reinnervation. The result was that the ectopic nerve was much more effective than was direct stimulation in preventing the original nerve from re-establishing permanent synapses, although its synapses are a long distance form the motor endplate. That is a surprising observation.

Lowrie: That result could also arise through failure adequately to mimic the activity of the nerve by direct electrical stimulation, either in its pattern or in its degree of spread.

Brown: So that experimental result is not a good argument for a specific chemical released at the synapses and working at long range?

Lowrie: Not on its own.

Thesleff: Speaking of the control of the neuron by the muscle, we have been studying endocytotic activity in skeletal muscle. After denervation a lot of endocytotic activity is seen, but only in the endplate region (Libelius & Tågerud 1984). Because endocytosis and exocytosis are closely coupled, one would expect that the endplate would also have exocytotic activity, although we cannot demonstrate this. My feeling is therefore that the endplate region of the nerve might secrete a neurite- or growth-promoting factor. You are suggesting that the innervated muscle fibre normally secretes a neurite-blocking factor and, when the muscle is crushed, the nerve starts to grow. I believe that it is the denervated muscle which secretes a growth-promoting factor.

Brown: I said that at least two factors are required at the endplate—a growth-promoting factor which is presumably taken up by the neuron and enables its normal metabolic characteristics to exist (its dendritic tree, receptors for transmitter on dendrites, and so on), and, secondly, something that stops the neuron growing. You need two factors, as a minimum.

Sanes: Perhaps the nerve, at some point, is ready for the putative growth-inhibiting factor. When neuritic processes grow in culture, they extend past many cells with which they make functional synapses. At a certain point the growth cone may be able to interpret such a signal, but not before that. So it is, in some sense, a two-way conversation.

My other point has to do with calcium itself. Kater's group has shown that a specific range of intracellular calcium concentrations appears to regulate process outgrowth *in vitro* (Mattson & Kater 1987). The calcium concentrations that promote elongation and growth cone motility are different. Therefore, it seems unlikely that inhibiting growth is a simple matter of activating a calcium-dependent protease.

Ribchester: I agree very much with the intuitive notion that there should be an inhibitory factor in addition to the growth-promoting factor to stabilize the nerve terminal at the synapse, but I am not sure that Michael Brown's experiment demonstrates the existence of that inhibitory factor. You raised the possibility earlier in this discussion that the difference between the effects of blocking activity and denervation on the development of post-denervation changes could be due to an inflammatory response. Could it not be something similar in your own experiment, and that when you injure the muscle you stimulate a non-specific inflammatory response and the sprouting stimulus is derived from that response?

Brown: I was suggesting that our experiments show that the distal end of the normal, non-growing neuron has material within it that it is capable of using to grow. It doesn't normally do so; the terminals remain static. Therefore growth (incorporation of these existing components) must be being stopped.

Henderson: You interpret your results in terms of a positive signal for axonal growth and sprouting and a 'stop' signal, but it seems to me that one could explain them in terms of a positive signal for axonal growth which is switched off once contact is made, and a different positive signal for sprouting. In adult sensory neurons, it has been shown (Diamond et al 1987) that endogenous NGF can play a role in promoting sprouting of collaterals but not in the regrowth of nociceptive axons. Perhaps you could also explain your results without invoking a real 'stop' signal, but just by lack of a positive sprouting signal.

Vrbová: In your muscle crush experiment, Dr Brown, what happens to the original contacts in the synapses where you see elongated branches coming out? Is the original contact broken or does it remain and just side-branches grow out? What happens to the endplate? If the crush injury causes damage to the axon terminal, the mechanism of tubulin assembly could account for the growth. And secondly, in a normal muscle, about 4% of endplates always show some sort of terminal sprouts, and that cannot be a stable situation; so this is a question of sampling. You may look at one endplate that is stable, but 4% may not be.

Buller: David Barker used to say that there was flux at all neuromuscular junctions, and I wondered whether your micrographs show the stability of terminals, rather than the stability of the clefts into which they are forced to run, if they are re-growing. You have shown a fixed disposition of the terminals; you have not shown that terminals don't re-form within that space.

Brown: Lichtman's work in living mice is more elegant in some ways than using fixed material; he can look at the same terminal over and over again in the living muscle and it always appears the same (Lichtman et al 1987). If it was undergoing change, you would expect to see some retraction. The fact that some terminals can be extremely constant implies that the control of growth can be very strong. In a normal mouse living its daily life—male mice fight a great deal—I would be surprised if you didn't get damage and therefore sprouting for that reason.

On Gerta Vrbová's question, in our 24h picture (Fig.1) the growing endplates don't look like normal ones with extra bits added on to the ends of existing branches; they look deformed. The initial growth is almost entirely restricted to extra branches and doesn't move out of the endplate area. At three days, when the nerve starts to grow fast, we see one or two branches which shoot away up the muscle fibre at a much faster rate. After 5–7 days the muscle fibre has repaired itself and the long branches retract right back into the original endplate area. So you revert to a more normal overall length of endplate. It is in the same place, rather as Jack McMahan showed for the reinnervated basal lamina in frogs (Anglister & McMahan 1984). But the nerve terminal now has many more branches, as if, because the cell body has been switched on to make extra material, it has to incorporate it somewhere; so it develops a fairly massive structure within limited confines. We haven't followed it long enough to see whether the normal structure is eventually re-formed.

Van Essen: On the stability of the normal endplate, Lichtman's stains do not stain the entire presynaptic terminal; they stain a central core. (One of their dyes stains mitochondria, for example.) It still remains possible (even plausible) that the margins of the nerve terminals are sending out small sprouts and that there is some dynamic interaction of the type that Barker & Ip originally proposed (1966). Lichtman's evidence shows that whatever oscillations there are seem to occur around a very stable general position, but it doesn't rule out some sprouting.

Carlson: Is there any information on the ageing neuromuscular junction that would shed any light on this? The nerve terminals seem to become unstable in older laboratory rodents. Is this due to changes at the muscular or the neural side of the junction?

Brown: Will Hopkins, looking at silver-stained nerve terminals, showed that while mice are growing and their muscles are getting bigger the endplates become more complicated and more branched. During adult life the endplates remain remarkably constant. In old age (some 400 days in mice) one begins to see changes in the shape of terminals; presumably some motoneurons die, and sprouting occurs.

Carlson: It is interesting that when the ageing animal is in theory going downhill, neuronal sprouting occurs. Is this due to a defect in the muscle fibres?

Brown: I don't think anybody knows whether the muscles or nerves are the culprits. A proportion of neurons used to be thought to die gradually throughout adult life, well after the embryonic period of cell death. But whether this has really been established in normal brains, I don't know.

Vrbová: To be optimistic, motoneurons don't seem to die in old age! So I think it is at the muscle end, not the motoneurons.

Pette: Dr Brown addressed the question as to whether muscle fibres can be truly transformed. Considering the data published so far (see Pette & Vrbová 1985), I doubt that a complete fibre transformation has been achieved with the use of chronic stimulation. According to our own results, the transformation process is, in addition, species specific. For example, stimulation-induced fast-to-slow transitions are more pronounced in the rabbit than in the rat.

In collaboration with Terje Lømo, we were able to show that direct stimulation of the slow rat soleus muscle with a fast motoneuron-like stimulus pattern resulted in a transformation of the type I fibres into type IIC fibres with pronounced increases in parvalbumin and sarcoplasmic reticulum Ca^{2+}-ATPase content, as well as specific changes in metabolic enzyme activities. However, the transformation process of the fast rat extensor digitorum longus muscle (EDL) was much less pronounced when it was subjected to a slow motoneuron-like stimulus pattern (Gundersen et al 1988). When compared to the fast-to-slow transformation of rabbit EDL, the resistance to transformation of rat EDL was especially pronounced. This resistance had previously been observed by Kwong & Vrbová (1981) who used indirect stimulation. Therefore, it is clear that species-specific differences have to be taken into account.

In summary, it is my impression that a complete transformation of fast into slow or, conversely, slow into fast fibres has not been observed with chronic stimulation. This, of course, may be due to the use of inappropriate stimulus patterns. However, it may also be that fast and slow fibre populations possess certain adaptive and species-specific ranges beyond which transformation cannot be achieved. This notion could also explain the commonly observed overlap of many properties in normal fast and slow fibre populations.

Mudge: Michael Brown has asked me to comment on the role of calcitonin gene-related peptide (CGRP) at the neuromuscular junction. This peptide is found only in a subpopulation of motoneurons in both the chick and rat. We have mapped the distribution of CGRP-positive motoneurons in the chick lumbar spinal cord (Tessier-Lavigne & Mudge 1987 and in preparation) and find that there are some motoneuron pools which contain only CGRP-positive neurons while other pools have only CGRP-negative neurons and some pools are mixed. One might expect that there is another, as yet unknown peptide in the CGRP-negative neurons. We could not find any correlation between CGRP-positive and CGRP-negative pools with known functional differences between muscles.

As I said earlier (p 71), CGRP causes a massive increase in cyclic AMP in muscle. In trying to fit this into a picture of trophic effects by CGRP on muscle, one obvious target is phosphorylation of the AChR, which leads to desensitization of the receptors. With respect to regulating the levels of AChR on muscle, you also have to consider that muscle responds well to β-adrenergic stimulation by increasing cyclic AMP. So, if the action of cyclic AMP in increasing the synthesis of AChR is important, it seems that you would have to confine the cyclic AMP stimulation of synthesis to the junctional region. You could think of ways of doing this, just as you could think of ways of confining the action of Ca^{2+} in down-regulating AChR synthesis to the extrajunctional region. However, it seems to me that cyclic AMP stimulation of AChR is unlikely to be the main action of CGRP on muscle; perhaps CGRP couples glycogenolysis (which is also stimulated by cyclic AMP) to activity. In any case, we first need to know if and when CGRP is released from motor nerve terminals.

References

Anglister L, McMahan UJ 1984 Extracellular matrix components involved in neuromuscular transmission and regeneration. In: Basement membranes and cell movement. Pitman, London (Ciba Found Symp 108) p 163–178

Barker D, Ip MC 1966 Sprouting and degeneration of mammalian motor axons in normal and de-afferentated skeletal muscle. Proc R Soc Lond B Biol Sci 163:538–554

Cangiano A, Lømo T, Lutzemberger L 1978 Do products of nerve degeneration contribute to the response of muscle to denervation? Neurosci Lett 10 suppl S24

Diamond J, Coughlin M, Macintyre L, Holmes M, Visheau B 1987 Evidence that endogenous β nerve growth factor is responsible for the collateral sprouting, but not the regeneration, of nociceptive axons in adult rats. Proc Natl Acad Sci USA 84:6596–6600

Gundersen K, Leberer E, Lømo T, Pette D, Staron RS 1988 Fibre types, calcium-sequestering proteins and metabolic enzymes in denervated and chronically stimulated muscles of the rat. J Physiol (Lond) 398:177–189

Hennig R 1987 Late reinnervation of the rat soleus muscle is differentially suppressed by chronic stimulation and by ectopic innervation. Acta Physiol Scand 130:153–160

Jones R, Vrbová G 1974 Two factors responsible for denervation hypersensitivity. J Physiol (Lond) 236:517–538
Kwong WH, Vrbová G 1981 Effects of low-frequency electrical stimulation on fast and slow muscles of the rat. Pfluegers Arch Eur J Physiol 391:200–207
Libelius R, Tågerud S 1984 Uptake of horseradish peroxidase in denervated skeletal muscle occurs primarily at the endplate region. J Neurol Sci 66:273–281
Lichtman JW, Magrassi L, Purves D 1987 Visualization of neuromuscular junctions over periods of several months in living mice. J Neurosci 7:1215–1222
Luco JV, Eyzaguirre C 1955 Fibrillation and hypersensitivity to acetycholine in denervated muscle: effect of length of degenerating nerve fibres. J Neurophysiol 18:65–73
Mattson MP, Kater SB 1987 Calcium regulation of neurite elongation and growth cone motility. J Neurosci 7:4034–4043
Peng HB, Ping-Chin Cheng, Luther PW 1981 Formation of ACh receptors induced by positively charged latex beads. Nature (Lond) 292:831–834
Pette D, Vrbová G 1985 Neural control of phenotypic expression in mammalian muscle fibers. Muscle & Nerve 8:676–689
Tessier-Lavigne M, Mudge AW 1987 Soc Neurosci Abstr no. 444.6

Development of neuromuscular connections: guidance of motoneuron axons to muscles in the embryonic chick hindlimb

Cynthia Lance-Jones

Department of Neurobiology, Anatomy and Cell Science and Center for Neuroscience, University of Pittsburgh, School of Medicine, Pittsburgh, Pennsylvania 15261, USA

Abstract. A striking feature of the outgrowth of motor axons from the embryonic chick spinal cord is the accuracy with which they grow to correct target muscles in the limb. Studies involving the surgical manipulation of the early neural tube or limb bud have led to the conclusion that motoneurons are specified or acquire distinctive properties early in development that enable them to detect and respond to specific cues in the periphery. We have recently carried out studies to identify the cells or tissues responsible for producing these cues. Major contributors to the outgrowing axon's environment are the lumbosacral (LS) somites which give rise to limb muscle cells and the LS somatopleural mesoderm which gives rise to limb connective tissues. To determine whether the LS somites contain specific guidance information at stages when motoneurons are known to be specified, motor projections were assessed in chick embryos after early somitic tissue reversals. The finding of normal projection patterns suggests that the LS somites do not contain such information at early stages. To assess the role of the LS somatopleural mesoderm in specific axon guidance, we examined motor projections in embryos where the LS somatopleural mesoderm had been shifted anteriorly before limb bud formation. In these embryos the limb developed in an anteriorly shifted position. LS motoneuron axons entered the posterior part of the limb but then extended anteriorly along aberrant paths to innervate appropriate muscles. These observations suggest that it is the precursor of limb connective tissues that is the source of specific axon guidance cues.

1988 Plasticity of the neuromuscular system. Wiley, Chichester (Ciba Foundation Symposium 138) p 97–115

In the development of the neuromuscular system one of the first tasks that a motoneuron faces is to find its way to an appropriate target region. In order to understand how this is accomplished, several investigators have studied the development of motoneuron projections to the chick limb. One approach taken has been to look closely at the course of motoneuron axons as they grow into the periphery. Another has been to assess the behaviour of motoneurons after altering the spatial relationships between the spinal cord and limb. The results of both types of studies suggest that limb-innervating motoneurons project quite accurately to correct target regions from the onset of axon outgrowth and that they do so by responding to specific environmental guidance cues. Similar conclusions have been drawn most recently from studies of axon outgrowth in a diversity of other neuronal systems (for review see Westerfield & Eisen 1988), thus suggesting that directed outgrowth and environmental cue recognition may be common features of the early stages in the development of precise neuronal connections.

An important step toward gaining an understanding of these processes is to identify the cellular origin of the environmental cues. The chick limb provides a unique system in which to address this question, for limb target tissue has been demonstrated to be of dual origin. Muscle cells are derived from the somites while limb connective tissues are derived from the somatopleural layer of the lateral plate. Since each component of limb target tissue is derived from a spatially discrete precursor population, one can separately manipulate them. Motoneuron connectivity patterns can then be compared to normal patterns in order to determine whether either of the precursor populations might be the primary source of specific environment guidance cues in the limb. We have used this approach to study motoneuron axon guidance to targets within the chick hindlimb. Specifically, we have focused on the guidance of motoneurons to different muscle targets on the anteroposterior axis of the embryonic thigh. This report describes the evidence for directed outgrowth and specific recognition on this axis and reviews ongoing studies on the roles of somitic and somatopleural tissues in axon pathway choice.

Evidence for directed outgrowth and specific cue recognition

Horseradish peroxidase labelling techniques have been used to map the location of motoneurons projecting to hindlimb muscles in 10–12-day chick embryos (stage 36–38 of Hamburger & Hamilton 1951) and in hatchlings (Landmesser 1978a, Hollyday 1980). Each individual hindlimb muscle is innervated by a discrete cluster or pool of motoneurons that occupies a characteristic position within the lateral motor column of the lumbosacral (LS) cord (Fig. 1). For thigh muscles each pool extends through two to four of the eight lumbosacral cord segments, there being a general topographical correspondence between segmental motoneuron pool position and anteroposterior target position. Retrograde horseradish peroxidase (HRP) labelling studies (Landmesser 1978b) have also shown that similar patterns exist at seven days (stage 30) of development, which is before the peak period of naturally occurring motoneuron cell death (Hamburger 1975). This conclusion was extended to even earlier stages of development through the use of

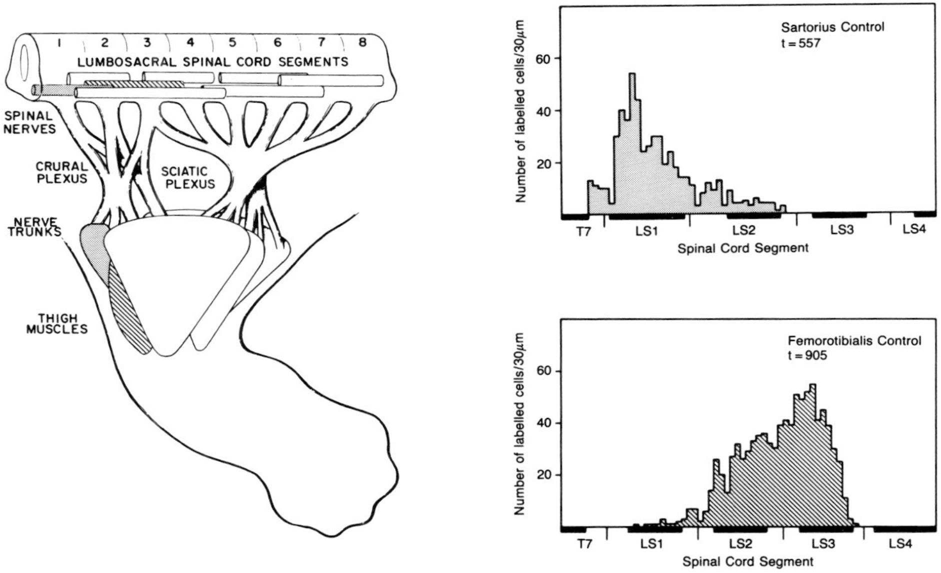

FIG. 1. Motoneuron pools to the chick hindlimb. The schematic on the left illustrates the general topographical relationship between motoneuron pool position within the LS spinal cord and target position within the thigh. The sartorius muscle (stippled) is the most anterior muscle of the thigh and is innervated by the most anterior LS motoneuron pool. The femorotibialis muscle (cross-hatched) is a slightly more posterior muscle and is innervated by a slightly more posteriorly positioned pool. Each of these muscles and most other anterior thigh muscles are innervated by muscle nerves originating from the crural or anterior limb plexus. This plexus is formed by spinal nerves LS 1–3. The sciatic or posterior limb plexus is formed by spinal nerves LS 3–8. Its nerve trunks and muscle nerves project to posterior thigh muscles as well as more distal limb regions. Representative histograms of the distribution of labelled cells following HRP injections of stage 35–36 sartorius and femorotibialis muscles are shown at right (modified from Lance-Jones 1988b). Most sartorius motoneurons lie in LS 1; most femorotibialis motoneurons in LS 2 and 3.

anterograde HRP labelling (Lance-Jones & Landmesser 1981a, Tosney & Landmesser 1985). After injecting a discrete cord region with HRP, one can trace the peripheral course of a small population of labelled motoneurons originating from an identified cord segment. At all stages of axon growth to the target region (3½–7 days, stage 24–30) segmental projection patterns were found to be like the mature patterns.

One cannot conclude that projection errors are never made. For example, the studies described do not allow one to discern errors made by axons at pool boundaries. In addition, some errors have been reported prior to the cell death period in a few different vertebrate limb systems (see for example Lamb 1979). However, the important point to be made is that most chick hindlimb innervating motoneurons make no major projection errors and appear to be precisely guided to a specific target region from the onset of outgrowth (see also Landmesser 1988).

Do motoneuron axons actively recognize specific target regions or guidance cues that will lead them there? Three lines of evidence have indicated that they do. The first evidence came from experiments in which 3–4 anterior LS cord segments had been reversed about the anteroposterior axis at 2½ days (stage 15), before axon outgrowth (Lance-Jones & Landmesser 1980). Motoneurons consistently innervated the correct targets and did so by altering their position within the plexus region at the base of the limb. Second, anterior shifts of the limb bud (Lance-Jones & Landmesser 1981b) suggested that motoneurons could also correct for spatial shifts by altering their course within the limb proper. Following this manipulation, anterior LS motoneurons grew initially into posterior limb regions but then took totally novel or aberrant paths to the correct anterior targets. Last, tracing of nerve pathways in normal embryos indicated that motoneuron axons do not project in a tightly ordered array as they might if they were being precisely but passively guided. Both anterograde (Tosney & Landmesser 1985) and retrograde (Lance-Jones & Landmesser 1981a) HRP labelling studies indicate that motoneuron axons coursing between two specific points can take diverse routes. It was concluded from these four studies that subpopulations of limb-innervating motoneurons must acquire distinctive properties, perhaps specific surface labels, before axon outgrowth. During outgrowth, these prelabelled or specified motoneurons appear to detect and respond to specific environmental guidance cues which are located at the base of the limb and perhaps also in the limb proper.

The somites as a source of guidance

What is the origin or source of specific environmental guidance cues? The first possibility that I chose to consider was that the somites or their derivatives provide such cues. By transplanting groups of quail somites into chick embryos prior to limb formation and later mapping the distribution of quail cells, Chevallier and Christ and their colleagues (see Chevallier 1979, Jacob et al 1979) had shown that the somites adjacent to the prospective limb region give rise to all limb muscle cells. Muscle cell precursors migrate into the developing limb bud beginning at 2½ days (stage 16), before motoneuron axons enter the limb (Jacob et al 1979). Perhaps these somitic cells bear distinctive labels prior to their migration. Remembering from the cord reversal experiments that subpopulations of LS motoneurons are specified by this early stage, a simple model one might propose is that motoneurons and muscle cell precursors acquire matching labels as a result of an early shared segmental position. Much as Sperry (1963) proposed in his chemoaffinity model, motoneurons from a single cord segment might then recognize and innervate muscle cells from a corresponding segmental level. Alternatively, to chose the correct target regions, motoneurons might follow a trail left by migrating muscle cells.

There is reason to think this idea likely. First, the general correspondence between anteroposterior motor pool position and target position in the limb (see Fig. 1) is certainly compatible with the idea of matching segmental labels. Second, a study of the reinnervation of adult mammalian trunk muscles suggests that target selectivity may be based on a positional quality shared by neurons and their targets (Wigston & Sanes 1982). To test this in the embryonic chick hindlimb, one needs to know first whether cells that populate a specific muscle arise from somites at a discrete position along the neuroaxis. Studies in which single quail somites have been transplanted into chick embryos indicate that this is true in the brachial region (Beresford 1983). Recent studies carried out in our laboratory indicate that it is also true in the LS region (Lance-Jones 1988a).

As Beresford (1983) had done for the wing, I mapped the somitic level of origin of individual limb muscles using quail/chick grafting techniques. Single somites or equivalent lengths of unsegmented somitic tissue from LS axial levels were removed from 2½–3 day (stages 15–18) quail embryos and transplanted to an equivalent position in similar-staged chick hosts. At 6–8 days (stages 28–34), when muscle cell migration is over and specific muscle primordia can be readily recognized, paraffin sections of chimeras were examined and limb muscles scored for the presence of quail cells using a quail cell-specific antiserum (Lance-Jones & Lagenaur 1987).

Quail cells from a specific somitic or axial level consistently made major contributions to a discrete subset of limb muscles (Fig. 2). In thigh regions, the position of this subset corresponded to the axial level of the transplant. That is, as one mapped somitic contributions from somitic level 26 to somitic level 33 (those levels which yielded limb muscle labelling), progressively more posterior and posterodorsal muscles were labelled (see Figs. 2 and 3). In addition, the pattern of somitic contribution to thigh as well as shank muscles

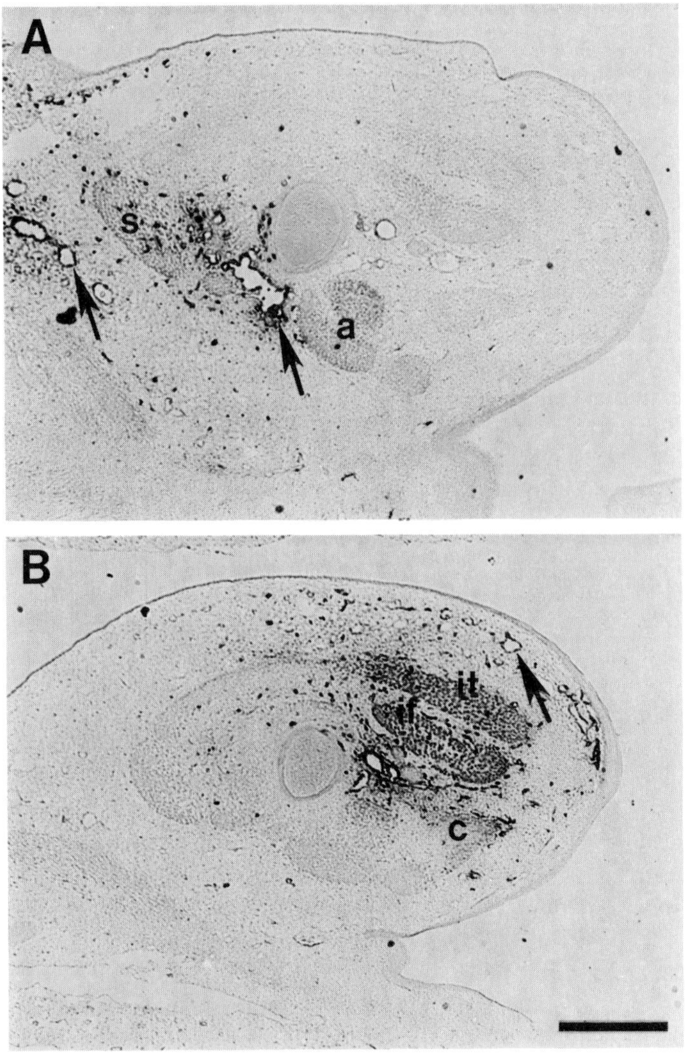

FIG. 2. Photomicrographs of thigh cross-sections from stage 30 quail/chick somitic mesoderm chimeras. The sections have been stained with a quail-specific antiserum (Lance-Jones & Lagenaur 1987). A peroxidase-coupled second antibody has been used to visualize quail-derived tissues. (A) From a chimera that had received a transplant of an anterior LS quail somite (somite 27) at stage 15. Only anterior thigh muscles such as the sartorius (s) and the adductor (a) show substantial quail cell labelling. (B) From a chimera that had received a transplant of quail somitic tissue from a posterior LS level (somitic mesoderm segment 32) at stage 17. Posterior thigh muscles such as the iliofibularis (if), the posterior iliotibialis (it) and the caudilioflexorius (c) show heavy quail cell labelling. The extremely darkly stained tissues in both sections (arrows) are quail-derived blood vessels. Calibration bar, 500 μm.

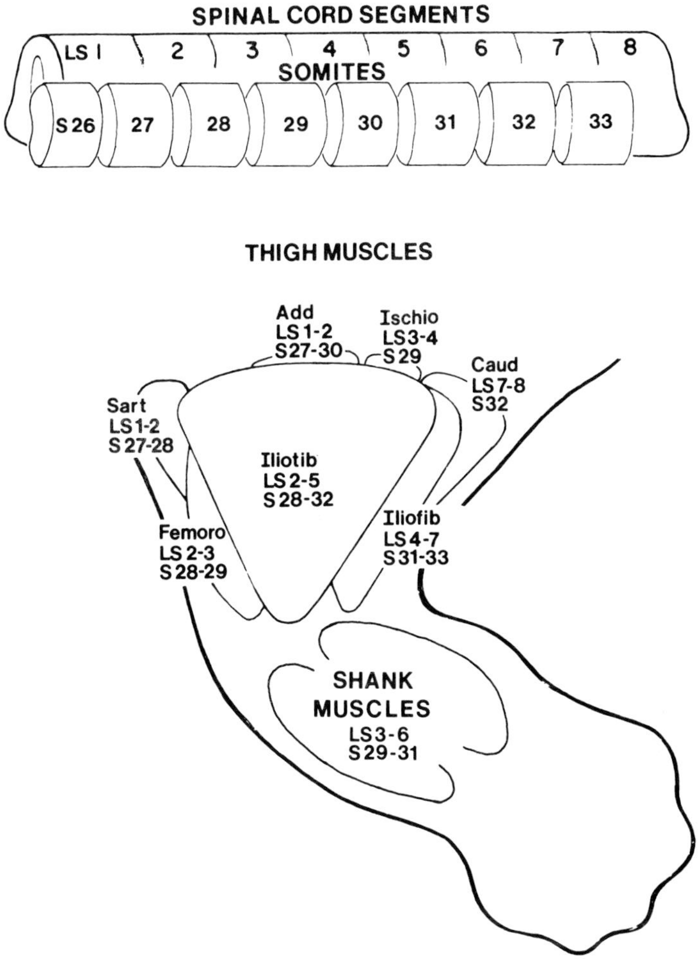

FIG. 3. Somitic levels of origin and segmental motor innervation of chick hindlimb muscles. Each somite (S) that contributed to limb musculature in quail/chick chimeras is identified by axial level at top. Each is positioned along the cord according to the vertebral level that it also contributed to. For example, somite 27 is positioned adjacent to LS1–2 because it normally contributed to vertebra LS1 and 2 as well as to limb musculature. For selected thigh muscles and the shank muscle complex, major contributing somites and the major segments providing motor innervation are given. Most muscles received major somitic contributions from those somites adjacent and slightly posterior to their cord segments of innervation (see, for example, the sartorius and iliofibularis). For a few thigh muscles, somitic levels of origin were quite broad when compared to segmental innervation levels (see, for example, the iliotibialis). However, differences between quail and chick muscle migration patterns or the fact that unsegmented somitic mesoderm was sometimes transplanted may account for these exceptions (see Lance-Jones 1988a). *Abbreviations:* Sart, sartorius; Femoro, femorotibialis; Iliotib, iliotibialis; Iliofib, iliofibularis; Caud, caudilioflexorius; Add, adductor; Ischio, ischioflexorius.

generally paralleled innervation patterns. This can be seen in Fig. 3, where the major somitic derivation and motor innervation patterns for selected muscles are shown. This finding is compatible with the hypothesis that motoneurons recognize muscle cells or somitic tissue from a corresponding axial level.

To directly test this hypothesis, one would like to alter a muscle's somitic level of origin by manipulating somites and then ask whether motor innervation patterns are correspondingly altered. However, this requires ascertaining first that the manipulation indeed alters the muscle's somitic level of origin. To do this, I asked how somitic cells behave when somitic tissue is positioned in a new axial level. Specifically, single segments of quail somitic tissues were transplanted to a site in a chick 2–3 segments away from their site of origin. In all of these heterotopic transplants, the limbs appeared morphologically normal and quail cells contributed to muscles in accord with their new position in the host, not with their original position in the donor. This finding suggests that LS somitic cells are not predetermined to follow local cues in the hindlimb or to migrate to a specific muscle region.

The above results indicated that a muscle's somitic origin could be altered by moving somites prior to muscle cell migration and thus provided a model in which to test the hypothesis that somitic cells are the source of guidance cues (Lance-Jones 1988b). At stage 15, 3–4 segments of somitic tissue (plus overlying ectoderm) adjacent to prospective anterior limb regions (somitic levels 26–29) were removed from one side of a chick embryo and positioned in reversed order on the opposite side of the same embryo (Fig. 4). This operation is comparable, with respect to size and axial level, to that performed in the cord reversal experiments where axons projected only to correct targets. Since the results of heterotopic somite transplants indicate that somitic contributions to anterior limb muscles should be reversed, the finding of reversed motoneuron projection patterns would suggest that the motoneuron axons were responding to somitically derived guidance cues.

In fact the opposite was found. Retrograde HRP labelling studies of 7–11 day (stage 30–37) somite reversal embryos indicated that the positions of motor pools to anterior thigh muscles were quite normal. Histograms of the distribution of labelled motoneurons following injections of a sartorius and a femorotibialis muscle in two stage 36 somite reversal embryos are shown in Fig. 4. The pools to both of these muscles are spatially discrete within anterior LS cord segments and each pool occupies an anteroposterior position which is similar to normal (compare to Fig. 1). The overall normality of pools in somite reversal embryos suggested that guidance cues were not affected by the operation. To verify this conclusion, I assessed projection patterns in a few embryos in which both anterior LS somites and anterior LS cord segments were reversed. Thus, I combined the cord reversal and somite reversal paradigms. In these embryos, like the cord reversal experiments described

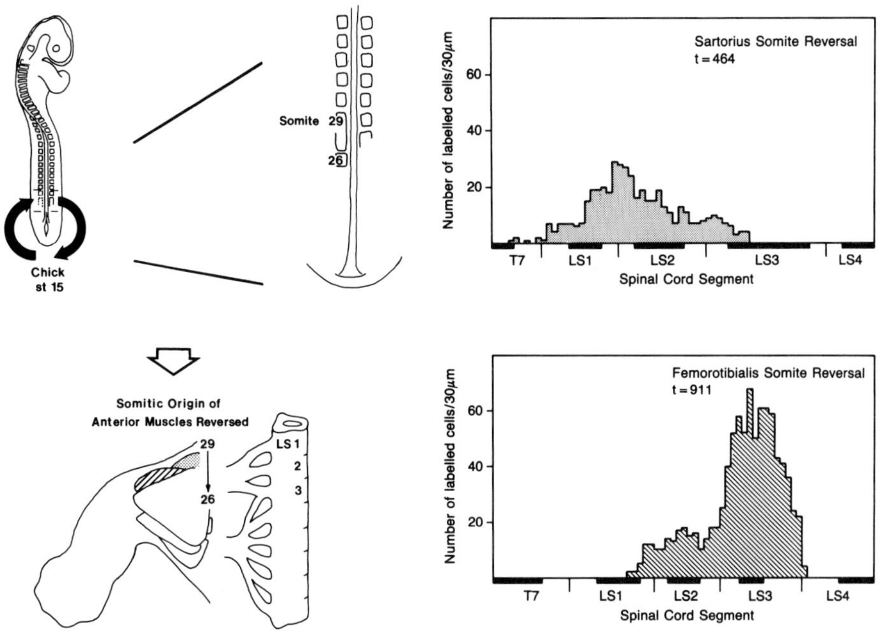

FIG. 4. The effect of LS somitic mesoderm reversal on the development of motoneuron projection patterns. On the left are shown diagrams of the surgery and its effect on limb muscle origins. Previous experiments (see Lance-Jones 1988a) had shown that somites contribute to muscles in accord with their anteroposterior position just prior to migration. Thus an early reversal of the somitic mesoderm that normally contributes to anterior thigh muscles (levels 26–29) reverses the somitic origins of these muscles. Two muscles normally derived from specific subsets of somites 26–29 are the sartorius (somites 27–28) and the femorotibialis (somites 28–29). Histograms of the distribution of labelled motoneurons after HRP injections of each muscle in two stage 35–36 embryos are shown on the right (modified from Lance-Jones 1988b). Despite the reversal of the somites contributing to each muscle, the motoneuron pool to each muscle is positioned quite normally within anterior LS spinal cord segments. In the cases illustrated as well as others examined there is a slight posterior shift in the segmental position of the individual pools (compare to Fig. 1). This is likely to reflect a posterior shift in spinal nerve position, not the alteration of muscle somite origin. Keynes & Stern (1984) have shown that spinal nerves form opposite the anterior half of the somite under both normal and experimental conditions. Thus, a reversal of a block of somites will have caused a local posterior shift in the site of spinal nerve formation. A corresponding shift in motoneuron access to the limb may have slightly shifted individual motor pools (see Lance-Jones 1988b).

earlier (Lance-Jones & Landmesser 1980), pool positions were reversed. Motoneurons consistently appeared to alter their course to project to correct targets. These results suggest that specific guidance cues remain intact after somite reversals. Thus, one is led to the conclusion that the LS somites or their derivatives, limb muscle cells, do not bear guidance cues before muscle cell migration and axon outgrowth. Similar conclusions were reached by Keynes et al (1987) after slightly different somite manipulations at the wing level. It is important to note that neither my experiments (Lance-Jones 1988b) nor those of Keynes et al (1987) rule out the possibility that muscle cells acquire labels or cues after they migrate. However, they do direct one's attention to other embryonic tissues as a source of guidance cues.

The somatopleure as a source of guidance cues

In a pre-limb bud stage embryo, the somatopleural mesoderm is that mesoderm that lies lateral to the somites and the intermediate mesoderm. It is the dorsal layer of the lateral plate and, in conjunction with the overlying ectoderm, forms the somatopleure. The somatopleure in the hindlimb region gives rise to the limb bud, the somatopleural mesoderm giving rise specifically to all the connective tissues in the limb (see Chevallier 1979, Jacob et al 1979). Somatopleural mesoderm deserves attention as a possible source of specific guidance cues not only because of its contribution to the local environment encountered by the outgrowing axons, but also because it is thought to define the sites of major nerve trunks in the limb (Lewis et al 1981). To address this issue, Dr Mark Dias and I have assessed motoneuron connectivity patterns in chick embryos after manipulation of the somatopleure (Dias & Lance-Jones 1987). Since limb and muscle patterning is believed to be governed by components of the somatopleure (Chevallier & Kieny 1982), we chose to manipulate the whole limb somatopleure rather than to carry out reversals of a few segments which would have disrupted the normal pattern of limb development. At stage 14–15, posterior thoracic somatopleure from axial levels 22–25 was removed. LS somatopleure from axial levels 26–33 was excised and shifted anteriorly. The posterior thoracic somatopleure was then placed posteriorly (Fig. 5). When sacrificed at stages 27–36, all embryos had anteriorly shifted limbs.

This manipulation was similar to the limb shift manipulation mentioned earlier but differs in one important way. The somatopleure shifts were made prior to muscle cell migration into the limb. Since heterotopic somite manipulations suggest that limb muscle cell migration is passive, we would expect that the set of somites contributing to the limb would now be shifted anteriorly and include thoracic somites. This was shown to be the case by replacing chick posterior thoracic somites with equivalent quail tissue in a few embryos with somatopleure shifts. Anterior thigh muscles showed both nor-

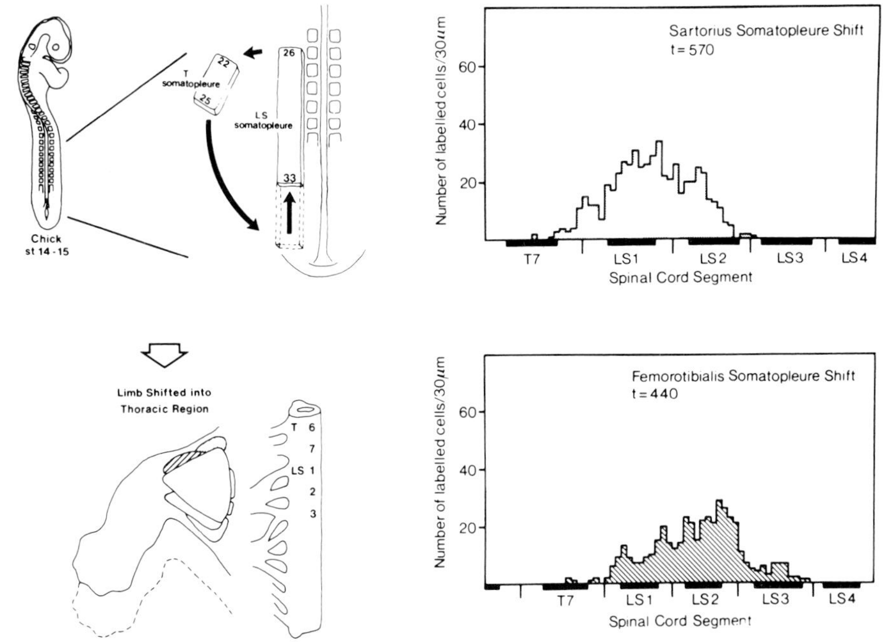

FIG. 5. The motor innervation of limb muscles following anterior shifts of the LS somatopleure. On the left are shown diagrams of the surgery and the effect on limb and plexus development. Shifts of the somatopleure from axial levels 26–33 into the thoracic (T) region gave rise to limbs shifted 1–3 segments anteriorly. In such limbs, all or some of the anterior LS spinal nerves (LS1–3) entered more posterior limb regions than normal and contributed to the posterior sciatic rather than the anterior crural plexus. Shown on the right are representative histograms of the distribution of labelled cells after injections of the sartorius and femorotibialis in stage 36 embryos. These two anterior muscles are both innervated by LS motoneurons despite the shifts of the limb somatopleure and the resultant limbs into the thoracic region. While not as discrete as normal, each pool appears to be generally positioned correctly on the anteroposterior cord axis. For example, the sartorius pool is located primarily in LS1 as in a normal embryo. The femorotibialis pool is not positioned as far posterior as normal (compare to Fig. 1) but is clearly more posterior than that of the sartorius. These observations suggest that LS motoneurons have shifted their projections anteriorly to innervate anteriorly shifted target regions. Since only the somatopleure has been shifted anteriorly, these observations also suggest that LS motoneuron axons may be recognizing specific guidance cues associated with the LS somatopleure.

mal morphology and quail cell labelling. This finding indicates that anterior thigh muscles are of composite origin in somatopleure-shifted limbs: the connective tissue from anterior LS levels, the muscle cells from thoracic levels.

We then defined the positions of motoneurons projecting to anterior thigh muscles in stage 34–36 somatopleure-shifted embryos using retrograde HRP labelling. Representative sartorius and femorotibialis pools are shown in Fig. 5. First, one can see that only a few thoracic (T) motoneurons are labelled. Virtually all labelled cells are in anterior LS segments. Even though these muscles were shifted anteriorly into thoracic regions, it is motoneurons in LS cord segments that innervate them. Further, the positions of cells projecting to each individual muscle suggest a specificity of projection among LS motoneurons. As in a normal embryo, the sartorius motoneuron pool lies anterior to the femorotibialis pool. These results suggest that after a somatopleure shift, motoneurons from anterior LS cord segments shift their projections anteriorly and innervate targets correct with respect to connective tissue origin.

To define how this may have been accomplished, peripheral nerve patterns were reconstructed in younger embryos (stage 27–31) where it is easiest to visualize gross nerve morphology. In the shifted limbs, 1–3 anterior LS spinal nerves entered posterior thigh regions and contributed to the sciatic plexus (see Fig. 5). As in limb shift embryos, aberrant muscle nerves were observed projecting from this plexus to anterior thigh muscles. Since anterior thigh muscles are normally innervated by anterior LS spinal nerves via anterior or anterior crural nerve trunks (see Fig. 1), this observation shows that axons channelled into the sciatic plexus have altered their course well within the limb proper to project to anterior thigh regions. While preliminary, these experiments implicate the LS somatopleure in the specific guidance of motoneurons.

Summary

In this paper I have described motoneuron axon guidance to different target regions on the anteroposterior axis of the limb. Here, precise target choice appears to be preceded by a prelabelling of motoneurons and their specific recognition of environmental cues during outgrowth. One of the simplest models by which to explain these phenomena is that motoneuron axons and their target muscle cells acquire axial position labels early in development and that a matching process later occurs. This hypothesis appears to be ruled out by the results of somite manipulation experiments that I have performed. While the pattern of somitic cell contributions to muscles in the limb is orderly and shows parallels to motoneuron projection patterns, specific recognition appears to be independent of the segmental level of origin of muscle cell precursors at the time when motoneurons are known to be specified. Rather, our preliminary analyses of projections after shifts of the somatopleure suggest that it may be the connective tissue precursor populations of the limb that provide specific guidance cues.

If future experiments support these conclusions, it will be important to identify how such cues are provided by the somatopleure during axon outgrowth. The role of the somatopleure may be a direct one, the LS somatopleure actually being the tissue recognized by LS motoneurons. However, its role may also be an indirect one, related to the recognized role of the somatopleure in defining limb muscle patterns (see Chevallier & Kieny 1982 and above). Perhaps the somatopleure is responsible for setting up a molecular gradient which defines muscle cell position and also marks muscle cells with specific axon guidance cues.

I have focused on one event in the development of neuromuscular connections, namely guidance to targets. Yet, as discussed in several of the presentations made in this symposium, there are numerous stages of neuron–target interactions and clearly many types of interactions which can affect a developing motoneuron's behaviour. In the future, it may be of interest to examine the role of the connective tissues in later neuromuscular interactions and to explore the relationship between specific guidance factors and trophic factors.

Acknowledgements

I am indebted to Marcia Honig and Bruce Mendelson for their comments on the manuscript, to Joan Kumpfmiller for technical assistance, to Sharon Wesolowski for secretarial assistance and to the National Science Foundation for support (BNS 8518864).

References

Beresford B 1983 Brachial muscles in the chick embryo: the fate of individual somites. J Embryol Exp Morphol 77:99–116

Chevallier A 1979 Role of the somitic mesoderm in the development of the thorax in bird embryos. II. Origin of thoracic and appendicular musculature. J Embryol Exp Morphol 49:73–88

Chevallier A, Kieny M 1982 On the role of the connective tissue in the patterning of the chick limb musculature. Dev Biol 191:227–280

Dias M, Lance-Jones C 1987 A possible role for somatopleural tissue in specific motoneuron guidance in the embryonic chick hindlimb. Soc Neurosci Abstr 13:468

Hamburger V 1975 Cell death in the development of the lateral motor column of the chick embryo. J Comp Neurol 160:535–546

Hamburger V, Hamilton HL 1951 A series of normal stages in the development of the chick embryo. J Morphol 88:49–92

Hollyday M 1980 Organization of motor pools in the chick lateral motor column. J Comp Neurol 194:143–170

Jacob MB, Christ B, Jacob HJ 1979 The migration of myogenic cells from the somites into the leg region of avian embryos. Anat Embryol 157:291–309
Keynes RJ, Stern CD 1984 Segmentation in the vertebrate nervous system. Nature (Lond) 310:786–789
Keynes RJ, Stirling RV, Stern CD, Summerbell D 1987 The specificity of motor innervation of the chick wing does not depend upon the segmental origin of muscles. Development 99:565–575
Lamb AH 1979 Evidence that some developing limb motoneurons die for reasons other than peripheral competition. Dev Biol 71:8–21
Lance-Jones C 1988a The somitic level of origin of embryonic chick hindlimb muscles. Dev Biol 126:394–407
Lance-Jones C 1988b The effect of somite manipulation on the development of motoneuron projection patterns in the embryonic chick hindlimb. Dev Biol 126:408–419
Lance-Jones C, Lagenaur C 1987 A new marker for identifying quail cells in embryonic avian chimeras: a quail-specific antiserum. J Histochem Cytochem 35:771–780
Lance-Jones C, Landmesser LT 1980 Motoneurone projection patterns in the chick hind limb following early partial reversal of the spinal cord. J Physiol (Lond) 302:581–602
Lance-Jones C, Landmesser LT 1981a Pathway selection by chick lumbosacral motoneurons during normal development. Proc R Soc Lond B Biol Sci 214:1–18
Lance-Jones C, Landmesser LT 1981b Pathway selection by embryonic chick motoneurons in an experimentally altered environment. Proc R Soc Lond B Biol Sci 214:19–52
Landmesser LT 1978a The distribution of motoneurons supplying chick hindlimb muscles. J Physiol (Lond) 284:371–389
Landmesser LT 1978b The development of motor projection patterns in the chick hindlimb. J Physiol (Lond) 284:391–414
Landmesser LT 1988 Peripheral guidance cues and the formation of specific motor projections in the chick. In: Easter SS et al (eds) From message to mind. Sinaur Associates, Inc., Sunderland, Massachusetts, p 121–133
Lewis J, Chevallier A, Kieny M, Wolpert L 1981 Muscle nerve branches do not develop in chick wings devoid of muscle. J Embryol Exp Morphol 64: 211–232
Sperry R 1963 Chemoffinity in the orderly growth of nerve fiber patterns and connections. Proc Natl Acad Sci USA 50:703–710
Tosney KW, Landmesser LT 1985 Development of the major pathways for neurite outgrowth in the chick hindlimb. Dev Biol 109:193–214
Wigston DJ, Sanes JR 1982 Selective reinnervation of adult mammalian muscle by axons from different segmental levels. Nature (Lond) 299:464–467
Westerfield M, Eisen JS 1988 Common mechanisms of growth cone guidance during axon pathfinding. In: Easter SS et al (eds) From message to mind. Sinaur Associates, Inc., Sunderland, Massachusetts, p 110–120

DISCUSSION

Crowder: Could it be that the somatopleure induces the limb, with the muscle actually inducing the nerve, and that the chick nerve cannot recognize the quail muscle?

Lance-Jones: It is possible that information in the somatopleure determines limb position and morphology and that it also labels limb muscle cells. These labels, in turn, may be recognized by motoneuron axons. In our experiments, we only used quail/chick chimeras, to make certain that thoracic myogenic cells were migrating into the shifted lumbosacral somatopleure. The analyses of nerve patterns were all done in chick embryos in which the chick's own lumbosacral somatopleure was shifted anteriorly. Thus, we were not asking chick axons to recognize quail muscle. In fact, chick motoneuron axons can specifically recognize quail muscles. Dr Tanaka and Dr Landmesser (1986) have shown that when a quail limb bud is placed on a chick embryo or vice versa, specific motoneuron projections are found. This finding suggests that quail and chick species have common motoneuron axon guidance cues.

Van Essen: I am interested in how the innervation pattern within the sciatic and crural plexuses is set up. Is it known whether the somatopleure contributes to the connective tissue to that region?

Lance-Jones: We have begun a fate-mapping study of the somatopleure. At stages prior to muscle cell migration we injected discrete regions of the lumbosacral somatopleural mesoderm with a long-lasting dye (DiI; Honig & Hume 1986). Examinations of the limb at later stages indicate that the lumbosacral somatopleural mesoderm contributes to the plexus region and that it does so in a generally topographical manner. We don't know much about the disposition of the somatopleural mesoderm when axons first enter the limb. The electron microscopic studies of Al-Ghaith & Lewis (1982) and Tosney & Landmesser (1985a) indicate that axonal growth cones make close contact with mesenchymal cells. However, we don't know whether these mesenchymal cells are of somatopleural origin, somitic origin, or neural crest origin.

Henderson: One striking result of your experiments on the transplantation of the somatopleure is that initially the axons seem to go in the 'wrong' direction, towards the 'wrong' plexus. They then realize that they have done something wrong, and shoot out across country, to innervate the correct muscles. Does this choice correspond with the first time that the growth cones come in contact with mesenchymal cells?

Lance-Jones: No; growth cones may come into contact with mesenchymal cells quite far proximally. We do not know why motoneuron axons alter their course distally at midthigh levels in somatopleure shift experiments. One possibility is that the girdle and the proximal part of the femur act as mechanical barriers preventing axons from altering their course proximally.

Henderson: I am reminded of the results in the retinotectal system, where a certain number of experimentally displaced axons in the optic nerve reach their correct target, not by going across country, but by getting to the right position along one axis and then making a right-angle turn to grow along the other. In your system, I wondered if this 90degree turn occurs at different distances from the spinal cord in different transplants? This might suggest that the axon is 'measuring', first on one axis and then on another.

Lance-Jones: In the chick hindlimb, axons may be 'measuring' the dorsoventral and anteroposterior axes separately. Dorsoventral pathway choices appear to be made further proximally than anteroposterior choices and may be governed by separate mechanisms (see Lance-Jones 1986). In the somatopleure shift experiments, only the motoneuron's relationship to the anteroposterior axis has been disrupted. Sharp turns were generally made at midthigh levels, near the normal site of muscle nerve formation and anteroposterior pathway choice.

Eisenberg: It struck me too that the right-angled bend is an important moment in the decision of the nerve. It almost looks as if there are cues telling the nerve that it is headed in the wrong direction, and there must be some corrective mechanism. At that stage, is there any searching in other directions—perhaps sproutings or branchings in the wrong direction that then retract? Are false turns seen at that point?

Lance-Jones: Studies of normal development (Tosney & Landmesser 1985b,c) suggest that, within a limited range, axons do do a certain amount of searching. Axon trajectories appear to be quite straight as axons exit from the cord. However, when axons approach and enter the plexus region, they may make alterations in their course or show convoluted trajectories. As the axons reach the muscle nerve level, which is about at the level where the right-angle turns are made, axons again may show alterations in trajectory and may cross one another. Tosney & Landmesser (1985c) have also found that growth cone morphology is different at the plexus and muscle nerve level from what it is at spinal nerve and nerve trunk levels. These authors have termed these sites 'decision sites', to emphasize the idea that here axons may be 'searching' or discriminating between correct and incorrect paths.

Holder: What distance is involved in the formation of the aberrant branches? Is it likely to be 'feelable' by the growth cone, or is it more likely to be long range?

Lance-Jones: The muscles innervated by aberrant nerves frequently had components in central as well as anterior limb regions. Thus, axons mischannelled by the shift into the sciatic or posterior nerve trunks were potentially not far from parts of their correct targets. We do not know the exact distances involved at the time of axon outgrowth. However, it is likely that parts of correct targets were within the range that outgrowing axons may normally sample.

Holder: So the growth cone is perhaps feeling its way around, in making the decision to turn. On the question of where the decision regions actually are, in a regenerating system, like the newt or axolotl, if you cut and re-route a nerve, moving it away from its normal pathway, the axons re-grow towards their original target along their old pathway (Holder et al 1984). This suggests that guidance cues are located all the way along the limb. I would also agree with you that the plexus is an important decision region, but it's not the only one, because there are such regions in many places in the limb (see Wilson & Holder 1988 for a detailed discussion of decision regions in the axolotl limb).

Lance-Jones: Yes, I would agree that guidance cues could be located all along the limb. It is important to note that the presence of mechanical barriers rather than an absence of guidance cues could be the reason why axons do not alter their course at some sites in the limb.

Carlson: I am trying to dissociate the influences of muscle from those of connective tissue, to see whether the latter instructs the muscle and then the muscle tells the nerve what to do. You might look at this by somatopleure transplantation experiments in conjunction with somite removal. But would you have enough of the nerve pattern coming into the limb to be able to analyse any later deviations?

Lance-Jones: We haven't done that experiment. It might in fact be difficult, because muscle nerves may form infrequently after somite removal (see below).

Carlson: So you are taking away the potential secondary guidance cues that muscle might provide, if it does get secondarily instructed?

Lance-Jones: In the future we should like to address the question of whether motoneuron axons directly recognize cues associated with the somatopleure or whether they recognize cues on muscle cells which were imparted by the somatopleure. This might be done *in vivo* by transplanting small pieces of somatopleure with and without muscle tissue into a host limb bud. If host axons can be directed to the transplant, then analyses of when and which axons do so might answer this question. We also hope to do *in vitro* experiments in which neurite outgrowth on somatopleure explants is studied.

Carlson: This is something that Nigel Holder might do in his system, because you can get axolotl limbs to regenerate without muscle by irradiating the limb stump and then adding unirradiated skin.

Holder: This experiment has been done in chicks by Julian Lewis and his colleagues (Lewis et al 1981), and we also have some preliminary results with the axolotl.

Lance-Jones: Lewis et al (1981) examined nerve branching patterns in wings after the irradiation of the somites. They reported that nerve trunks and the plexus form normally despite the absence of muscle but that muscle nerves do not form. More recently, Phelan & Hollyday (1986) have examined branching patterns in wings after surgical excision of somites. In contrast to Lewis et al (1981), these investigators did find muscle nerves. Phelan & Hollyday (1986) also assessed the specificity of dorsoventral pathway choice in the plexus region and found it to be normal. They concluded that muscle tissue is not required for proximal dorsoventral pathway choice. They did not explicitly test for specificity on the anteroposterior axis. However, the segmental position of motoneurons projecting to selected target regions did appear to be normal (M. Hollyday, personal communication). Last, Tosney (1987) has reported that limb motoneuron projections are specific on both axes after the excision of the dermamyotome portion of one or more chick lumbosacral somites.

Holder: In the axolotl, we don't know about the specificity either, but we get many motoneurons growing down into muscleless limb regenerates. We don't know whether the neurons are in precisely the right place, but nerves are present in the flexor and extensor compartments. So the muscle is not essential for the motor nerves to grow into the limb, or for the positioning of the major pathways.

Carlson: Have you tried analysing supernumerary regenerates yet?

Holder: No, but that would be an interesting experiment.

Lance Jones: As I mentioned earlier, we hope to look at neurite outgrowth *in vitro* on somatopleure tissue in order to further address these issues.

Lowrie: Have you found any evidence in your reversal experiments of a difference between motoneurons in their ability to project axons back to their correct target? I am thinking of the possibility of the 'pathfinder' axons which other, later axons follow.

Lance-Jones: In our somite and cord reversal experiments the earliest assessments of projection patterns were generally made at 6–7 days of development, a couple of days after the first motoneuron axons enter the limb. Thus we do not know whether the very earliest outgrowing axons might behave differently from later ones. However, electron microscopic studies of early neurite outgrowth in the chick hindlimb (Tosney & Landmesser 1985a) suggest that there is not a distinct pathfinder population.

Lowrie: Does this mean that all the axons projecting to a particular part of the limb musculature innervate it at the same time?

Lance-Jones: Yes; I believe that most axons reach the target region at roughly the same time.

References

Al-Ghaith LK, Lewis JH 1982 Pioneer growth cones in virgin mesenchyme: an electron microscope study in the developing chick wing. J Embryol Exp Morphol 68:149–160

Honig MG, Hume RI 1986 Fluorescent carbocyanine dyes allow living neurons of identified origin to be studied in long term culture. J Cell Biol 103:171–187

Holder N, Tonge DA, Jesani P 1984 Directed regrowth of axons from a misrouted nerve to their target muscles in the limb of the adult newt. Proc R Soc Lond B Biol Sci 222:477–489

Lance-Jones C 1986 Motoneuron projection patterns in chick embryonic limbs with a double complement of dorsal thigh musculature. Dev Biol 116:387–406

Lewis JH, Chevallier A, Kieny M, Wolpert L 1981 Muscle nerve branches do not develop in chick wings devoid of muscle. J Embryol Exp Morphol 64:211–232

Phelan K, Hollyday M 1986 Pathway selection in muscleless chick limbs. Soc Neurosci Abstr 12:1210

Tanaka H, Landmesser LT 1986 Interspecies selective motoneuron projection patterns in chick-quail chimeras. J Neurosci 6:2880–2888

Tosney KW 1987 Proximal tissues and patterned neurite outgrowth at the lumbosacral level of the chick embryo. I. Deletion of the dermamyotome. Dev Biol 122:540–558

Tosney KW, Landmesser LT 1985a Development of the major pathways for neurite outgrowth in the chick hindlimb. Dev Biol 109:193–214

Tosney KW, Landmesser LT 1985b Specificity of early motoneuron growth cone outgrowth in the chick embryo. J Neurosci 5:2336–2344

Tosney KW, Landmesser LT 1985c Growth cone morphology and trajectory in the lumbosacral region of the chick embryo. J Neurosci 5:2345–2358

Wilson S, Holder N 1988 Evidence for axonal decision regions in the axolotl peripheral nervous system. Development, in press

In vitro analysis of specificity during nerve–muscle synaptogenesis

Dan H. Sanes* and Mu-ming Poo

Section of Molecular Neurobiology, Yale University School of Medicine, New Haven, Connecticut 06510, USA

Abstract. The early phase of synapse formation was studied in cultures of *Xenopus laevis* spinal neurons and myotomal muscle cells. Two early events are described: the pulsatile secretion of acetylcholine from the nerve terminal in response to myocytic or neuronal contacts, and the development of nerve–myocyte adhesion during the first few minutes of contact. The specificity in these early events in synaptogenesis was assessed with respect to the positional and clonal relationships of the neurons and myocytes. Axial position and lineage were determined by injecting embryos with a fluorescent dye, such that dissociated cells could subsequently be identified in culture. We examined the efficacy of spontaneous synaptic currents, and the relative preponderance of growth cone–myocyte associations, for neurite–myocyte pairs of the same or dissimilar origin. Neither of these two assays revealed a dependence on the axial position or the lineage of the cells. Although these studies indicate that early nerve–muscle interactions show little positional or clonal selectivity, myocytes clearly influence the onset of synaptic function.

1988 Plasticity of the neuromuscular system. Wiley, Chichester (Ciba Foundation Symposium 138) p 116–130

Upon contact, motoneuronal growth cones and muscle cells engage in a number of cellular interactions that result in the emergence of a mature synapse. Information must be transmitted between pre- and postsynaptic cells to activate fundamental events such as the clustering of acetylcholine receptors (AChR), the establishment of efficient transmitter release, and the maturation of synaptic specializations. For example, it has recently been found that postsynaptic membrane components can influence the transmitter phenotype of neurons (Adler & Black 1986).

Given that cholinergic neurons are generally capable of innervating all striated muscle fibres, a second class of interactions could participate in the correct matching of motoneurons with specific target muscle populations. There are two general ways in which a cell may come to express a unique

---

* *Present address:* Departments of Otolaryngology, and Physiology & Biophysics, New York University, Medical Center, 550 First Avenue, New York, NY 10016, USA.

molecular identity: through its position within the embryo, or from a unique lineage (Purves & Lichtman 1985). The positional cue is thought to be a gradient of soluble or cell surface molecules (Wolpert 1969), and the lineage cue could, in principle, be related to the cell cycle (e.g., number and pattern of divisions) (Sulston & Horvitz 1977, Lawrence 1975).

In the neuromuscular system there are several phenomenological findings that suggest that positional information may influence the final pattern of connections. For example, topographical projections from spinal motor centres to the periphery may initially exhibit a less restricted pattern of innervation (Stephens & Govind 1981, Brown & Booth 1983, Bennett & Ho 1988). Furthermore, it has been found that regenerating preganglionic sympathetic neurons show some preference for intercostal muscle fibres originating from a related axial position (Wigston & Sanes 1985). Although evidence for an influence of cell lineage on neuromuscular connectivity is scant, there is one report that notes an apparent association of clonally related nerve and muscle cells during the initial stages of development (Moody & Jacobson 1983). It has also been suggested that the innervation of primary fast and slow muscle fibres may arise from lineage-dependent selectivity (Sanes 1987).

The studies described in this chapter indicate that specific membrane–membrane interactions induce spontaneous transmitter release, and the rapid increase in adhesion. However, there seems to be little, if any, specificity for functional or physical association based upon the axial position or lineage of these cells during early development.

Methods

The neural tube and associated mesoderm from *Xenopus laevis* embryos were dissociated in a Ca^{2+}-free Ringer solution at Nieuwkoop and Faber (1967) stages 20–22 (approximately 24 hours after fertilization), and plated onto glass coverslips (Spitzer & Lamborghini 1976). After 24 hours at room temperature there are numerous myocytes and neurons in the culture dish, and 60–90% of the latter are known to be cholinergic (Xie & Poo 1986, Sun & Poo 1987). The cells that we study *in vitro* are known to give rise to the segmental neuromuscular system of the tail. The motoneurons normally exit the cord at approximately Nieuwkoop and Faber stage 19 (Nieuwkoop & Faber 1967; Kullberg et al 1977), and grow along the myotomal junctions (Lewis & Hughes 1960). Although nerve fibres rarely leave this pathway, a minority do exit the junction and grow across the segmental musculature (Chow & Cohen 1983, D.H. Sanes & M-m. Poo, unpublished observations).

The use of whole-cell and patch-clamp recording techniques allow one to monitor synaptic events with high spatial and temporal resolution (Xie & Poo 1986, Sun & Poo 1987). Round myocytes (myoballs) and membrane patches are manipulated into contact with neurites, and the release of transmitter is measured as whole-cell or single channel currents. The precise timing of nerve–myocyte interactions can be controlled by physical manipulation of the myocyte with the tip of an electrode. The adhesion between a myocyte and a neuron is qualitatively assessed by withdrawing the myocytes after a designated period of contact.

The original lineage or axial position of a cultured cell was assigned by monitoring the presence of a fluorophore that had been injected at an earlier stage of development (Weisblat et al 1978) (Fig. 4, below). To determine a cell's original axial position we injected fertilized eggs (i.e., one-cell stage) with a small volume of fluorescein- or rhodamine-conjugated dextran. Dissociated cell cultures were prepared from rostral pieces of rhodamine-labelled embryos, and caudal pieces of fluorescein-labelled embryos (or vice versa). The fluorescent colour, therefore, signified axial position. For the purposes of the present study, *rostral* was defined as the first $^1/_6$th of the neuraxis just caudal to the presumptive rhombencephalon, and *caudal* was defined as the fourth $^1/_6$th of the neuraxis. To follow the cell lineage, we injected one blastomere with rhodamine-conjugated dextran at the 64-cell stage, and dissociated an individual embryo for each cell culture dish.

Results and discussion

Induction of transmitter release

When a myoball was voltage clamped at its resting potential and manipulated into contact with the neuritic process of a co-cultured neuron, the immediate appearance of spontaneous synaptic currents was observed (Fig. 1). These current events commonly increased in frequency and average amplitude during the first 10–15 min of contact (Xie & Poo 1986). They were abolished in the presence of α-bungarotoxin or curare, which block the action of

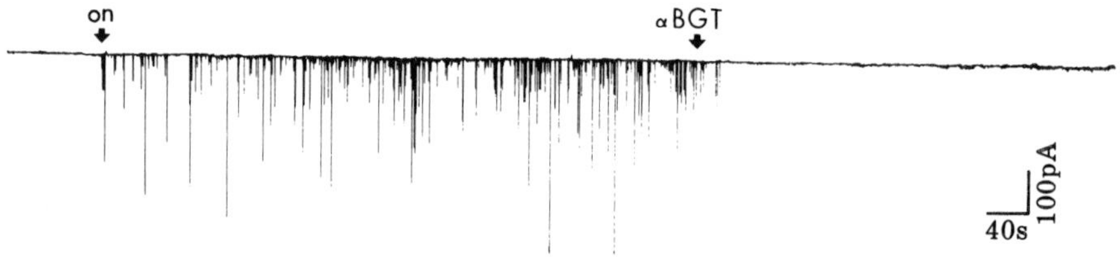

FIG. 1. The onset of electrical activity immediately following a manipulated association between neuron and myoball. Tracing is a continuous recording of a myoball membrane current (inward current downward) when the cell was voltage clamped to −70 mV. Note that the amplitude and frequency of spontaneous synaptic currents appear to increase somewhat during the first several minutes. The application of α-bungarotoxin (α-BGT) (20 μg/ml) eliminated all activity, indicating that the synaptic transmission was cholinergic.

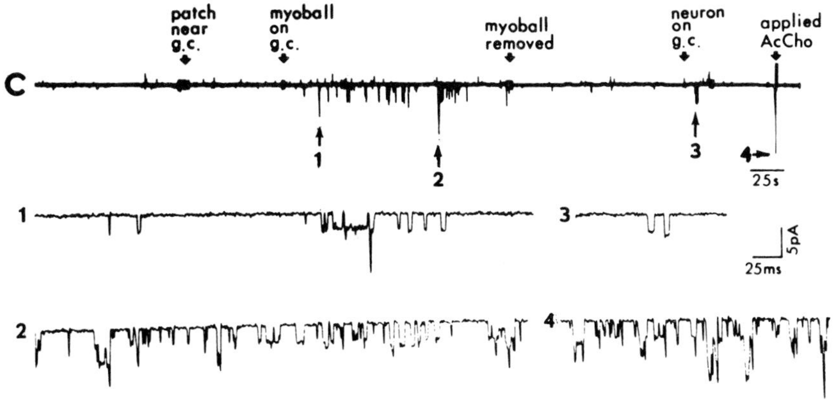

FIG. 2. The induction of acetylcholine (AcCho) release by a myoball and a neuron, as recorded with an excised patch of muscle membrane in outside-out configuration. The top trace shows the patch membrane current (inward current downward) recorded at a holding voltage of −80 mV (pipette inside negative). Four samples of single channel events are shown at higher resolution. It is clear that little activity is observed near the growth cone (g.c.) prior to myoball contact, and that the myoball induced a greater release of transmitter from the growth cone than did a neuron. 'Applied AcCho' indicates that a drop of acetylcholine-containing saline was added to the culture at the end of the experiment to confirm that the patch membrane was appropriately sensitive.

acetylcholine (ACh), and must, therefore, be due to the pulsatile release of ACh from the contacted neurite.

That the pulsatile secretion of ACh was induced by the muscle contact was demonstrated by experiments in which an excised patch of muscle membrane (outside-out configuration) was used as a probe to monitor the presence of extracellular ACh (Xie & Poo 1986). As depicted in Fig. 2, the patch membrane probe detected few ACh-like molecules near an isolated growth cone. Contact by a myoball markedly elevated the level of ACh near the growth cone, as indicated by the appearance of ACh-induced single channel events at the probe. This induction of transmitter release is relatively specific for the muscle membrane, since persistent, high level ACh release was observed when a myocyte, but not a neuron, was manipulated into contact with the growth cone (Fig. 2). Placing a myoball onto the growth cone resulted in an increase of the average frequency of current events from 4.9 to 19.9 per min (n = 46), while the frequency was only 8.4 per min for contacts by neuron cell bodies (n = 7). Furthermore, contact by the excised patch of muscle membrane (the ACh probe) itself was capable of inducing ACh release from the growth cone, suggesting that it was the direct physical contact, rather than diffusible factors released by the myocyte, that was responsible for the induction of ACh release (Xie & Poo 1986). These results indicate that specific membrane–membrane interactions are involved in triggering the onset of synaptic function.

Nerve-muscle adhesion

The strength of adhesion between neurites and myocytes was found to increase dramatically during the first 15 min after the cells were manipulated into contact (Fig. 3). The adhesion assay was a qualitative grading of the

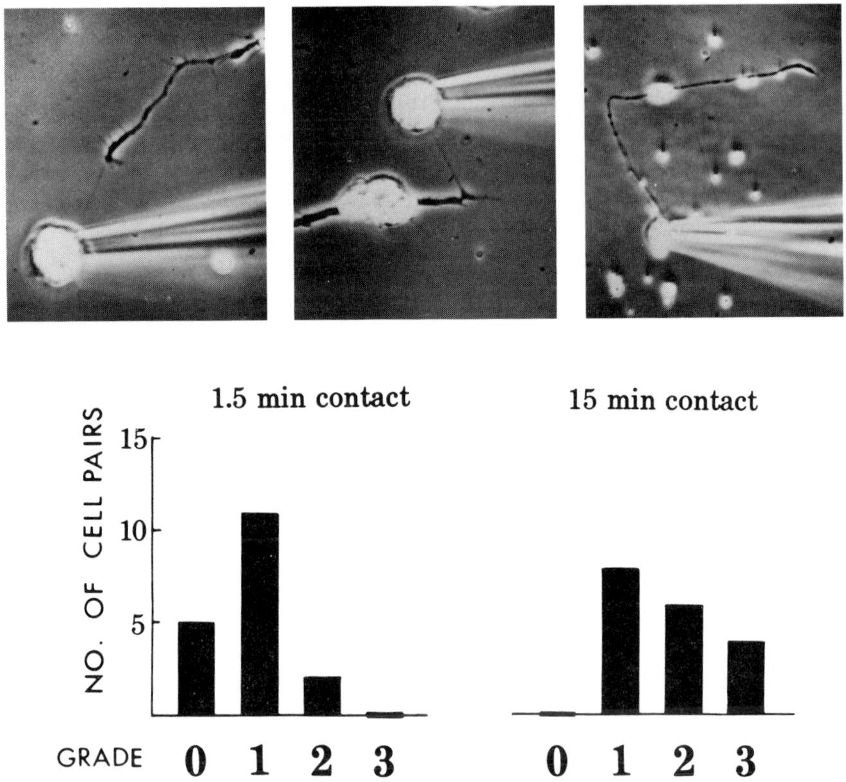

FIG. 3. The degree of adhesion between neurites and manipulated myoballs. The top three panels show the grades of adhesion between myoballs and neuritic growth cones: *Grade 1:* a filamentous connection; *Grade 2:* a deformation of the growth cone; or *Grade 3*: a detachment of the growth cone from the substratum. 'No visible adhesion' (*Grade 0*) is not illustrated. There is a rapid increase in the amount of adhesion between neurite and myoball over the first 15 min, as judged by the ability of the myoball to distort the neurite shapes when it is pulled away. At 1.5 min, most contacts show Grade 1–2 characteristics, while at 15 min all contacts show Grade 1–3 characteristics.

morphological features observed during the physical separation of the myocyte from the neurite. Fig. 3 shows three examples of myoball separation that illustrate the grades of neurite-muscle adhesion: *Grade 1:* membranous filament is seen connecting the myoball and the neurite but the main 'palm' of the growth cone shows no visible deformation; *Grade 2:* myoball removal leads to deformation of the growth cone; *Grade 3:* myoball removal leads to detachment of the growth cone or neurite from the substratum. (*Grade 0:* no visible neurite–muscle attachment seen during the separation, not shown.) Fig. 3 also depicts the histograms of adhesion grades from two groups of myoball–neurite contacts of duration 1.5 or 15 min, respectively. It is apparent from this qualitative study that low grade adhesion develops quickly after the contact, but high grade stable adhesion requires time (approximately 10 min) to become fully established. Many questions remain to be addressed: for example, how selective is the nerve–muscle adhesion? Does the stabilization of adhesion during the early phase of neurite–muscle contact play any important role in synaptogenesis? What are the molecules involved in the formation of adhesive bonds between the cells? What is the mechanism that accounts for the time course in the stabilization of adhesion?

Specificity in nerve–muscle interactions

It has been postulated that subpopulations of the pre- and postsynaptic cells could differentially express molecular markers for selective synaptogenesis by virtue of their lineage or axial position within the embryo. We have tested both these postulates in cell culture by examining the synaptic activity and the physical association of growth cones with the co-cultured myocytes. The original rostro-caudal position or cell lineage of each cell was tracked according to the presence of fluorescent labels (see Methods and Fig. 4). The short-term (approximately 20 min) synaptic activity was studied by manipulating myocytes into contact with neurites. We have also studied contacts resulting from the natural innervation of myocytes by growing neuritic processes after approximately two days in culture. Examples of manipulated and natural contacts are shown in Fig. 5.

The experimental procedure required that recordings be made at the contacts established by the same neuron with a related and an unrelated myocyte. The term 'related' means that the nerve and muscle came from the same axial position or were clonally related at the 64-cell stage. The term 'unrelated' means that the nerve and muscle came from different axial positions or were not clonally related at the 64-cell stage. These paired recordings greatly reduced the contribution of neuronal heterogeneity to the variability of synaptic transmission. We first compared the rate and the average amplitude of spontaneous synaptic currents (SSCs) at contacts between *positionally* related and unrelated neurons and myocytes. There was no reliable difference

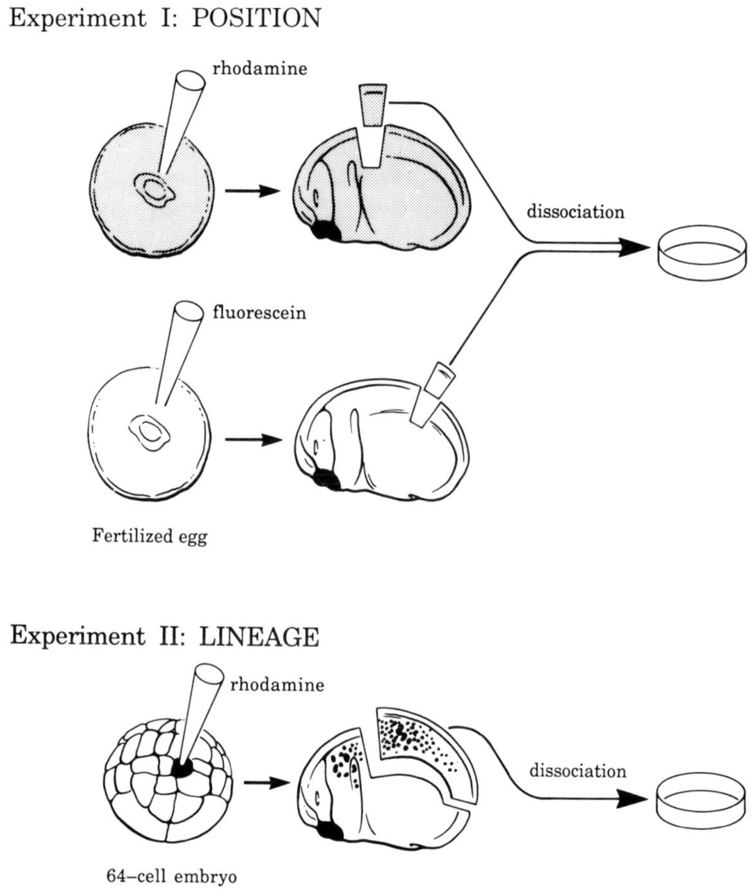

FIG. 4. Schematic diagram of the experimental procedures. In *Experiment I* (top) we injected fertilized eggs with either rhodamine- or fluorescein-conjugated dextran. The rostral (1/6) pieces of three rhodamine-labelled embryos, and the caudal (4/6) pieces of three fluorescein-labelled embryos were dissociated and co-cultured. In *Experiment II* (bottom) one blastomere of the 64-cell stage embryo, that is known to give rise to muscle and motoneurons (Jacobson & Hirose 1981), was injected with rhodamine-conjugated dextran. The neural tube and associated myotome of individual embryos were dissected free and cultured separately.

in these measures, either for manipulated or natural contacts (Fig. 6A and B). Moreover, there was no apparent difference in nerve-evoked synaptic currents (ESCs) for natural contacts between positionally related and unrelated neurons and myocytes (average amplitudes 562 and 555 pA, respectively; $n = 4$). For individual pairs of recordings, the ratio of average (ESC) amplitude for the related and unrelated contact was 0.51, 1.01, 1.26 and 1.37.

A similar analysis of *clonally* related versus unrelated neuron–myocyte

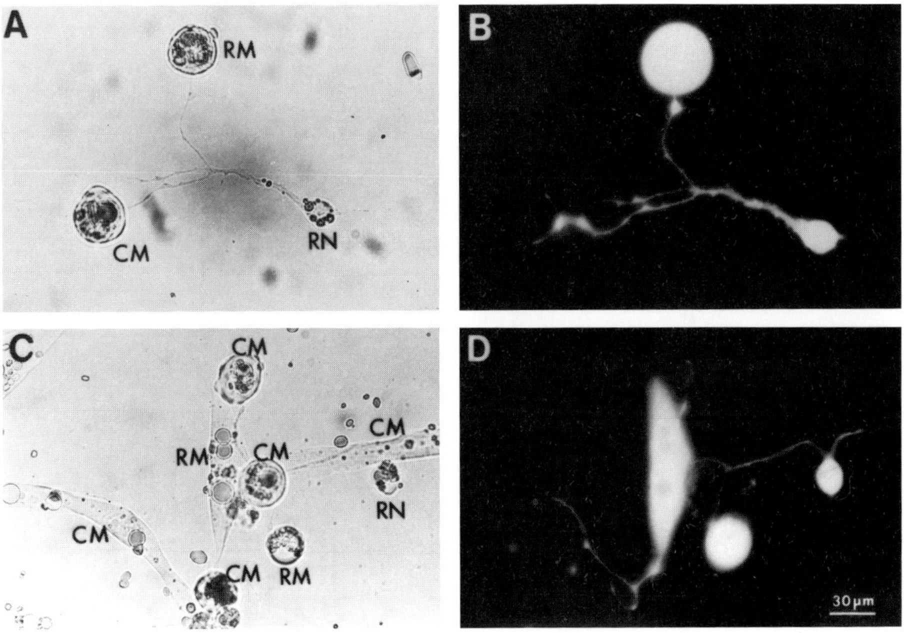

FIG. 5. Examples of manipulated (A and B) and natural (C and D) nerve–muscle contacts in cell culture from experiments testing the influence of position on nerve–muscle synaptogenesis. On the left are bright-field photomicrographs of the nerves and muscle cells in each culture. On the right are fluorescence photomicrographs of the same field. Note that the fluorescently labelled nerve cell (RN) contacts both a labelled (RM) and an unlabelled (CM) myocyte in each case, allowing us to assess the efficacy of synaptic transmission for a related and unrelated contact made by a single neuron. RN, rhodamine-labelled nerve; RM, rhodamine-labelled muscle; CM, clear muscle.

contacts yielded comparable results. As shown in Fig. 6A and B, there were no significant differences between related and unrelated contacts for either SSC rate or amplitude. We noted that natural contacts in two day cultures exhibited SSCs of higher rate and average amplitude than those obtained from manipulated contacts. However, there was considerable variability between cultures, as is exemplified by the relatively low rate of SSCs recorded from natural contacts in the lineage experiment.

Neurites in culture tend to grow best on substrata of greater adhesiveness (Letourneau 1975). As a qualitative assay for the physical affinity of the neurite for the myocyte, we counted the number of growth cones that bypassed related or unrelated myocytes to terminate on the culture dish substratum. Growth cones clearly preferred the myocyte membrane to the substratum, as approximately 80% remained on the myocyte surface after about two days in culture. However, neurons exhibited no selective affinity for positionally or clonally related myocytes (Fig. 6C).

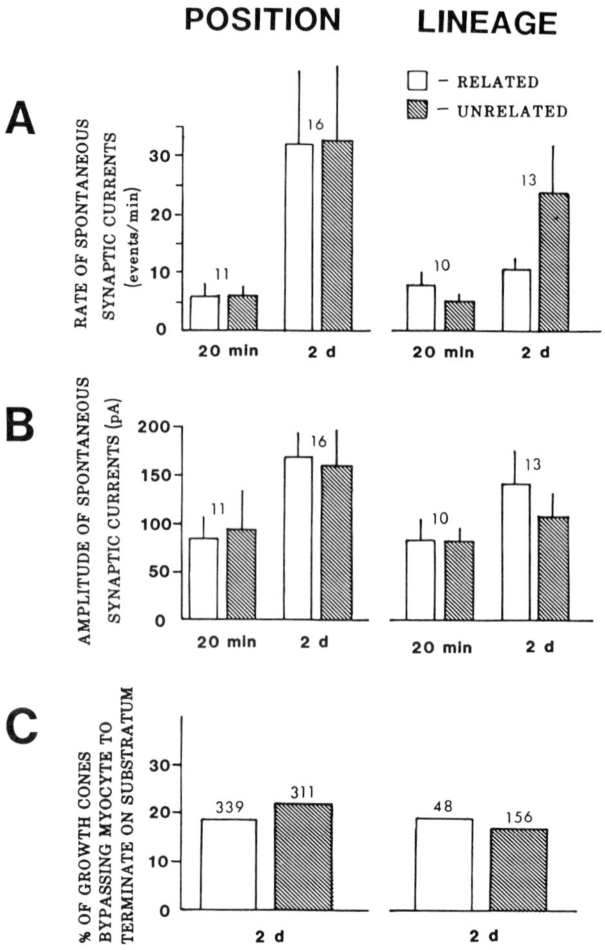

FIG. 6. Quantitative comparison of synaptic function and physical association between nerve–muscle contacts that were either related or unrelated for axial position (left) or cell lineage (right). (A) The rate of spontaneous synaptic currents for related (clear bars) vs. unrelated (cross-hatched bars) nerve–muscle contacts. (B) The amplitude of spontaneous synaptic currents for related vs. unrelated nerve–muscle contacts. (C) The percentage of growth cones that bypass related vs. unrelated myocytes to terminate on the culture of substratum. The number of observations is indicated above each set of bars. The approximate duration of the contact is shown beneath each set of bars. Mean and standard deviation are shown in A and B. No reliable difference was found between related and unrelated contacts for any of the parameters examined ($P > 0.05$; Wilcoxon signed ranks test and paired samples *t*-test).

The absence of selective interactions between nerve and muscle cells in culture could result from the dissociation of the embryonic tissue, even though the method was comparatively gentle (i.e., non-enzymic). A much harsher dissociation procedure (using trypsinization) does not interfere with the *in vitro* manifestation of the position-dependent affinity between chick retinal and tectal cells (Bonhoeffer & Huf 1982). These experiments showed that temporal retinal neurons prefer to grow on the surface of neurons isolated from anterior tectum. Although the intrinsic properties of a cell may be preserved after dissociation, the procedure does exclude the normal array of cellular interactions. For example, it is possible that the influence of non-target cells, which continuously impose a positional identity on nerve or muscle cells, was excluded from the present assay system (see chapter by C. Lance-Jones in this volume). Finally, the excision and transplantation of tissue to a heterotypic location apparently does not interfere with the position-dependent innervation patterns (Walthall & Murphey 1984, Wigston & Sanes 1985). Taken together, these observations suggest that the unselective innervation of myocytes *in vitro* is an accurate reflection of their interactions in the developing embryo.

The present results are consistent with a mechanism in which extracellular cues are used to guide outgrowing motor axons towards the appropriate target cells. In the case of the *Xenopus laevis* embryo, there may be segmentally repeated cues that direct the motor axon to the myotomal junction, and constrain it to this pathway. There are several experimental studies on chick limb innervation indicating that explicit cues along the growing motor axon's pathway are responsible for the adult pattern of connectivity (Lance-Jones & Landmesser 1981a, 1981b, Summerbell & Stirling 1981, Whitelaw & Hollyday 1983, Lance-Jones 1986). The elegant descriptions of primary motoneuron outgrowth in the zebrafish demonstrate that axonal arbors are invariably confined to a single myotomal segment throughout development (Myers et al 1986). Therefore, the developing primary motor axon appears to be initially capable of innervating all myocytes, although its choice of target is restricted by the exigencies of pathway choice. This does not exclude the possibility that secondary or regenerating motor axons make use of cues present on the muscle cell surface.

Summary

Very rapid physicochemical changes occur at the initial site of nerve-muscle contact. One of these, the induction of presynaptic transmitter release by the myocyte, is due to membrane–membrane interactions. The signal from the myocytic membrane is somewhat specific in that neural membrane is far less effective at inducing the spontaneous release of ACh. The dynamics of nerve–myocyte contacts are also evidenced by enhanced adhesion of the contact site over the initial 15 minutes. Since the muscle cell membrane appears to have some unique characteristics during synaptogenesis, we assessed the degree to which selective innervation occurs. Nerve–muscle contacts *in vitro* do not show any selective function or adhesion based on either their original axial position or their clonal relationships within the embryo.

Acknowledgements

This work was supported by grants from the National Institutes of Health (NS 17558, NS 12961) and the National Science Foundation (BNS-8543366).

References

Adler JE, Black IB 1986 Membrane contact regulates transmitter phenotypic expression. Dev Brain Res 30:237–241

Bennett MR, Ho S 1988 The formation of topographical maps in developing rat gastrocnemius muscle during synapse elimination. J Physiol (Lond) 396:471–496

Bonhoeffer F, Huf J 1982 *In vitro* experiments on axon guidance demonstrating an anterior-posterior gradient on the tectum. EMBO (Eur Mol Biol Organ) J 1:427–431

Brown MC, Booth CM 1983 Postnatal development of the adult pattern of motor axon distribution in rat muscle. Nature (Lond) 304:741–742

Chow I, Cohen MW 1983 Developmental changes in the distribution of acetylcholine receptors in the myotomes of *Xenopus laevis*. J Physiol (Lond) 339:553–571

Jacobson M, Hirose G 1981 Clonal organization of the central nervous system of the frog. II. Clones stemming from individual blastomeres of the 32- and 64-cell stages. J Neurosci 1:271–285

Kullberg RW, Lentz TL, Cohen MW 1977 Development of the myotomal neuromuscular junction in *Xenopus laevis:* an electrophysiological and fine-structural study. Dev Biol 60:101–129

Lance-Jones CC 1986 Motoneuron projection patterns in chick embryonic limbs with a double complement of dorsal thigh musculature. Dev Biol 116:387–406

Lance-Jones C 1988 Development of neuromuscular connections: guidance of motoneuron axons to muscles in the embryonic chick hindlimb. In: Plasticity of the neuromuscular system. Wiley, Chichester (Ciba Found Symp 138) p 97–115

Lance-Jones C, Landmesser L 1981a Pathway selection by chick lumbosacral motoneurons during normal development. Proc R Soc Lond B Biol Sci 214:1–18

Lance-Jones C, Landmesser L 1981b Pathway selection by embryonic chick motoneurons in an experimentally altered environment. Proc R Soc Lond B Biol Sci 214:19–52

Lawrence PA 1975 The structure and properties of a compartment border: the intersegmental boundary in *Oncopeltus*. In: Cell patterning. Elsevier/Excerpta Medica/North-Holland, Amsterdam (Ciba Found Symp 29) 3–23

Letourneau PC 1975 Cell to substratum adhesion and guidance of axonal elongation. Dev Biol 44:92–101

Lewis PR, Hughes AFW 1960 Patterns of myo-neural junctions and cholinesterase activity in the muscles of tadpoles of *Xenopus laevis*. Q J Microsc Sci 101:55–67

Moody SA, Jacobson M 1983 Compartmental relationships between anuran primary spinal motoneurons and somitic muscle fibers that they first innervate. J Neurosci 3:1670–1682

Myers PZ, Eisen JS, Westerfield M 1986 Development and axonal outgrowth of identified motoneurons in the zebrafish. J Neurosci 6:2278–2289

Nieuwkoop PD, Faber J 1967 Normal table of *Xenopus laevis* (Daudin), 2nd edn, Elsevier/North-Holland Publishing, Amsterdam
Purves D, Lichtman JW 1985 Principles of neural development. Sinauer Associates, Sunderland
Sanes JR 1987 Cell lineage and the origin of muscle fiber types. Trends Neurosci 10:219–221
Spitzer NC, Lamborghini JE 1976 The development of the action potential mechanism of amphibian neurons isolated in culture. Proc Natl Acad Sci USA 73:1641–1645
Stephens PJ, Govind CK 1981 Peripheral innervation fields of single lobster motoneurons defined by synapse elimination during development. Brain Res 212:476–480
Sulston JE, Horvitz HR 1977 Post-embryonic cell lineages of the nematode *Caenorhabditis elegans*. Dev Biol 56:110–156
Summerbell D, Stirling RV 1981 The innervation of dorsoventrally reversed chick wings: evidence that motor axons do not actively seek out their appropriate targets. J Embryol Exp Morphol 61:233–247
Sun Y-a, Poo M-m 1987 Evoked release of acetylcholine from the growing embryonic neuron. Proc Natl Acad Sci USA 84:2540–2544
Walthall WW, Murphey RK 1984 Rules for neural development revealed by chimaeric sensory systems in crickets. Nature (Lond) 311:57–59
Weisblat DA, Sawyer RT, Stent GS 1978 Cell lineage analysis by intracellular injection of a tracer enzyme. Science (Wash DC) 202:1295–1298
Whitelaw V, Hollyday M 1983 Neural pathway constraints in the motor innervation of the chick hindlimb following dorsoventral rotations of distal limb segments. J Neurosci 3:1226–1233
Wigston DJ, Sanes JR 1985 Selective reinnervation of intercostal muscles transplanted from different segmental levels to a common site. J Neurosci 5:1208–1221
Wolpert L 1969 Postitional information and the spatial pattern of cellular differentiation. J Theor Biol 25:1–47
Xie Z-p, Poo M-m 1986 Initial events in the formation of neuromuscular synapse: rapid induction of acetylcholine release from embryonic neuron. Proc Natl Acad Sci USA 83:7069–7073

DISCUSSION

Thesleff: When you place a myoball or membrane patch close to a growth cone you record potentials produced by acetylcholine and you call it a functional contact. Is that ACh release similar to the release occurring during development? That is, is it calcium dependent? Is it enhanced by the depolarization? Or is it just a spontaneous release, unrelated to what you see physiologically?

Sanes: The nerve-evoked ACh release is calcium dependent. The spontaneous release events are mostly calcium and TTX independent in culture. Moreover, from the initial moment of contact there appears to be the ability to sensitize that release (Sun & Poo 1987). If one stimulates at 0.2 per second and looks at the baseline level of endplate currents, and then stimulates at 10 per second for five seconds, the endplate currents are enhanced. Therefore, one can observe post-tetanic potentiation, even on initial contact, in these culture systems.

Dubowitz: You have done these elegant experiments with a relatively small change in position between the cells. At the other extreme, one obtains functional contact between neurites and myocytes of different species. Have you done anything at more crude levels, such as across species, or with some other gross differences, and found any obvious preferential contacts?

Sanes: We haven't, but Nirenberg's group looked at the ability of chick retinal neurons to form functional synapses with striated muscle fibres (Puro et al 1977). Functional contacts were made rapidly, but were not maintained. Spinal cord neurons did, however, form stable synapses with these myocytes.

Kernell: You described two important interactions between muscle cells and nerve endings. Firstly, you obtained an increasing amount of adhesion which was different in different cases. Secondly, there was a release of ACh from the nerve ending. Were these two effects related? Was there more adhesion in cases with a more striking increase of ACh release?

Sanes: A thesis that Dr Poo is testing is that ACh receptors may endow cells with adhesivity, and that where one does see larger potentials or endplate currents one would see more adhesion.

Henderson: Moody & Jacobson (1983) seemed to have far fewer neurons and muscles clonally related in the parts of the *Xenopus* embryo that they were studying, compared to your micrographs (not reproduced in text).

Sanes: This is because they published photographs of embryos that had been injected at the 128- or 256-cell stage. For the purpose of showing you plenty of motoneuronal pathways, I showed animals that had been injected at the 16- or 8-cell stage. However, Moody & Jacobson did see preferential association between clonally related nerves and muscles down to 16-cell stage injections.

Henderson: So it's not possible that in your studies a clonal relationship is diluted out by labelling early? If the determination of a nerve–muscle pair occurred later, you might have too many labelled cells to allow detection of a significant correlation.

Sanes: We injected at the stage (64 cell) when Moody & Jacobson observed highly significant relationships.

Holder: In the *Xenopus* spinal cord, only one particular group of motoneurons innervates the axial muscles, as distinct from motoneurons that innervate limb muscles. Also, in the rat experiment by Wigston & Sanes (1985) that you referred to, there is an unusual set-up, in that sympathetic axons are innervating skeletal muscle. Would that make the interpretation of the results difficult, if unusual motor neurons are innervating unusual muscles?

Sanes: In our culture experiments we do not even know that they are motor neurons, which does lead to interpretative difficulties. This could be why we fail to pick up something that's there, or see only marginal effects.

Holder: Have you tried *Xenopus* motoneurons and myoballs in culture?

Sanes: No. We would have to retrogradely label the motoneurons with a fluorophore and cell-sort before plating.

Buller: Can you say what proportion of your neurites are motor neurons? You might get good adhesion between two neurites that were interneurons rather than motoneurons.

Sanes: Roughly 66% of our cultured neurons show electrically induced release of ACh. In 87% of the naturally occurring contacts we see spontaneous, TTX-independent release of ACh. So a large percentage of the neurons are cholinergic, but they are not necessarily motoneurons.

Thesleff: Do you see non-quantal release of acetylcholine?

Sanes: We do see that. These phenomena have been documented by Sun & Poo (1985) and by Young (1986). Non-quantal events were observed after several days in culture, as revealed by a DC shift in the membrane potential (hyperpolarization) after curare block. Such DC shifts were not observed during the first few minutes of nerve–muscle contact, suggesting that early spontaneous release of ACh was mostly pulsatile.

Thesleff: Would the non-quantal release of ACh affect the membrane potential of the cell?

Sanes: Yes, and this non-quantal release appears to affect membrane potential for longer than brief synaptic currents do.

Thesleff: Does the non-quantal release play a role in the assembly of ACh receptors?

Sanes: It is possible. Dr Poo's thesis is that there may be an 'electrophoresis' of ACh receptors or other molecules towards the synaptic site, although this would only enhance the normal accumulation of ACh receptor that occurs in the presence of ACh receptor antagonists. With an on-going DC shift, instead of brief pulses of current, one might obtain a disproportionate physical effect on molecules in the area in that way.

Kernell: How variable were your results in relation to how much of the nerve terminal came in contact with the muscle cell? You could have a long stretch of a nerve fibre curving around a muscle cell, or you could just touch it with the axon tip. Does this matter?

Sanes: It absolutely matters! When one manipulates a myoball into contact with a neurite and puts increasing pressure on it (which presumably corresponds to a greater surface area of contact), endplate current amplitude increases.

Kernell: Could you influence your results by pushing the muscle cell and the nerve ending closer together?

Sanes: That is so for the endplate currents, but possibly not for the adhesion, which may have more to do with the ability of a growth cone to put out small filamentous processes and adhere to the myocyte surface.

Kernell: So there was a certain lack of correlation, or a degree of independence, between these two phenomena (transmitter release and adhesion) here?

Sanes: There has not yet been a quantitative comparison.

Henderson: Mu-ming Poo has suggested that electrical currents flowing through the ACh receptors might be important in attracting the growth cone, leading to the initial steps in nerve–muscle adhesion. I wondered whether, in these *in vitro* experiments, the electrical activity of the system affects the efficiency of the formation of contacts. What happens if you block transmission, using α-bungarotoxin or curare?

Sanes: I don't think that has been done. In culture, if one blocks activity, a large amount of activity is observed as soon as the block is removed.

Lance-Jones: Have you noticed in culture whether any of the *neurons* that originate from the same position, or that were clonally related, tend to selectively adhere to one another?

Sanes: They didn't seem to, although we did not quantify this parameter.

References

Moody SA, Jacobson M 1983 Compartmental relationships between anuran primary spinal motoneurons and somitic muscle fibers that they first innervate. J Neurosci 3:1670–1682

Puro DG, DeMello FG, Nirenberg M 1977 Synapse turnover: the formation and termination of transient synapses. Proc Natl Acad Sci USA 74:4977–4981

Sun Y-a, Poo M-m 1985 Non-quantal release of ACh at a developing neuromuscular synapse in culture. J Neurosci 5:634–642

Sun Y-a, Poo M-m 1987 Evoked release of acetylcholine from the growing embryonic neuron. Proc Natl Acad Sci USA 84:2540–2544

Wigston DJ, Sanes JR 1985 Selective reinnervation of intercostal muscles transplanted from different segmental levels to a common site. J Neurosci 5:1208–1221

Young SH 1986 Spontaneous release of transmitter from the growth cones of *Xenopus* neurons *in vitro*: the influence of Ca^{2+} and Mg^{2+} ions. Dev Biol 113:373–380

Reorganization of synaptic inputs to developing skeletal muscle fibres

G. Vrbová, M.B. Lowrie and *J. Evers

Department of Anatomy & Developmental Biology, University College London, Gower Street, London WC1E 6BT, UK and **Department of Molecular Neurobiology, School of Medicine, New Haven, Connecticut 06510, USA*

Abstract. During early stages of postnatal development skeletal muscle fibres of mammals are contacted by several axons. The transition from poly- to mononeuronal innervation has been extensively studied on the rat soleus. The role of activity in this process has been acknowledged but the mechanisms leading to synapse remodelling are not understood. The participation of the muscle has to be taken into account; if muscles are paralysed by α-bungarotoxin, the elimination of terminals is arrested. Changes in Ca^{2+} also influence the rate of removal of terminals. Calcium seems to act through a calcium-activated neutral protease (CANP) present in nerve endings. If CANP is inhibited, elimination fails to take place. Thus Ca^{2+} enters the terminal and activates the CANP. Release of K^+ ions from active muscle could link muscle activity and synapse elimination. Excess K^+ was found to reduce nerve–muscle contacts, by depolarizing terminals and allowing Ca^{2+} entry. A greater increase of Ca^{2+} concentration in smaller terminals would be expected, because of their surface-to-volume ratio, and they are preferentially eliminated. Thus elimination depends on the unequal size of terminals at the endplate. Therefore the 'survivability' of individual nerve endings may already be determined at the time of synapse elimination.

1988 Plasticity of the neuromuscular system. Wiley, Chichester (Ciba Foundation Symposium 138) p 131–151

In the developing neuromuscular system, motoneurons and muscles develop independently and establish connections with each other at a certain stage of development. Later, at a time when the animal is beginning to display neurogenic movement, a proportion of the motoneurons die, while the surviving cells form a large number of synaptic contacts with the skeletal muscle fibres that they supply (see Oppenheim & Wu Chuwang 1983). During further maturation in mammals, all the synapses but one on each muscle fibre are eliminated (Redfern 1970, Brown et al 1976). This reorganization determines the size distribution of motor units within each muscle.

The elimination of the superfluous nerve endings is brought about by the increased activity of the developing animal. It coincides in time with the increase in locomotor activity at later stages of development (Navarrete &

Vrbová 1983), it is delayed when nerve activity is reduced by tenotomy, tetrodotoxin or spinal cord transection (Benoit & Changeux 1975, Thompson et al 1979, Caldwell & Ridge 1983), and it can be speeded up by increasing neuromuscular activity (O'Brien et al 1978, Zelená et al 1979). The participation of the target muscle in this regulation of synaptic inputs has been demonstrated, because postsynaptic neuromuscular blocking agents, such as curare or α-bungarotoxin, which prevent activity, retard or even prevent the elimination of superfluous nerve endings (Srihari & Vrbová 1978, Duxson 1982). Thus the activity of the muscle fibres is essential for the synaptic reorganization of this system. The mechanism by which muscle activity leads to a reduction of its synaptic inputs is not understood, but from our previous work it appears that a Ca^{2+}-activated neutral protease (CANP), known to be present in nerves (Kamakura et al 1983), plays an important part in this event.

This paper will present evidence to show that muscle activity can lead to the activation of this protease in nerve terminals. In addition we shall argue that the unequal sizes of nerve terminals present at the time of synapse elimination are important for the selective destruction of nerve endings. Our results indicate, however, that at the time of synapse elimination the fate or 'survivability' of the nerve terminal is already determined, and while we can shift the time course of the developmental process we cannot influence its outcome. Thus the size distribution of motor units is likely to be determined earlier in development, before the onset of rapid synapse elimination.

Methods

The experiments were carried out on soleus muscles of 8–14-day-old Wistar rats. The soleus muscles were removed under ether anaesthetic and the level of polyneuronal innervation was examined either electrophysiologically, histologically, or using ultrastructural criteria (see O'Brien et al 1978, 1980, 1984). Before the innervation pattern was assessed the soleus muscles were incubated in solutions containing acetylcholine (ACh), raised Ca^{2+} concentration, and inhibitors of CANP, while the other muscle was kept in normal Krebs–Henseleit solution. In another group of experiments nine-day-old rats were anaesthetized with ether and, using sterile precautions, a strip of silicone rubber containing either BAPTA (1, 2-bis (2-aminophenoxy)ethane-N,N,N′,N′-tetraacetic acid, a calcium-chelating agent), or an inhibitor of CANP was placed along the soleus muscle. Three days later the muscles were removed and processed so that the extent of polyneuronal innervation could be assessed (see O'Brien et al 1978).

In a group of rats aged 8–10 days electrodes were implanted so that the sciatic nerve could be stimulated (see O'Brien et al 1978). The animals were stimulated at 8Hz for three hours a day for one to two weeks. Within a week after stimulation had stopped the rats were anaesthetized (4.5% chloral hydrate, 1 ml/100 g of body weight) and their extensor digitorum longus muscles prepared for tension recording (see Lowrie & Vrbová 1984). The spinal cord was then exposed and ventral root filaments were teased out so that individual axons could be stimulated. Contractions were elicited using square-wave pulses of 0.01 ms duration and supramaximal intensity.

Results and discussion

Modifications of synaptic contacts by ACh treatment or neuromuscular activity

Soleus muscles of nine-day-old rats were incubated for two hours in solutions

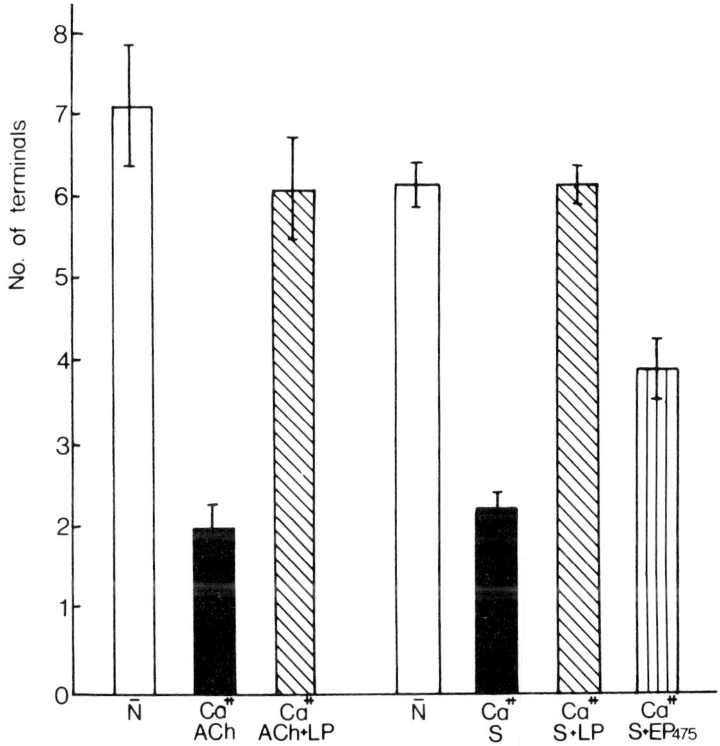

FIG. 1. The columns show the number ± SD of axon profiles per endplate in soleus muscle fibres from nine-day-old rats. N̄, incubated for 2 h in normal Krebs Henseleit solution; ACh + Ca^{++}, incubated in solutions containing ACh (10^{-3} g/ml) and 12 mM Ca^{2+}; + LP, incubated in solution as before with added leupeptin (0.1 mM) and pepstatin (0.1 mM); S, after 2 h stimulation at 8 Hz in 12 mM Ca^{2+}; S + LP, after 2 h stimulation in 12 mM Ca^{2+} with leupeptin and pepstatin included in the bathing solution; S + EP_{475}, after 2h stimulation in 12 mM Ca^{2+} with EP_{475} included in the bathing solution. (Data from O'Brien et al 1980 and 1984.)

containing ACh (10^{-3} g/ml). They were processed for electron microscopy and the number of axon profiles in contact with the muscle fibre on each endplate was counted. The results showed that exposure to ACh significantly reduced the number of synaptic contacts.

A further reduction in the number of connections was seen when the Ca^{2+} concentration was increased from 1.9 to 12 mM (see Fig. 1). This treatment also produced complete denervation of some endplates. The possibility that a Ca^{2+}-activated protease could be involved in this rapid reduction of synaptic contacts was suggested by results in which protease inhibitors (leupeptin, 0.1 mg/ml and pepstatin, 0.1 mg/ml) were included in the incubation medium. As shown in Fig. 1, these inhibitors prevented the reduction of neuromuscular contacts seen in the previous group of experiments.

Diffuse application of ACh to skeletal muscles is not a physiological stimulus, and it was desirable to apply naturally released ACh to the neuromuscular junction. This was achieved by stimulating the motor nerve to skeletal muscle fibres for two hours at 8 Hz. Stimulation alone did not lead to a reduction of synaptic contacts in this relatively short span of time, but when the muscles were activated in solutions containing an increased concentration of Ca^{2+}, a reduction in synaptic inputs occurred within the two hours of stimulation (see Fig. 1). As in the previous group of experiments the addition of inhibitors of CANP prevented the reduction of neuromuscular contacts induced by neuromuscular activity (see Fig. 1).

These results taken together suggest that the regulation of neuromuscular contacts is mediated by a Ca^{2+}-sensitive protease present in the nerve terminal. In order to establish whether such a mechanism could operate *in vivo* during the time when synapse elimination usually takes place, we attempted to alter the time course of synapse elmination *in vivo* by interfering (a) with the Ca^{2+} concentration around the muscle fibres, and (b) with the protease activity of the soleus neuromuscular junctions.

Effect of protease inhibitors and Ca^{2+} concentration on the time course of neuronal elimination

In a series of experiments the concentration of Ca^{2+} around soleus muscle fibres was reduced by locally applying a chelating agent (BAPTA) to the soleus muscle of nine-day-old rats. Three days later the muscles were removed and the extent of polyneuronal innervation was assessed histologically and electrophysiologically. Fig. 2 shows examples of endplates from control and treated muscles. The block diagram shows that in treated muscles a significantly higher proportion of the endplates was supplied by more than one axon. These neuromuscular contacts were functional, for electrophysiological studies showed a similarly higher proportion of endplates with polyneuronal innervation (see Connold et al 1986). Lowering the Ca^{2+} concentration around the soleus muscle fibres was not sufficient to decrease

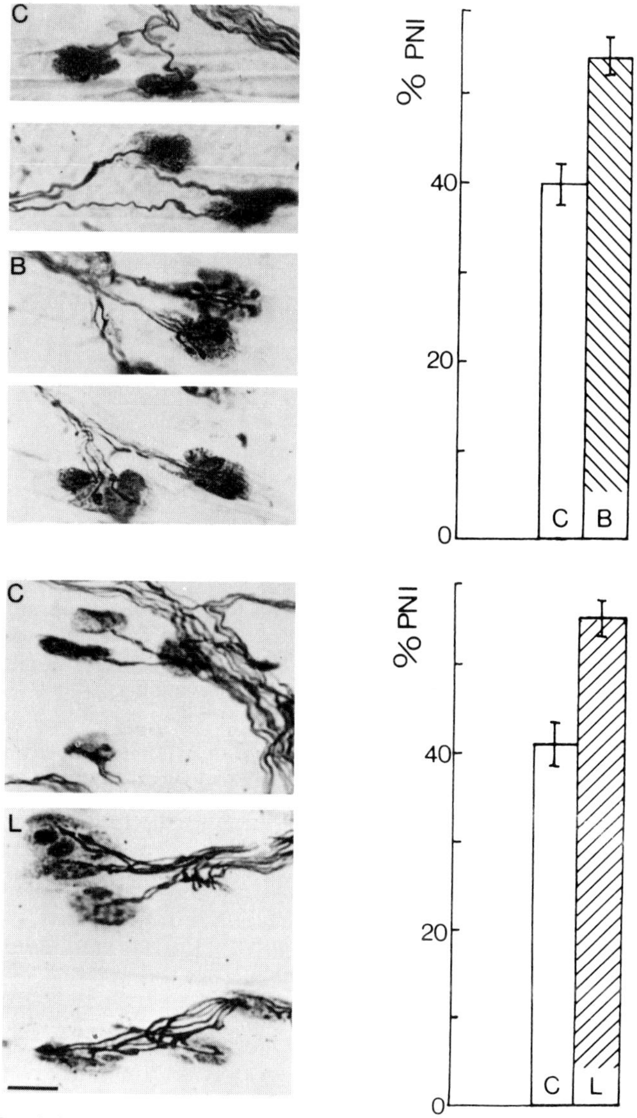

FIG. 2. On left, cholinesterase–silver-stained endplates from control 12-day-old soleus muscles and muscles treated with a chelating agent (BAPTA) or leupeptin for three days *in vivo*. The block diagram compares the percentage of polyneuronal innervation (PNI) in control 12-day-old soleus muscles to that in muscles treated with a chelating agent (BAPTA) or leupeptin. C, control; B, BAPTA treated; L, leupeptin treated. (Results from Connold et al 1986.)

muscle activity (see Connold et al 1986), yet it prevented the normally occurring reduction in polyneuronal innervation.

Whether the reduction of Ca^{2+} concentration was exerting its effect by

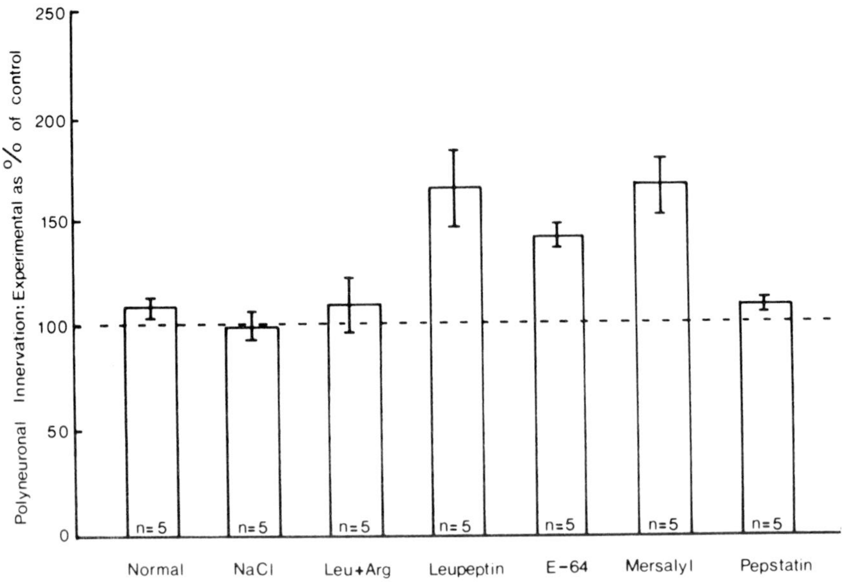

FIG. 3. The block diagram shows the percentage of polyneuronal innervation of soleus muscle fibres of 12-day-old rats pretreated for three days with various agents, some of which (leupeptin, E-64 and mersalyl) are inhibitors of CANP. Note that treatment with inhibitors preserves polyneuronal innervation.

preventing the activation of CANP was studied in experiments where, instead of a chelating agent, various inhibitors of CANP were applied to the muscle. Fig. 2 shows examples from muscles treated with leupeptin, a CANP inhibitor. Histological examination (left) indicated that, as in muscles treated with chelating agents, treatment with leupeptin preserved the neuromuscular contacts. The histogram on the right confirms this impression. The effect of a number of compounds on the normally occurring reduction of neuromuscular contacts was studied using electrophysiological criteria. Because of the relatively big variation in polyneuronal innervation between animals, and the small difference between the two soleus muscles in the same animal, the results were calculated so that the degree of polyneuronal innervation of the treated muscles was expressed as a percentage of the control (see Connold et al 1986). Fig. 3 summarizes the results and shows that all muscles treated with a compound that inhibited the CANP maintained the high level of neuromuscular connections, whereas compounds that did not affect the CANP (NaCl, leucine and arginine, pepstatin) had little or no effect.

Thus Ca^{2+} appears to affect nerve–muscle contacts by activating CANP, which in turn degrades the cytoskeleton of the terminals and leads to their retraction. If this were the case it could be expected that small nerve terminals would be preferentially eliminated, for, because of an unfavourable surface-to-volume ratio, the concentration of Ca^{2+} would rise more steeply in these small nerve terminals than in large ones.

Synapse elimination and nerve terminal size

During normal development of the rat soleus the number of axon profiles contacting the skeletal muscle fibres decreases (Korneliussen & Jansen 1976). At 10 days each endplate has an average of 5.4 ± 0.27 axon profiles, and at 12 days this decreases to 4.3 ± 0.22 axon profiles (see Duxson 1982, 1983). During this period the size distribution of the nerve terminals changes in that there is a decrease in small and medium-sized ones, and an increase in large terminals. From results obtained by Duxson (1982) there is a loss of one small and 0.3 medium-sized terminal per endplate and an increase of 0.2 large terminals. Thus the loss of 1.1 terminal per endplate that occurs during this time is due to the loss of small terminals.

Elimination of polyneuronal innervation can be speeded up by prolonging the action of ACh by injecting muscles *in vivo* with an anticholinesterase agent, diisopropyl fluorophosphate (DFP). This leads to a rapid reduction in nerve–muscle connections within 24 hours and, in this case also, the small and medium-sized terminals are removed, while the large ones survive the treatment (Duxson & Vrbová 1985).

In these experiments the change took place over a period of one to two days during the animal's normal development. We also examined muscles in which the number of nerve–muscle contacts was reduced by stimulating the muscles *in vitro* for two hours in solutions containing high Ca^{2+} concentrations. In this case nerve terminals of all sizes were affected by the treatment, but the small and medium-sized terminals suffered greater reduction than the large ones (O'Brien et al 1984).

Although these results show that Ca^{2+} and CANP are important for the elimination of polyneuronal innervation, they do not explain the link between muscle activity and elimination of synapses.

The effect of K^+ concentration on the number of neuromuscular contacts

It is known that skeletal muscle fibres on depolarization release K^+. This in turn could open voltage-dependent Ca^{2+} channels in the nerve terminal, thus allowing an increase of the Ca^{2+} concentration in the terminal. We tested the possibility that increased K^+ concentration could change the degree of polyneuronal innervation.

Soleus muscles from 8–11-day-old rats were removed and one muscle was placed in normal Krebs–Henseleit solutions while the other was kept in solutions containing 20 mM K^+.

Polyneuronal innervation was assessed electrophysiologically and, as

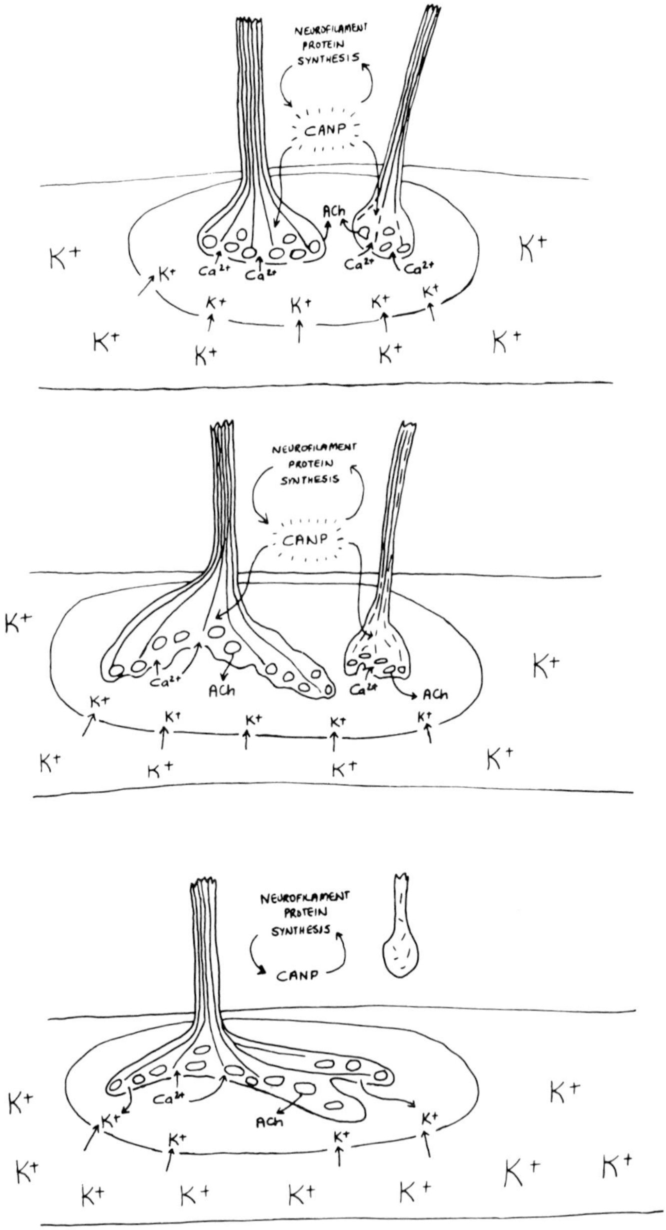

FIG. 4. Schematic diagram illustrating the proposed mechanism of elimination of polyneuronal innervation during postnatal development.

before, was expressed as a percentage of the contralateral muscle kept in normal Krebs solution. In comparison to such controls, polyneuronal innervation of soleus muscles incubated in 20 mM K^+ for only two hours was significantly reduced to 80 ± 3% (n = 11). To test whether this decrease can be prevented by inhibiting the CANP, leupeptin was included in the solution containing raised [K^+]. This not only prevented the decrease induced by 20 mM K^+, but increased the percentage of muscle fibres supplied by more than one axon to 118 ± 4% (n = 8). This apparent increase was however not significant.

Model of neuronal elimination

A satisfactory model of the elimination of excess terminals has to take into account the following facts:

(a) The process is dependent on the activity of the postsynaptic membrane — that is, the muscle fibre;

(b) Ca^{2+} concentration plays an important role, in that increased Ca^{2+} accelerates and decreased Ca^{2+} reduces the rate of elimination;

(c) Inhibitors of CANP preserve neuromuscular contacts;

(d) Increased K^+ concentration rapidly reduces neuromuscular contacts, and this is prevented by inhibitors of CANP; and

(e) The smaller terminals are preferentially eliminated.

Fig. 4 illustrates schematically the mechanism that we consider accommodates these facts.

Neuromuscular activity leads to the release of K^+ concentration in the synaptic cleft. This event is already well documented (Hohlfeld et al 1981). This increase in K^+ concentration will open voltage-dependent Ca^{2+} channels, which are inactivated slowly, and therefore would remain open for a relatively long period of time. The Ca^{2+} that enters the nerve terminals will then activate the CANP, leading to the destruction of cytoskeletal elements of the terminal and its removal. Small terminals should be more vulnerable, for in view of their surface-to-volume ratio the concentration of Ca^{2+} should increase more in these than in the larger terminals. Thus this mechanism would account for all the observations presented here.

If our proposal is correct, it follows that at the time of synapse elimination the fate of the terminal is already determined, for, at eight to 10 days of age, each endplate is contacted by axon profiles of variable size. Therefore at this time it may be possible to change the time course of the elimination of polyneuronal innervation but not its final outcome. In muscle this outcome is represented by the distribution of motor unit sizes. The next experiments were therefore designed to test this suggested mechanism.

The effect of chronic nerve stimulation on motor unit distribution in the rat EDL muscle

As in soleus, virtually all EDL muscle fibres from six-day-old rats are polyneuronally innervated and this declines with age to a residual level of 5% at 16–17 days of age. The sciatic nerve was stimulated during this period as

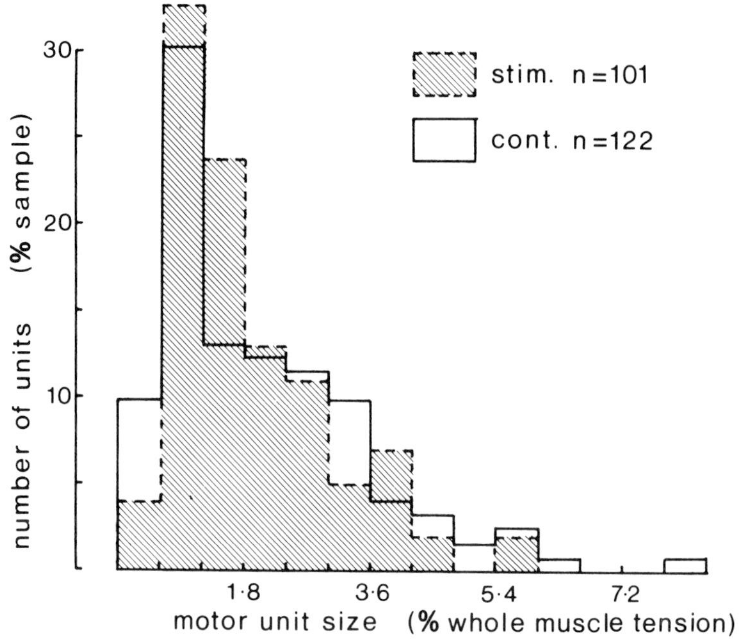

FIG. 5. Distribution of motor unit sizes in control and chronically stimulated EDL muscles of 3–5-day-old rats. On the abscissa the size of the unit is expressed as a percentage of total muscle tension. The ordinate shows the number of units expressed as a percentage of the total within each muscle.

described by O'Brien et al (1978) and the distribution of motor unit sizes was assessed in rats aged three to five weeks, by stimulating single ventral root filaments and recording tetanic tension in individual motor units. The stimulation had no appreciable effect on the tension developed by the EDL muscles, their weight or contractile speeds. The histochemical examination revealed that the muscle fibres of the stimulated EDL stained more darkly for succinate dehydrogenase (SDH), and there were no lightly stained fibres present. Therefore it can be assumed that activity had reached the muscle fibres. Fig. 5 shows the motor unit size distribution of control and stimulated EDL muscles. It is clear that stimulation of the sciatic nerve during the time when elimination of polyneuronal innervation normally takes place did not alter the development of the size distribution of motor units. These results therefore indicate that the 'survivability' of the nerve terminal is decided before the onset of the elimination of polyneuronal innervation.

Conclusion

The reorganization of synaptic inputs to skeletal muscle fibres that takes place during postnatal development depends on activity. We are proposing a simple model in which K^+ ions released from active muscles open voltage-dependent Ca^{2+} channels of nerve terminals. The entry of Ca^{2+} can in small terminals reach concentrations high enough to activate an enzyme that destroys cytoskeletal proteins (CANP), and can in this way lead to the disruption of many neuromuscular contacts. We provide evidence to show that if terminals of different sizes are present at the same synapse, the small terminals are preferentially lost. Thus at the time of synapse elimination the 'survivability' of the terminal is already determined by its relative size.

Acknowledgements

We are grateful to the Medical Research Council, the Muscular Dystrophy Group and The Wellcome Trust for supporting this work.

References

Benoit P, Changeux JP 1975 Consequences of tenotomy on the evolution of multi-innervation in the developing rat soleus muscle. Brain Res 99:354–358

Brown MC, Jansen JK, Van Essen D 1976 Polyneuronal innervation of skeletal muscle in new-born rats and its elimination during maturation. J Physiol (Lond) 261:387–422

Caldwell JH, Ridge RMAP 1983 Effect of deafferentation and spinal cord transection on synapse elimination in developing rat muscles. J Physiol (Lond) 339:145–159

Connold AL, Evers JV, Vrbová G 1986 Effect of low calcium and protease inhibitors on synapse elimination during postnatal development in the rat soleus muscle. Dev Brain Res 28:99–107

Duxson MJ 1982 The effect of postsynaptic block on development of the neuromuscular junction of postnatal rats. J Neurocytol 11:395–408

Duxson MJ 1983 Mechanisms controlling maturation of the mammalian neuromuscular junction. PhD thesis, University of London

Duxson MJ, Vrbová G 1985 Inhibition of acetylcholinesterase accelerates axon terminal withdrawal at the developing neuromuscular junction. J Neurocytol 14:337–363

Hohlfeld R, Sterz R, Peper K 1981 Prejunctional effects of anticholinesterase drugs at the endplate mediated by presynaptic acetylcholine receptors or by postsynaptic potassium efflux? Pfluegers Arch 391:213–218

Kamakura K, Ishiura S, Sugita H, Toyokura Y 1983 Identification of Ca^{2+} activated neutral protease in the peripheral nerve and its effect on neurofilament degeneration. J Neurochem 40:908–913

Korneliussen H, Jansen JKS 1976 Morphological aspects of the elimination of polyneuronal innervation of skeletal muscle fibres in newborn rats. J Neurocytol 5:591–604

Lowrie MB, Vrbová G 1984 Different patterns of recovery of fast and slow muscles following nerve injury in the rat. J Physiol (Lond) 349:397–410

Navarrete R, Vrbová G 1983 Changes of activity patterns in slow and fast muscles during postnatal development. Dev Brain Res 8:11–19

O'Brien RAD, Ostberg AJC, Vrbová G 1978 Observations on the elimination of

polyneuronal innervation in developing mammalian skeletal muscle. J Physiol (Lond) 282:571–572
O'Brien RAD, Ostberg AJC, Vrbová G 1980 The effect of acetylcholine on the function and structure of the developing mammalian neuromuscular junction. Neuroscience 5:1367–1379
O'Brien RAD, Ostberg AJC, Vrbová G 1984 Protease inhibitors reduce the loss of nerve terminals induced by activity and calcium in developing rat soleus muscles in vitro. Neuroscience 12:637–646
Oppenheim RW, Wu Chuwang 1983 Aspects of naturally occurring motoneurone death in the chick spinal cord during embryonic development. In: Burnstock G et al (eds) Somatic and autonomic nerve–muscle interactions. Elsevier/North-Holland, Amsterdam p 57–109
Redfern PA 1970 Neuromuscular transmission in newborn rats. J Physiol (Lond) 209: 701–709
Srihari T, Vrbová G 1978 The role of muscle activity in the differentiation of neuromuscular junctions in slow and fast chick muscles. J Neurocytol 7:529–540
Thompson W, Kuffler DP, Jansen JKS 1979 The effect of prolonged, reversible block of nerve impulses on the elimination of polyneuronal innervation of new-born rat skeletal muscle fibres. Neuroscience 4:271–281
Zelená J, Vyskocil F, Jirmanova I 1979 The elimination of polyneuronal innervation of end-plates in developing rat muscles with altered function. Prog Brain Res 49:365–372

DISCUSSION

Sanes: Does the proposed mechanism explain why all active nerve terminals, whatever their size, do not get chewed up by the calcium-activated neutral protease (CNAP)? Is there any preferential localization of CNAP at a certain time in development?

Vrbová: I would not think so, because, as Michael Brown put it earlier, something has to stop the nerve from growing. Your experiments and those of Dr Poo do show that you are transforming a neuron from a growing cell into a secretory cell, and it would be interesting to see what happens in the cell body when this transformation occurs. This protease mechanism may be important for that, because it probably destroys some cytoskeletal elements and that will stop the neuron from growing further. In addition, the Ca^{2+} entry that we suggest may also replenish calcium stores, which will enable the terminal to be a more efficient secretory structure. So the 'protease' type of mechanism could explain both events.

Buller: Presumably all motor terminals are small at some stage; do you envisage a lot of wastage, until one terminal reaches a critical size and can outlive the rest?

Vrbová: No. Each terminal is initially small, and has very little transmitter and therefore little influence on the effector; perhaps muscle maturation at this stage depends on the whole complement of terminals and only when terminals become bigger will some become redundant. It is also possible that the terminals support each other. The event of polyneuronal innervation and its elimination can therefore have other morphogenetic functions, in addition to refining the circuitry.

Thesleff: Perhaps your results could be explained by an influx of calcium into the muscle caused by the transmitter–receptor interaction. Calcium would activate muscle proteases which damage the postsynaptic endplate membrane and thereby cause the withdrawal of the nerve.

Vrbová: If we block the postsynaptic membrane with curare we do not get increased withdrawal, so this seems unlikely as a mechanism.

Kernell: Do relatively large axons tend to produce relatively large nerve endings? If so, the preferential disappearance of thin endings would explain why thick axons would tend to catch larger motor units with a greater number of fibres than those of the thinner axons.

Vrbová: At this stage of development the distribution of axon sizes is really quite uniform, so I cannot answer that question.

Ribchester: Did you count the numbers of motor units in the muscles after incubation in high levels of potassium, to rule out the possibility that high $[K^+]$ was producing a block of motor axon collaterals within the muscle? You describe the high potassium concentrations that you used as modest, but threefold is a lot, and one would expect osmotic effects on the very small diameter nerve endings and terminals within the muscle. If they had been blocked by K^+-induced depolarization, the reduced polyinnervation that you see would have been artefactual.

Vrbová: We didn't count the motor units, but the histological study done subsequently also shows a reduction in contacts which could not be explained in this way.

Ribchester: Dan Sanes raised the issue of the persistence of terminals. Where it has been possible to measure intracellular calcium, as in squid giant synapses, activity in nerve terminals raises the level from 10^{-7} to 10^{-5} M. So your proposed mechanism has to provide an explanation of why normal, active terminals persist throughout life, particularly in view of experiments by Lichtman et al (1987) showing that not only is there no sprouting of the terminals at the endplate in the intact animal, but there is no regression of those terminals either.

Vrbová: If the nerve terminal is large enough, the increase in calcium concentration caused by the opening of the slowly inactivating calcium channels will not be sufficient to activate the CANP. There are also other mechanisms in nerve terminals that regulate the calcium concentration and might be acquired later: one is the calcium-gated potassium channels, which produce an outward current that doesn't allow further calcium to enter the nerve terminal.

Ribchester: What level of calcium is needed inside the terminal to activate the calcium-activated neutral peptidase?

Vrbová: I don't know.

Eisenberg: The role of potassium that you postulate worries me. As the neuromuscular postsynaptic region continues to develop in the adult, the folds become deeper, so the accumulation of potassium in the clefts could be expected to be greater, which seems to be working in the wrong direction. The more clefts there are, the more potassium accumulates and therefore the more depolarization would result. How do you reconcile that?

Vrbová: If the terminal is bigger, it may be able to cope with the larger shifts of Ca^{2+}. In addition, with development the terminal acquires properties that stop Ca^{2+} entry by activating Ca^{2+}-controlled K^+ channels. Also, I wouldn't agree about the stability of the adult terminal. The sternocostalis muscle of the adult rat is a flat sheet of muscle that has two types of endplates, one being complex and the other being simple. The normal sternocostalis muscle of the six-week-old rat has about half of each type. When the protease is inhibited with leupeptin or a calcium chelator for two days, the number of complex endplates increases (Swanson & Vrbová 1987). This doesn't show such great stability.

Hoffman: Your observation suggests that calcium-activated protease activity plays a role in the elimination of excess axon terminals. Degradation of the cytoskeleton at the axon terminals may also provide a mechanism for maintaining a constant axonal length, with inhibition of protease activity leading to terminal sprouting (Lasek & Hoffman 1976). Did you find terminal sprouting after leupeptin treatment?

Vrbová: If we partially denervate the muscle, leupeptin does not increase the amount of sprouting, but the number of endplates that remain functional is increased, as if it protects terminals to endplates that normally would withdraw. If we section the L5 ventral root in the rat, for example, normally the axons from L4 sprout to occupy 60% of the soleus muscle fibres. If at the same time the muscle is treated with leupeptin, L4 axons supply all soleus muscle fibres. We think that leupeptin does act on the nerve ending.

Crowder: I can see how this mechanism accounts for synapse elimination in general, but is it by itself enough to account for all phenomena of synapse elimination, such as eliminating down to only one synapse per muscle fibre, in almost all cases?

Vrbová: As you say, it isn't all cases; and a mathematical model was proposed by Wilshaw (1981) to show how this could be a random process. Also, we and others have shown that many muscle fibres won't make it during development; the EDL loses a proportion of muscle fibres, presumably due to loss of neuromuscular contacts.

Crowder: You are saying that in such cases all the synapses are eliminated and then the muscle fibres die?

Vrbová: Yes; if there is a random process, one would expect this would happen occasionally.

Nadal-Ginard: Is it possible that there is some feedback among the different terminals? This would fit with the model you have published. One can imagine that it works in a manner similar to the way pacemaker activity is established in a variety of systems, from electrical pacemakers to *Dictyostelium* aggregation with cyclic AMP release. In all these systems, the structure or cell that generates the strongest signal becomes the pacemaker and eliminates the activity of the others. The calcium-activating protease may be the physical means for the elimination of the terminals that fail to become these dominant ones, but perhaps that is a later stage—just a way of cleaning up the mess after one of the terminals has become functionally established as the dominant one.

Vrbová: It is quite possible that the terminals that have remained smaller are those where the axon has undergone most branching. You are asking why some of them remain small?

Nadal-Ginard: Yes, and explaining why generally only one of them remains per neuromuscular plate. If there were some type of feedback inhibition in the endplate, that could explain why one and only one is always left there.

Vrbová: You still have to explain the role of the target in the elimination of the smaller terminals, which would not fit in with your model; but your suggestion could explain why one stays smaller.

Hoffman: Do we know whether the big primary endings are related anatomically to the smaller ones on the same muscle fibre? Are they branches of separate axons, or could they be branches of the same one?

Vrbová: We can't tell this from the electron microscope studies, but from the silver preparations we can see the different terminals originating from separate axons.

Pette: Which biochemical process could be responsible for synapse elimination? A protease could be something secondary in the entire process.

Vrbová: In the presence of the protease inhibitor, the terminal remains functional, so the protease doesn't just clean up the mess; it must kill the terminal or in some way make it non-functional. The protease inhibitor maintains the terminals as functional entities, because you can measure their endplate potentials caused by their activity.

Gordon: One neglected issue in this discussion is the role of the extracellular matrix in synaptogenesis. With the elaboration of the adult endplate, adhesion between the nerve terminal and the extracellular matrix may be important. Certainly in peripheral nerve regeneration, synapses can form even without muscle. Do you get polyneuronal innervation on an irradiated muscle, which retains its extracellular matrix? And, secondly, do the activated proteases that you refer to in fact disrupt the bond between the terminal and the extracellular matrix, rather than acting as calcium-activated proteases within the terminal?

Vrbová: We have never looked at irradiated muscle. We have tried to see whether new synapses are formed on muscle fibres poisoned with α-bungarotoxin, in young rats. We wanted to see whether sprouting terminals were able to make contact with muscle fibres when the postsynaptic membrane doesn't respond. They were not able to do so. So if you paralyse the muscle at this stage of development you don't get functional contacts with the vacated muscle fibres; they will never be colonized.

Carlson: In adult rats, if we eliminate muscle fibres with Marcaine (an anaesthetic drug), the nerve terminals seem to be perfectly happy at the electron microscopic level. Yet one would expect the motor endplate region to be awash in a sea of potassium and calcium and other juices that muscle fibres are releasing. Perhaps there is a totally different mechanism for conferring instability at this point?

Vrbová: I think the extracellular matrix in the adult plays an important role in synapse formation, as has been shown by Jack McMahan and Joshua Sanes and many others (see Sanes 1983), and it is certainly remarkable how stable the subneural apparatus is, once it is formed. In the young animal, such as the rat, in the first 10 days after birth, the myofibrils lie right under the sarcolemma. If a nerve contacts such a structure, the shearing force that it has to cope with, when trying to establish contact, must be phenomenal. Whereas, two or three weeks after nerve–muscle contact, there is an empty space beneath the sarcolemma and, presumably, no shearing force. In the young animal it may therefore be true that the extracellular matrix plays an important role in later events of synapse stabilization, but not in the initial formation of contacts.

Van Essen: In relation to the problem of why only one terminal survives, there are muscles in non-mammalian species (such as amphibians) with polyneuronal innervation which is perfectly stable. An attractive hypothesis is that the critical difference has to do with the sort of size principle that Dr Vrbová elaborated. If for any reason larger terminals have an advantage over smaller ones, this will inexorably lead to synapse elimination; if for any other reason smaller terminals are completely effective, polyneuronal innervation should be stable. The protease hypothesis is an attractive way of mediating that process, but there are other possible explanations as well.

With regard to the role of activity, there are two important issues here. One aspect, already discussed, is that changes in nerve or muscle activity can affect the overall rate of synapse elimination, either accelerating it or slowing it down. A second issue, conceptually quite distinct, concerns the effect of differential activity at a single endplate: does activity confer an advantage or not? Experiments with Ed Callaway and Jim Soha have produced a surprising result (Callaway et al 1987). We placed a tetrodotoxin-laden Silastic plug into one of the three spinal nerves innervating the soleus muscle of the newborn rabbit, using one that contributes only a minority of fibres. As a result, a subset of muscle fibres receive at least one active input from the major root (S1) and at least one inactive input from the blocked root (L7). We leave the block for 4–5 days postnatally and then take out the plug, wash the tetrodotoxin out, and measure motor unit sizes with a tension gauge.

The results immediately after this 4–5-day period of block were that motor unit sizes are about 50% larger than they would have been if normally active. That was true for both fast and slow motor unit populations.

We also allowed the activity to recover over an additional 4–5-day period to see whether there was a general delay in the synapse elimination process. We found the same effect, namely that in both fast and slow motor units the once-inactivated motor units are substantially larger than the control, normally active ones. This experiment argues strongly against the hypothesis that this is a Hebb-like competition; it seems to be the opposite, in which inactivity confers an advantage, rather than putting inactive synapses at a disadvantage.

Sanes: These results are surprising. It might be equally compatible with your results that tetrodotoxin actually enhances the growth of axons, independent of the nerve terminal's interactions with the postsynaptic target. Two groups have shown that TTX enhances neurite outgrowth *in vitro* (Bixby & Spitzer 1984, Van Huizen & Romijn 1987), and in the retinotectal system TTX causes an exuberance of growth of retinal ganglion cell axons (Reh & Constantine-Paton 1985). So is it conceivable that the motor axons sprouted, innervated a large number of muscle fibres, and, when you removed the block, they were able to activate a relatively large motor unit.

Van Essen: It is quite possible that TTX inactivation causes a general shift in the bias for growth, so that the inactive terminals are more inclined to sprout. We do not know whether this is a simple presynaptic set of interactions, or whether the state of affairs with the muscle fibres is significant. In our paradigm, the muscle remains active, and so it is not releasing sprouting influences, but whether it is providing some other kind of feedback is not known.

Sanes: Drs Magchielse and Meeter (1986) did a similar experiment with the opposite result to yours. They found that the more active nerve cells retained functional contacts with muscle cells.

Van Essen: This is not the only study in which activity has been said to confer an advantage. We certainly do not suggest that our result will be generalizable to all synapses in all parts of the nervous system at all developmental ages; there may be interesting differences, according to age and location. However, we would like to introduce the idea that the developmental role of activity can be important, but of either sign, with regard to the strengthening of connections.

Gordon: On this question of activity and the end result, namely the motor unit size, in adults the force developed by the motor unit, which is the functional measure of unit size, is related to the size of the motor neuron. This relationship is due, at least in part, to the fact that more muscle fibres are innervated by the largest motor neurons and these are the least active, in terms of recruitment. So Dr Van Essen's result is what actually happens; the largest motor neurons supply the largest number of muscle fibres. Various experiments have looked at whether activity is important in the success of reinnervation. Dr Ribchester has done elegant experiments in which he argued that the active motor neurons were at an advantage. But I think Dr Van Essen's experiments, as well as the experiments which demonstrate that the active motoneurons are at an advantage, tend to be those in which all motoneurons show asynchronous rather than synchronous activity.

Vrbová: Certainly in the adult it is so that the largest and least active motor units have the largest number of muscle fibres, but in development we don't know whether the largest units are the most active ones; moreover, motoneurons are electrically coupled, so we don't really know what an individual motoneuron does.

Eisenberg: Could we define 'least' and 'most' active? Because the factor seen by the muscle needn't be conveyed by how many nerve stimuli were delivered per day, but by how the volleys were packaged, and by how many ACh vesicles were actually released at any one volley of impulses.

Vrbová: Certainly at this stage (9–12 days after birth), the rat soleus is no more active than the EDL with the same number of impulses per hour. Later, when the rat starts supporting itself on its hind legs and is walking, there is a 30-fold difference.

Gordon: That is an issue of different muscles with different functions. One is considering the motor units within a muscle and the corresponding differences in activity of the motoneurons.

Lowrie: Dr Van Essen, the unit sizes you presented here were standardized to the median unit of the middle root, which was normally active. I recall from your paper (Callaway et al 1987) that when you measured the size of units by a different method, as a percentage of total muscle tension, the only statistically significant difference was for the slow units of the middle root in that they were smaller after TTX treatment of the extreme root. Can the increase in size of the inactive units, supplied by the extreme root, which you described here, be explained by the reduction in size of the normally active units of the middle root?

Van Essen: We have looked at this more extensively since then and find that in terms of the percentage of total muscle tension, the inactive extreme motor units are enhanced relative to what they would have been; the active middle roots which are competing against them are slightly smaller, so the degree of overlap has not changed noticeably. We believe this supports the hypothesis that the inactive ones are competing successfully against their active middle unit competitors.

Lowrie: In your paper (Callaway et al 1987) the activity of those muscle fibres partially innervated by inactive axons was reduced; therefore the activity of the muscle as a whole must have been affected. This might have a feedback effect on the units which are not being directly inactivated, and their activity might be enhanced, thus accelerating withdrawal of terminals and reducing unit size.

Van Essen: We have no direct measure of overall changes in activity in this preparation. We would emphasize that we inactivate only a small percentage of the motor units to the soleus muscle. When we start the inactivation, the entire muscle (97% or more) is innervated through normally active roots, so total muscle activity should be changed only rather slightly.

Ribchester: We have had some (published) correspondence (Callaway et al 1988, Ribchester 1988a) over this issue, but I want to stress a couple of points. First, the conclusions are made on the basis of a single indirect measure of motor unit size (the number of terminals which are supplied by a motor neuron), namely twitch contractions made in response to stimulation of individual motor axons. There are technical difficulties with the interpretation of those measurements with respect to motor unit size. Nevertheless, it is an interesting finding and I would like to see results that would confirm it, using an independent method.

The second point has been raised already, that different results were obtained under different conditions which ostensibly are designed to examine the same question—namely, when you challenge active with inactive nerve terminals, converging on muscle fibres at a single motor endplate, which one is most effective, and which wins and which loses? Our recent experiments are similar in design to these experiments; they involve reinnervation of a partially denervated adult skeletal muscle in which the muscle, through processes of nerve sprouting at the intact motor unit, becomes completely innervated by collaterals from intact and active motoneurons. If we challenge that completely innervated, active muscle with in-growing motor axons which are blocked through chronic application of tetrodotoxin, the regenerating inactive motor units are smaller than they would be in the absence of block (Ribchester 1988b).

So qualitatively different results are obtained under different experimental conditions (reinnervation versus development) but in which the questions and experimental design are otherwise similar. It would be nice to know what these apparently different results are telling us about the mechanisms that developing and regenerating terminals have in common.

Gordon: In the natural process of reinnervation of a totally or partially denervated muscle we always find a re-establishment of the size relationship, of small motoneurons to small motor units. We have recently looked at the factors which determine force, including fibre size, fibre number, and contractility. We find that the relationship is determined partly by the number of muscle fibres innervated. Under that natural situation of graded activity (because within the spinal cord there is still recruitment from the smallest to the largest motoneurons, with the smallest being the most active and largest motoneurons the least active), reinnervation re-establishes that the smallest motoneurons reinnervate fewer fibres than the largest inactive motoneurons (so the most active motoneurons will be innervating fewest muscle fibres). The competition experiments are interesting, but in natural situations you obtain the real picture.

Vrbová: I am concerned like Richard Ribchester about the fact that if you remove one input in your developing muscle, the development of those muscle fibres may be a little slower. This will give a bigger twitch response, because the twitch:tetanus ratio declines with age, so in a less mature group of muscle fibres you will have a bigger proportion of the twitch in relation to the tetanus. The fact that there are other terminals doesn't mean that it matures at the same rate; at least if you stained them for succinate dehydrogenase you would see whether you get the sort of maturation that you would expect in this preparation.

Van Essen: The experiments in which we allow a recovery period should address that, because the initially blocked axons have a longer period of recovery of activity than they had of the initial block; we wouldn't expect that a maturational difference could account for our findings.

Oppenheim: An idea that has often been used to explain both the elimination and the maintenance of polyneuronal innervation is the neurotrophic or competition theory. In the present context, this theory would predict that the target releases trophic factors that are preferentially taken up by the larger terminals, and thereby maintains them to the detriment of the smaller ones. I cannot see that Dr Vrbová's results are inconsistent with that idea.

Vrbová: We have no evidence for it, however.

Crowder: What evidence do you have against it?

Vrbová: The evidence against it is the fact that you can maintain the terminals if you inhibit a very specific enzyme. You would have to argue that the trophic substance is acting through this enzyme, and that you can produce it by changing the microenvironment of the terminal very subtly.

Crowder: It could be that the muscular activity results in the exocytosis of subsynaptic vesicles from muscle that are releasing this factor, and that activity by the neuron will allow it to take up the putative trophic factor.

Vrbová: You need to identify 'the substance', in that case.

Crowder: Of course, that is the difficulty, because there is no good assay.

Henderson: I don't think the effects of protease inhibitors argue against a trophic theory, since any trophic factor must work against the background of some regressive influence—perhaps a spontaneous tendency to regress in the neuron, or perhaps an external factor. If one removes the regressive influence and sees more polyneuronal innervation, it is still possible that in the normal situation terminals are stabilized by a positively acting trophic factor.

References

Bixby JL, Spitzer NC 1984 Early differentiation of vertebrate spinal neurons in the absence of voltage-dependent Ca^{2+} and Na^{+} influx. Dev Biol 106:89–96

Callaway EM, Soha JM, Van Essen DC 1987 Competition favouring inactive over active motor neurons during synapse elimination. Nature (Lond) 328:422-426

Callaway EM, Soha JM, Van Essen DC 1988 Competitive elimination of neuromuscular synapses. (Reply to RR Ribchester) Nature (Lond) 331:21-22

Lasek RJ, Hoffman PN 1976 The neuronal cytoskeleton, axonal transport and axonal growth. Cold Spring Harbor Conf Cell Proliferation 3:1021–1049

Lichtman J, Magrani L, Purves D 1987 Visualization of neuromuscular junctions over periods of several months in living animals. J Neurosci 7:1215–1222

Magchielse T, Meeter E 1986 The effect of neuronal activity on the competitive elimination of neuromuscular junctions in tissue culture. Dev Brain Res 25:211–220

Reh T, Constantine-Paton M 1985 Eye specific segregation requires neural activity in three-eyed *Rana pipiens*. J Neurosci 5:1132–1143

Ribchester RR 1988a Competitive elimination of neuromuscular synapses. Nature (Lond) 331:21

Ribchester RR 1988b Activity-dependent and-independent synaptic interactions during reinnervation of partially denervated rat muscle. J Physiol (Lond) 401:53–77

Sanes JR 1983 Role of extracellular matrix in neuronal development. Annu Rev Physiol 45:581–600

Swanson GJ, Vrbová G 1987 Effects of low calcium and inhibition of calcium activated neutral protease (CANP) on mature nerve terminal structure in the rat sternocostalis muscle. Dev Brain Res 33:199–203

Van Huizen F, Romijn HJ 1987 Tetrodotoxin enhances initial neurite outgrowth from fetal rat cerebral cortex cells *in vitro*. Brain Res 408:271–274

Wilshaw DJ 1981 The establishment and subsequent elimination of polyneuronal innervation of developing muscle: theoretical considerations. Proc R Soc Lond B Biol Sci 212:233–252

Neurotrophic interactions in the development of spinal cord motoneurons

Ronald W. Oppenheim and *Lanny J. Haverkamp

Department of Anatomy, Bowman Gray School of Medicine, Wake Forest University, Winston-Salem, North Carolina 27103 and **Department of Neurology, Baylor College of Medicine, Houston, Texas 77030, USA*

Abstract. The final number of spinal cord motoneurons is attained by a two-step process involving the proliferation of precursor cells and the loss by cell death of a proportion (≈50%) of the post-mitotic neurons. Although the mechanisms responsible for the proliferation of stereotyped numbers of motoneurons are not understood, considerable evidence from *in vitro* as well as *in vivo* studies indicates that the second step in attaining population size (cell death) is controlled by the interaction of motoneurons with both their efferent targets and their afferent inputs. Target influences on motoneuron survival are thought to be regulated by muscular activity and by competition for limited amounts of neurotrophic factors derived from striated skeletal muscles. However, evidence that such putative neurotrophic factors actually modulate motoneuron survival *in vivo* has been lacking. Using crude and partially purified extracts from embryonic hindlimbs (Days 8–9) we have found that the treatment of chick embryos *in ovo* with these agents during the normal cell death period (Days 5–10) rescues a significant number of motoneurons from degeneration. Kidney or lung extracts and heat-inactivated hindlimb extracts were ineffective. The survival-inducing activity of partially purified extract was dose dependent and developmentally regulated. The survival of sensory, sympathetic and a population of cholinergic sympathetic preganglionic neurons was unaffected by treatment with hindlimb extract. The massive motoneuron death that occurs after early target (hindlimb) removal was partially ameliorated by daily treatment with the hindlimb extract. Survival-inducing activity of the extract is lost after trypsin treatment. Taken collectively these results indicate that a target-derived protein or polypeptide neurotrophic factor is involved in the regulation of motoneuron survival *in vivo*.

1988 Plasticity of the neuromuscular system. Wiley, Chichester (Ciba Foundation Symposium 138) p 152–171

During the past 25 years, investigators of neuronal development have formulated a conceptual framework for explaining a large number of older as well as more recent findings concerning the development of neuron–target interactions. This framework has been most commonly referred to as the *competi-*

tion, trophic or neurotrophic theory (Hamburger & Oppenheim 1982, Purves 1986). This theory has been developed in large measure to explain how targets (neuronal and non-neuronal) influence the survival, differentiation and maintenance of neurons and their properties (although its usefulness is by no means restricted to these events). Briefly, the neurotrophic theory states that in many parts of the developing nervous system, neurons and their processes compete for target-derived or target-associated molecules (neurotrophic factors) that are in limited supply relative to the numbers of inceptive neurons and processes. Because of either intrinsic (i.e. genetic) or extrinsically imposed differences between neurons, some cells are thought to be at a competitive advantage in gaining access to sufficient quantities of neurotrophic agents and thereby survive and differentiate normally, whereas the remainder succumb or fail to differentiate normally (Hamburger & Oppenheim 1982, Oppenheim 1985).

Although the general notion that neurons are dependent on cell–cell interactions for their normal development and differentiation has been a hallmark of experimental embryology since the turn of this century, the specific tenets of the neurotrophic hypothesis concerning this interaction are of relatively recent origin. This is not to say that the general notion that neuron–target interactions may involve chemical signals was inconceivable to early investigators. Ramón y Cajal (1929) as well as others (e.g. Parker 1932, Detwiler 1936) were certainly aware of this possibility. However, for most of this century, chemical or molecular explanations of these phenomena were simply not a central part of the conceptual framework of neuroembryologists. This is all changed after the discovery and subsequent characterization of Nerve Growth Factor (NGF) in the 1950s and 1960s (Levi-Montalcini 1982).

Experiments begun by Hamburger and Levi-Montalcini in the 1930s and 1940s, designed to explore the role of peripheral targets in the regulation of the population size of spinal cord motoneurons and sensory neurons, ultimately led to the discovery of NGF (Oppenheim 1981, Levi-Montalcini 1986). Despite the fact that from the time of its discovery NGF was shown to have profound effects on the differentiation of sensory and sympathetic neurons, the unusual sites at which NGF was found to be stored (tumours, snake venom, salivary glands), together with the inability of most embryologists to break loose from the constraints that precluded a role for neuron-specific growth agents, led initially to a general denial of the significance of NGF for normal development (e.g. see the discussion of the paper by Cohen 1958). Only when specific antibodies to NGF were produced and shown to have deleterious effects on the survival and differentiation of neurons *in vivo* did it become obvious that endogenous sources of NGF must exist that very likely play an important role in development (Cohen 1960, Levi-Montalcini & Booker 1960). Nevertheless, it still required two more decades of intensive investigation before the full significance of NGF and the means by which it acts during development were clarified (Thoenen & Barde 1980, Harper & Thoenen 1981).

To focus for the moment on the survival-inducing properties of NGF, it is now clear that NGF is synthesized and stored in limited amounts in sensory and sympathetically innervated target tissues (Heumann 1987) and that it acts on presynaptic neurons by receptor-mediated uptake from nerve terminals and retrograde transport to the cell body, where it very likely alters gene expression, resulting in neuronal survival and differentiation. Thus in virtually all respects NGF has now been shown to meet the criteria for a target-derived neurotrophic factor that regulates the survival of specific classes of neurons during development. In this regard, NGF completes the search begun almost 40 years ago by Hamburger and Levi-Montalcini for the then only dimly understood 'peripheral influence' that regulates the population size of developing sensory and sympathetic neurons. Seldom, if ever, in neurobiology has such a long and dogged search for the ultimate causal events underlying a fundamental biological phenomenon been rewarded so handsomely.

The series of events associated with the discovery and characterization of NGF have to a great extent provided the factual basis for the neurotrophic hypothesis. For this hypothesis to be considered a major part of the conceptual framework of neurobiology, however, it must be applicable to a wider variety of cell types than the sensory and sympathetic ganglia. It has long been proposed that a variety of non-NGF neurotrophic agents that act on other types of neurons must exist, but only in the past 10 years or so has significant progress been made in identifying and characterizing other putative neurotrophic factors. Several putative neurotrophic or 'survival' factors have now been isolated, characterized *in vitro* and partially or completely purified (Berg 1984, Barde et al 1983). However, only one has been shown to meet the most important criterion for classification as a *bona fide* neurotropic factor, namely ability to act *in vivo* (Hofer & Barde 1988). Nonetheless, it was the success of these *in vitro* studies, together with the realization that the promotion of neuronal survival *in vitro* may reflect effects that are never operative during normal development *in vivo*, that provided the impetus for my colleagues and I to attempt to determine whether the targets of developing spinal cord motoneurons (skeletal muscle) contain factors that will rescue motoneurons from naturally occurring cell death, *in vivo*.

Identification of a putative motoneuron survival factor

In the chick embryo, 50% or more of the somatic motoneurons innervating skeletal muscle, in limb as well as non-limb regions, degenerate between embryonic day (E) 5½ and E12 (Hamburger 1975, Oppenheim 1985). Be-

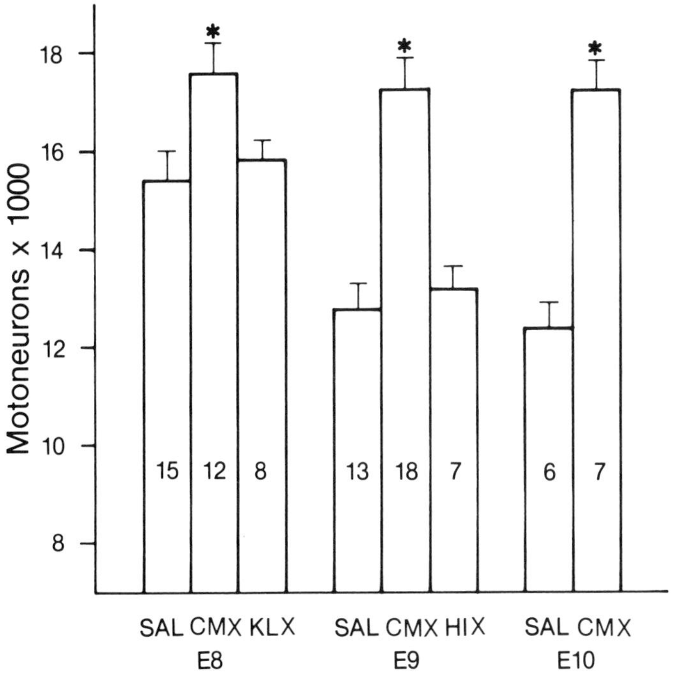

FIG. 1. The number of surviving lumbar motoneurons on E8, E9 and E10 after daily treatment from E5 or E6 with 250 μl of chick muscle extract (CMX), saline (SAL), kidney or lung extract (KLX) or heat-inactivated extract (HIX). Chick muscle extract, unless otherwise indicated, was prepared from the legs of embryonic day 8–9 white leghorns. Lung and kidney extracts were prepared from adult chickens. The numbers in the bars are the sample sizes (i.e., numbers of embryos). * $P < 0.01$, *t*-test.

cause these spinal motoneurons are thought to compete for some entity that is associated with primary myotubes, we began our studies by attempting to isolate a soluble factor from hindlimbs of E8–9 embryos when the hindlimb musculature is composed primarily, if not exclusively, of primary myotubes, and when motoneuron death is still occurring (McLennan 1982, Tanaka & Landmesser 1986).

As previously described (Oppenheim et al 1988), crude hindlimb (or control lung and kidney) extracts were prepared and administered daily through a small window in the shell onto the vascularized chorioallantoic membrane, beginning either on E5 or E6. Control embryos received equal volumes of a physiological saline solution. Because motoneuron survival is known to be dependent on neuromuscular activity (Oppenheim 1985), we also monitored the embryonic motility of embryos to determine whether these extracts alter neuromuscular activity. The data summarized in Fig. 1 are the results of several independent experiments in which embryos were sacrificed for histo-

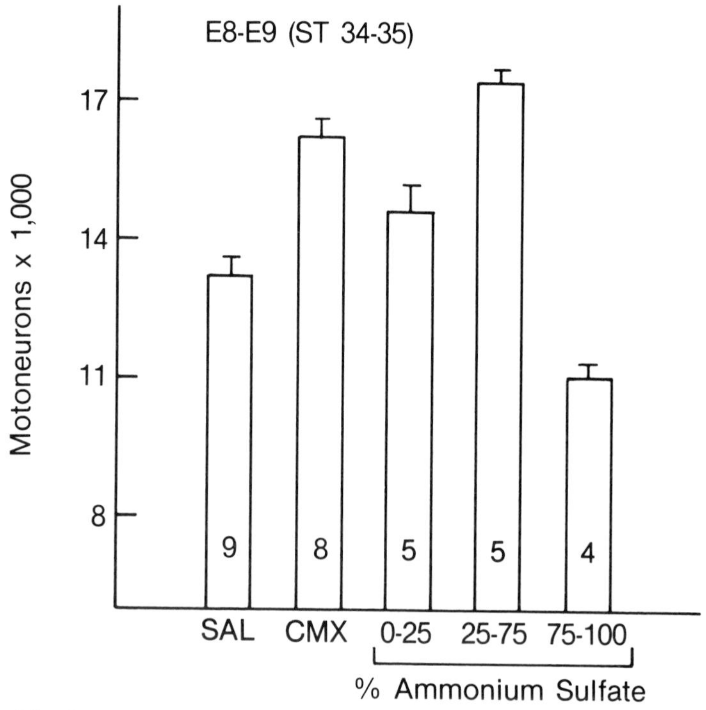

FIG. 2. Numbers of lumbar motoneurons in control (saline), chick muscle extract (CMX), and three ammonium sulphate fraction groups on E8–9. All three groups are significantly different from saline controls ($P < 0.05$ to $P < 0.01$, Mann-Whitney test). The 25–75% group was also significantly different from the CMX group ($P < 0.05$). The numbers in the bars are the sample sizes (i.e., numbers of embryos).

logical processing and cell counts at E8, E9 or E10. In all cases, embryos treated with crude hindlimb extracts had significantly more motoneurons surviving in the lumbar spinal cord than controls. Kidney or lung extracts were ineffective in rescuing motoneurons. Heat inactivation (60 °C for 45 min) of crude hindlimb extracts eliminated the motoneuron survival activity. Although the degree of motoneuron survival varied from experiment to experiment, treatment with hindlimb extracts always resulted in a significant increase in motoneuron numbers. The morphology of motoneurons appears normal after treatment with the extract. Embryos treated with hindlimb extract also displayed normal amounts of motility, indicating that increased motoneuron survival is not secondary to altered neuromuscular activity.

To further characterize the active components of the crude extract that mediate motoneuron survival we did ammonium sulphate ($AmSO_4$) precipitations, obtaining three fractions, 0–25%, 25–75% and 75–100%. The 25–75%

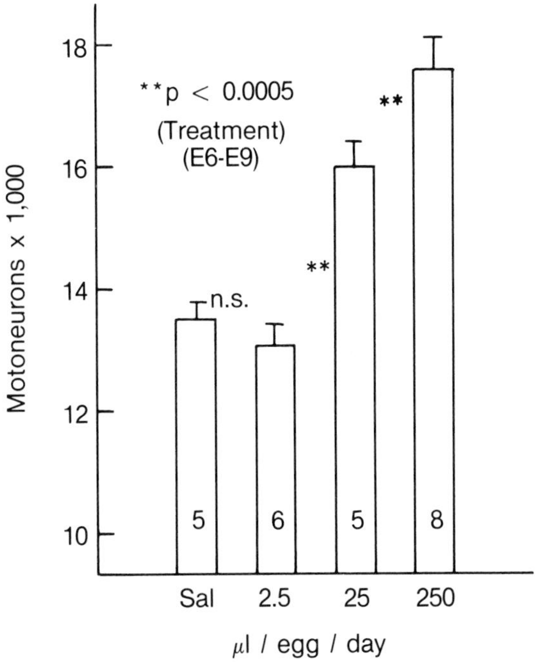

FIG. 3. The number of surviving lumbar motoneurons on E10 after daily treatment with 2.5, 25 or 250 µl of the ammonium sulphate CMX fraction. The groups receiving 25 and 250 µl also differ significantly ($P < 0.05$).

$AmSO_4$ fraction contained significantly more survival activity than either of the other fractions as well as more activity than the crude extract (Fig. 2). The 0–25% fraction contained a small but statistically significant amount of survival-inducing activity when compared to the saline control group. Interestingly, the 75–100% fraction had a significant negative effect on motoneuron survival. Because the maximum amount of motoneuron survival activity was contained in the 25–75% fraction, only this partially purified fraction was used in all subsequent experiments reported here. The motoneuron survival effects mediated by the 25–75% $AmSO_4$ fraction appears to be dose related, at least over the range of doses examined here (Fig. 3). Doses in excess of 250 µl were generally lethal to the embryo after only one or two injections.

The partially purified hindlimb extract acts only on a subpopulation of spinal cord neurons. Neurons in the dorsal root ganglia, sympathetic ganglia and sympathetic preganglionic column, all of which are also undergoing normal cell death during these stages, are unaffected by the hindlimb extract (Fig. 4). By contrast, NGF treatment alters the survival of these neuronal populations but has no effect on motoneuron survival. The hindlimb extract

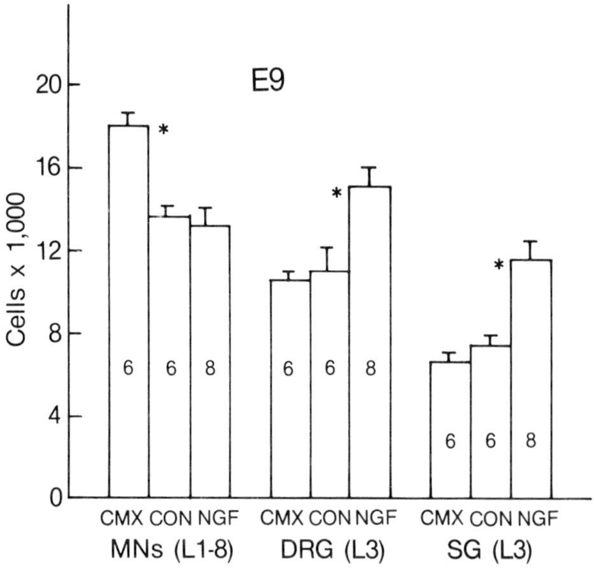

FIG. 4. The number of surviving lumbar motoneurons (MNs), dorsal root ganglion cells (DRG) and sympathetic ganglion cells (SG) in lumbar segment 25 (L3) after treatment with the 25–75% ammonium sulphate fraction (CMX), saline (control), or nerve growth factor (NGF, 20 μg per day). DRG and SG cells were counted in every 5th section. Although the data are not shown, cholinergic sympathetic preganglionic neurons were also counted in thoracic segments 5 and 6 (T5–6) in the same embryos. Controls had 4610 ± 435 cells *vs* 4663 ± 484 for the CMX-treated group. This difference was not statistically significant.

appears to be effective in rescuing motoneurons throughout virtually the entire extent of the lumbar spinal cord as well as non-limb (thoracic) motoneurons (Oppenheim et al 1988). Furthermore, the increased number of motoneurons after extract treatment is associated with a significant decrease in the numbers of degenerating motoneurons (data not shown, but see Oppenheim et al 1988). Thus, the partially purified hindlimb extract appears to act by maintaining or promoting the survival of motoneurons that would otherwise die.

Our rationale for using hindlimbs from E8–9 embryos as starting material for the crude and partially purified hindlimb extracts is the belief that in normal cell death motoneurons compete for some factor associated with the type of myotubes present at this time. Consequently, it seems plausible that the levels, source, or kind of putative survival factors present during these stages might differ from later (or earlier) stages when motoneurons are not dying. In accord with this possibility we have found that partially purified

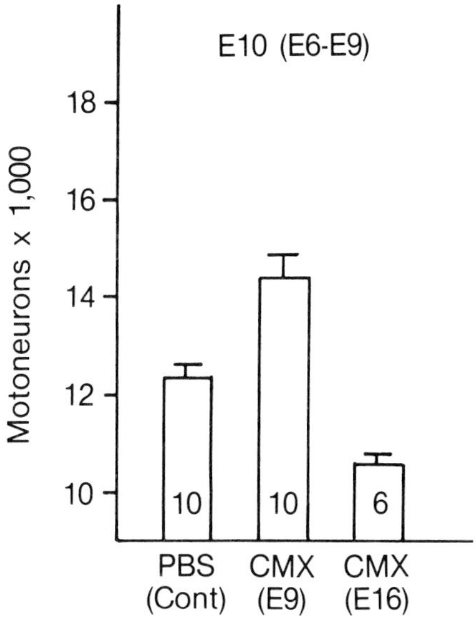

FIG. 5. Embryos were treated daily beginning on E6 with partially purified hindlimb extract (25–75% ammonium sulphate, CMX) prepared from either E8–9 or E16 embryos. The E16 group was significantly different ($P < 0.01$, *t*-test) from both control and E9 extract-treated groups. Control and E9 extract-treated groups were also significantly different ($P < 0.01$).

hindlimb extract from E16 embryos does not promote the survival of motoneurons (Fig. 5).

Removal of motoneuron targets by deleting hindlimb buds at stages before neuromuscular connectivity is established results in the subsequent death of over 90% (as compared to the normal 50%) of these limb-innervating motoneurons (Oppenheim et al 1978). We wished to determine whether treatment with partially purified hindlimb extract could rescue motoneurons destined to die because of early target removal. Limb buds were removed on E2 and treatment was begun on E4. Although embryos from which limb-buds had been removed before the treatment with extract continued to show increased cell loss, the extent of the loss was significantly ameliorated (Fig. 6).

The failure to rescue all, or at least a greater amount, of the induced cell loss could be due to a number of different factors. The simplest explanation is that this is a dose-dependent phenomenon. This possibility receives some support from the fact that the absolute number of motoneurons rescued by extract treatment ($\approx$3000) is similar on E8–9 on both control and operated sides of the spinal cord. It is also possible that, as with NGF, the most effective and normal means by which neurotrophic factors exert their effect

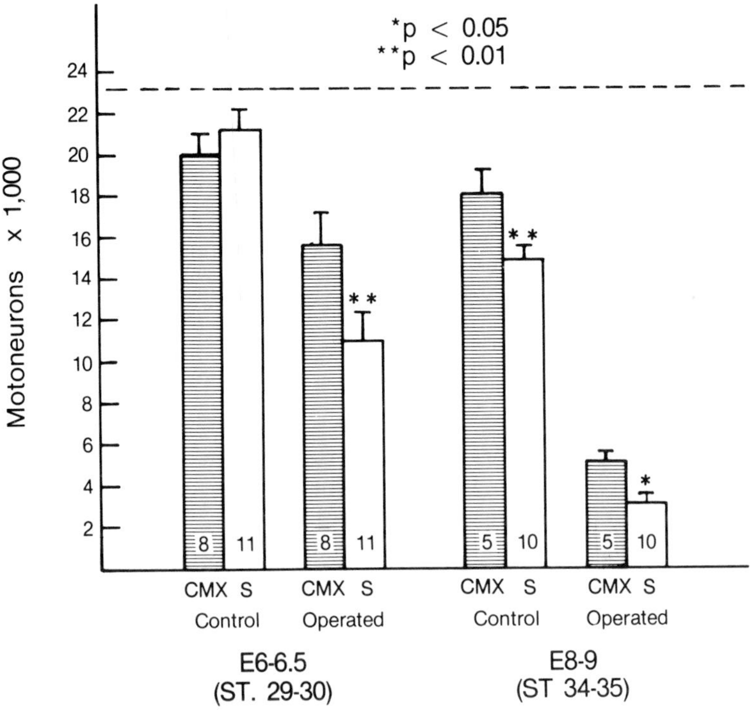

FIG. 6. The number of surviving lumbar motoneurons on E6–6½ and E8–9 on the side of the spinal cord ipsilateral (operated) and contralateral (control) to hindlimb removal after daily treatment with saline (S) or partially purified extract (CMX), beginning on E4. The dashed line at the top indicates the number of motoneurons present on E5, before the onset of normal cell death. Only embryos lacking all hindlimb musculature were retained for analysis.

on motoneurons is via terminal uptake and retrograde transport from neuromuscular contacts. Limb-bud-removed embryos lack such contacts. It is also possible, but less likely, that a proportion of the cell death that follows limb removal is due to a different mechanism from that operative during normal cell death (Lanser & Fallon 1987).

The apparent failure to rescue any motoneurons on the contralateral (control) side of the spinal cord on E6–6½ by extract treatment from E4 indicates: (a) that the small amount of normal cell death occurring at this time may be independent of limb-derived factors; and (b) that extract treatment does not alter the final stages of the proliferation or migration of motoneurons. Studies *in vitro* suggest that dissociated motoneurons from E5 (but not E6 or older) chick embryos are more dependent for their survival on immature astrocytes than on muscle-derived factors (Eagleson & Bennett 1986). Therefore, be-

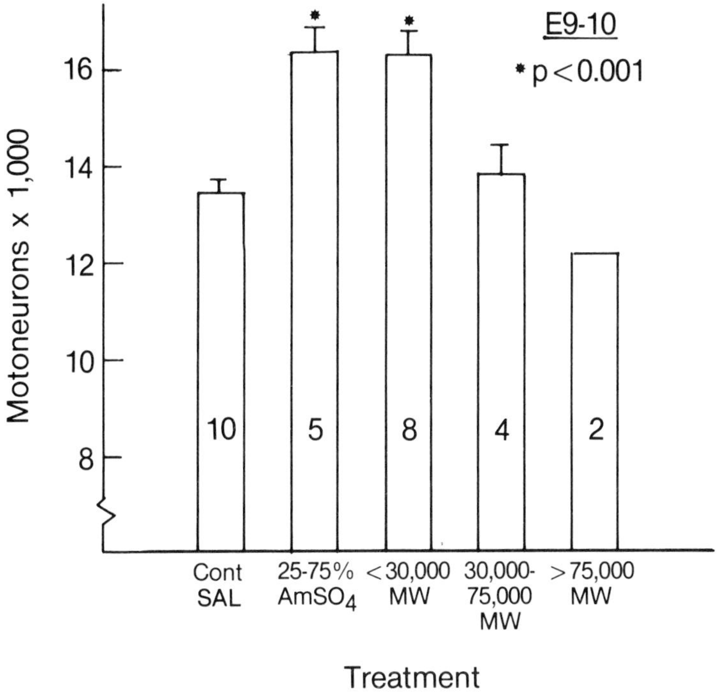

FIG. 7. The number of surviving lumbar motoneurons on E9–10 in embryos treated with saline, 25–75% $AmSO_4$ fraction, or three molecular weight fractions. Embryos were treated daily, beginning on E6. *P* values are for experimental *vs* control groups.

fore nerve–muscle contacts are initiated, motoneurons appear to go through a stage of limb-independent differentiation.

Molecular characterization

Putative neuronal survival factors include agents with a rather wide range of molecular weights. In an initial attempt to further characterize the hindlimb factor(s) mediating motoneuron survival we have used gel filtration chromatography to estimate the molecular weight range of the active factor(s) contained in the 25–75% $AmSO_4$ fraction. These initial experiments indicate that virtually all the survival activity can be recovered in a fraction composed of agents <30 000 Da (Fig.7). The amount of survival-inducing activity in this fraction is roughly equivalent to that contained in the 25–75% $AmSO_4$ starting material. A possibly related putative motoneuron survival factor of ≈20 000 Da derived from postnatal rat skeletal muscle also rescues a significant number of motoneurons when administered to chick embryos *in vivo* (data not shown).

The fact that the motoneuron survival activity in crude hindlimb extracts is heat labile suggests that the active factor may be a protein. Further support for this comes from experiments in which the activity in the 25–75% $AmSO_4$ fraction was shown to be trypsin sensitive. Pretreatment of this fraction with trypsin eliminated virtually all the motoneuron survival activity, whereas a number of control conditions were ineffective (data not shown, but see Oppenheim et al 1988).

Although there have been studies using *in vitro* methods demonstrating the presence of putative target tissue-derived survival factors for motoneurons (Bennett et al 1980, Calof & Reichardt 1984, Dohrmann et al 1986, O'Brien & Fischbach 1986), these suffer from the shortcomings discussed previously that are inherent to *in vitro* approaches. Nonetheless, several of these studies have been important in demonstrating that crude or partially purified muscle extracts or muscle-conditioned media can promote the survival of a population of identified motoneurons. A common theme in many of these studies is that factors associated with developing myotubes appear to be especially efficacious in promoting motoneuron survival. Furthermore, previously identified neurotrophic factors, such as NGF, brain-derived neurotrophic factor and other non-target-derived factors (e.g. laminin, a basement membrane glycoprotein), when administered alone are ineffective in promoting motoneuron survival *in vitro*. The putative motoneuron survival factor described here is also likely to be different from the neurotrophic factor, neuroleukin, described by Gurney et al (1986). Neuroleukin promotes the *in vitro* survival of an unidentified population of spinal cord neurons, as well as the survival of a subpopulation of spinal sensory ganglion neurons, whereas the factor described here has no apparent effect on sensory neurons.

In further accord with our *in vivo* results, Dohrmann et al (1986) have reported that maximum motoneuron survival activity *in vitro* is retained in a 25–75% $AmSO_4$ fraction of chick embryo muscle extract. *In vivo* studies of motoneurons and other neuronal types indicate that in addition to target-derived factors, afferent or CNS-derived influences are important for neuronal survival (Oppenheim 1988). *In vitro* studies also indicate that non-target, CNS-derived factors are important for motoneuron survival (e.g. Dohrmann et al 1987). Finally, both *in vivo* and *in vitro* studies have suggested that muscle-derived factors affecting motoneuron survival are influenced by activity and/or innervation (Hill & Bennett 1986, Oppenheim 1985). Thus, taken collectively, the results from both *in vitro* and *in vivo* approaches are in agreement on several essential points regarding the presence of a target-derived survival factor involved in the regulation of motoneuron numbers.

The most completely characterized neurotrophic factor, NGF, is known to be present in target tissues and to influence neuronal survival primarily by receptor-mediated terminal uptake and retrograde transport to the cell body. Although we have no direct evidence indicating that the putative motoneuron survival factor identified here acts in a similar manner to NGF, the results from the limb-bud removal experiment suggest that in order for the motoneuron factor to be maximally effective, intact nerve–muscle contacts may be required. We also do not know whether this factor is acting wholly or partly by influencing neurite growth. However, a number of *in vitro* studies have isolated two independent factors from muscle, a neurite-promoting and a survival-promoting factor, only the latter of which prevents motoneuron death. Consequently, it seems likely that the factor we have described here is, in fact, acting directly on motoneuron survival. It is also conceivable that the factor we have identified is acting directly on the muscle by increasing myotube numbers, size or type, all of which are characteristics that could conceivably indirectly influence motoneuron survival. However, preliminary studies of the medial adductor muscle from E9–10 extract-treated embryos indicate that this is unlikely to be the case. Muscle volume, myotube numbers and myotube size appear to be comparable to control hindlimb muscles. The distribution of muscle fibre types (fast *vs* slow) also appears normal in the hindlimbs of extract-treated chick embryos.

In summary, we have identified a putative protein motoneuron neurotrophic factor derived from normal target tissue that acts *in vivo* to rescue motoneurons in a manner consistent with its proposed role in the regulation of naturally occurring neuronal death in the developing chick embryo. The *in vivo* demonstration of motoneuron survival-promoting activity allows us to rule out a number of artifactual explanations for these results that are difficult or impossible to exclude from *in vitro* studies. Nonetheless, considerably more work is required before we can be certain that we are dealing with a real biological phenomenon and not merely a pharmacological effect. Of all the questions that remain to be answered, perhaps the chief one is whether antibodies against this putative factor (when they become available) when administered to embryos *in vivo* result in decreased motoneuron survival. Remembering once again the critical events leading to the acceptance of NGF as a biologically relevant molecule (Levi-Montalcini 1982, 1987), resolution of this single issue for the putative motoneuron neurotrophic factor described here would go a long way towards removing any nagging doubts about its role in naturally occurring motoneuron death *in vivo*.

Acknowledgements

The research described here on a putative motoneuron survival factor was supported by NIH grants NS 20402, NS 23058, the Muscular Dystrophy Association and California Biotechnology, Inc. We wish to acknowledge the advice, assistance and suggestions of Stan Appel, Jim McManaman, David Prevette and Sharen McKay.

References

Barde YA, Edgar D, Thoenen H 1983 New neurotrophic factors. Annu Rev Physiol 45:601–612

Bennett MR, Lai K, Nurcombe V 1980 Identification of embryonic motoneurons in vitro: their survival is dependent on skeletal muscle. Brain Res 190:537–542

Berg DK 1984 New neuronal growth factors. Annu Rev Neurosci 7:149–170

Calof AL, Reichardt LF 1984 Motoneurons purified by cell sorting respond to two distinct activities in myotube-conditioned medium. Dev Biol 106:194–210

Cohen S 1958 A nerve growth promoting protein. In: McElroy WD, Glass B (eds) The chemical basis of development. Johns Hopkins University Press, Baltimore, p 665–679

Cohen S 1960 Purification of a nerve growth promoting protein from the mouse salivary gland and its neuro-cytotoxic antiserum. Proc Natl Acad Sci USA 46:302–311

Detwiler SR 1936 Neuroembryology: an experimental study. Macmillan, New York

Dohrmann U, Edgar D, Sendtner M, Thoenen H 1986 Muscle-derived factors that support survival and promote fiber outgrowth from embryonic chick spinal motor neurons in culture. Dev Biol 118:202–221

Dohrmann U, Edgar D, Thoenen H 1987 Distinct neurotrophic factors from skeletal muscle and the central nervous system interact synergistically to support the survival of cultured embryonic spinal motor neurons. Dev Biol 124:145–152

Eagleson KL, Bennett MR 1986 Motoneurone survival requirements during development: the change from immature astrocyte dependence to myotube dependence. Dev Brain Res 29:161–172

Gurney ME, Heinrich SP, Lee MR, Yin HS 1986 Molecular cloning and expression of neuroleukin, a neurotrophic factor for spinal and sensory neurons. Science (Wash DC) 234:566–574

Hamburger V 1975 Cell death in the development of the lateral motor column of the chick embryo. J Comp Neurol 160:535–546.

Hamburger V, Oppenheim RW 1982 Naturally occurring neuronal death in vertebrates. Neurosci Comment 1:39–55

Harper GP, Thoenen H 1981 Target cells, biological effects, and mechanism of action of nerve growth factor and its antibodies. Annu Rev Pharmacol 21:205–229

Heumann R 1987 Regulation of the synthesis of nerve growth factor. J Exp Biol 132:133–150

Hill MA, Bennett MR 1986 Motoneurone survival activity in extracts of denervated muscle reduced by prior stimulation of the muscle. Dev Brain Res 24:305–308

Hofer MM, Barde Y-A 1988 Brain-derived neurotrophic factor prevents neuronal death *in vivo*. Nature (Lond) 331:261–262

Lanser ME, Fallon JF 1987 Development of the brachial lateral motor column in the *wingless* mutant chick embryo: motoneuron survival under varying degrees of peripheral load. J Comp Neurol 261:423–434

Levi-Montalcini R 1982 Developmental neurobiology and the natural history of nerve growth factor. Annu Rev Neurosci 5:341–362

Levi-Montalcini R 1987 The nerve growth factor 35 years later. Science (Wash DC) 237:1154–1162

Levi-Montalcini R, Booker B 1960 Destruction of the sympathetic ganglia in mammals by an antiserum to a nerve-growth protein. Proc Natl Acad Sci USA 46:384–391

McLennan I 1982 Size of motoneuron pool may be related to number of myotubes in developing muscle. Dev Biol 92:263–265

O'Brien RJ, Fischbach GD 1986 Isolation of embryonic chick motoneurons and their survival in vitro. J Neurosci 6:3265–3274
Oppenheim RW 1981 Neuronal cell death and some related regressive phenomena during neurogenesis: a selective historical review and progress report. In: Cowan WM (ed) Studies in developmental neurobiology: essays in honor of Viktor Hamburger. Oxford University Press, New York, p 74–133
Oppenheim RW 1985 Naturally occurring cell death during neural development. Trends Neurosci 8:487–493
Oppenheim RW 1988 Evidence for the role of afferents in the regulation of neuronal survival during normal periods of developmental cell death: motoneurons and ciliary ganglion. In: Ferrendelli J et al (eds) Neurobiology of amino acids, peptides, and trophic factors. Martinus Nijhoff, Boston, in press
Oppenheim RW, Haverkamp LJ, Prevette D, McManaman J, Appel SH 1988 Reduction of naturally occurring motoneuron death *in vivo* by a target-derived neurotrophic factor. Science (Wash DC) 240:919–922
Oppenheim RW, Chu-Wang IW, Maderdrut JL 1978 Cell death of motoneurons in the chick embryo spinal cord. III. The differentiation of motoneurons prior to their induced degeneration following limb-bud removal. J Comp Neurol 177:87–112
Parker GH 1932 On the tropic impulse, so-called, its rate and nature. Am Nat 66:147–158
Purves D 1986 The trophic theory of neuronal connections. Trends Neurosci 9:486–489
Ramón y Cajal S 1929 Studies on vertebrate neurogenesis. Thomas, Springfield, Illinois
Tanaka H, Landmesser L 1986 Cell death of lumbosacral motoneurons in chick, quail and chick–quail chimera embryos: a test of the quantitative matching hypothesis of neuronal cell death. J Neurosci 6:2289–2899
Thoenen H, Barde YA 1980 Physiology of nerve growth factor. Physiol Rev 60:1284–1335

DISCUSSION

Thesleff: Have you tried to isolate this trophic factor from chronically denervated adult muscle? Perhaps it would be releasing the same kind of factor.

Oppenheim: Although we haven't yet tried this, I would not be surprised if it did. Dr Henderson's work (see his chapter) would imply that this may be the case.

Gordon: I am interested in the parallels with nerve growth factor, because it has now been shown that NGF doesn't appear in the target tissues until the nerves get there and establish functional contacts. So in fact NGF is induced by nerves. Perhaps the NGF is coming from the Schwann cells. In the ablation experiments you were taking away the whole limb. To what extent are your extracts entirely muscle extracts, and could they contain a trophic substance which doesn't necessarily come from muscle, or is induced in muscle at a later time?

Oppenheim: Connective tissue, fibroblasts, Schwann cells and nerve tissue are also present in the crude homogenate. At present we are examining the effects of myotube-conditioned media and of Schwann cell-conditioned media on motoneuron survival *in vivo*. Although it seems unlikely, we cannot rule out the possibility that non-muscle tissues are the source of the active factor. Whether or not this or other neurotrophic factors are up-regulated in the targets after innervation, as happens with NGF, remains to be determined.

Gordon: The emphasis on the effect of NGF in promoting neurite outgrowth distorted the story a little, by suggesting that because the target made the nerves survive, the growing neurites were in fact searching out the growth factor. You could anticipate that by, for example, looking for a possible inductive effect by nerve on muscle.

Oppenheim: Yes; these recent data appear to rule out the role of NGF as a *tropic* factor for attracting growing axons to their target.

Dubowitz: Your factor is present in muscle of the nine-day-old chick embryo but not in the 16-day-old embryo. Can you explain why the survival factor should be there at a time when nerve death is occurring (5–10 days), and have you looked for it in younger embryos, and before cell death starts?

Oppenheim: It follows from the neurotrophic theory that target-derived survival factors are present during the cell death period when competition for these factors is greatest. We are now attempting to determine whether survival activity is present in muscle prior to the onset of cell death (i.e., on E5–6).

Dubowitz: When the rat neonatal muscle factor was studied, was there a critical period when the factor was active in the chick embryo?

Oppenheim: That hasn't been looked at. We have only examined muscle extract from postnatal rats. Cell death continues for a few days after birth in the sexually dimorphic nucleus of the rat spinal cord, so one might expect motoneuron survival factors to continue to be expressed at high levels until 7–10 days postnatally. But without systematically studying older, as well as younger, ages it is difficult to make firm statements about developmental regulation.

Van Essen: Is the level of this factor increased in bungarotoxin-treated animals?

Oppenheim: We are doing that experiment. In a recent paper by Tanaka (1987) it is reported that muscle extract from chronically paralysed embryos does not contain higher levels of motoneuron survival activity. However, this was an *in vitro* study and until comparable studies are done *in vivo* we won't know whether this also holds for normal development.

Sanes: What is the distribution of the factor in the chick embryo? Is it found in the central nervous system?

Oppenheim: Recent *in vitro* studies by Dohrmann et al (1987) indicate that there is a separate CNS-derived motoneuron survival factor that is distinct from and additive with the muscle-derived factor. We are now examining this in our *in vivo* model.

Sanes: Is there any evidence that it affects either vascularization of the nervous system or glial processes?

Oppenheim: We have not examined vascularization or glial development after treatment with muscle extract. However, one might expect that if the putative factor acts in this way it would also affect many other neurons that are undergoing cell death, but it doesn't.

Zak: In cardiac morphogenesis a substantial number of muscle cells die during the development. In this case the topography of cell death indicates that certain regions are affected preferentially. This results in the change in shape of the developing heart (Manasek 1969). So cell death is not unique to the nervous system.

Is anything known about the trigger for the death of the cell? In many types of cell, an influx of calcium seems to be associated with cell death.

Oppenheim: The precise triggers for embryonic cell death are not known for any population of neurons. There is an increasing amount of evidence, however, that in order for cell death to occur, active protein or RNA synthesis is required. Treatment with protein synthesis inhibitors blocks the death of muscle (L. Schwartz, personal communication) and neurons (Fahrbach & Truman 1987) in insects, suppresses the anti-NGF- induced death of dorsal root ganglion cells *in vitro* (Martin et al 1987) and blocks motoneuron death in the chick spinal cord *in vivo* (our own unpublished work).

Zak: Could a calcium ionophore be included in the experimental system, to test for a possible role of calcium?

Oppenheim: Yes. This would be a very interesting experiment.

Vrbová: Where do you think this factor acts? If you take the target away, the factor still rescues motoneurons. That suggests a central effect on the cell bodies themselves. I also wonder whether the molecule that Chris Henderson studies is similar, or whether it has been tested in this system. It is also derived from embryonic muscle.

Henderson: We haven't tried our neurite-promoting factors in Ron Oppenheims's system, but that should be done when we have enough of the pure molecule.

Pette: At what concentration does the protein act? In other words, is it a limiting factor, like calcium-binding protein or transferrin, or is it a receptor-linked factor?

Oppenheim: We inject 250μl of our 3mg/ml protein solution into a volume of about 50ml (the entire egg volume). At present we have no way of knowing if it is acting as you propose.

Henderson: How much do you inject of the 21K McManaman factor purified from rat skeletal muscle, to obtain similar effects?

Oppenheim: Those experiments were not done with the completely purified factor, but with a 50–100-fold purified preparation. We injected 100 μl per egg per day of a 3mg/ml protein solution.

Henderson: So the effects of the 21K preparation could be due to other muscle proteins which make up probably the great majority of that preparation?

Oppenheim: That is possible. Dr McManaman now has enough of the purified rat muscle factor to carry out a small *in vivo* experiment. This is currently being done in my laboratory.

Mudge: Could you convert your 250 μl of extract into chick-equivalents for me? Since you can dilute your extract and still see effects, this suggests that the rescuing factor is rather abundant. How many chicks are used to make the 250 μl injected?

Oppenheim: To do an experiment injecting about 50 embryos over four or five days would require hindlimb muscles from about 300 embryos. The dilution of the extract varies between experiments, because we always try to make the final concentration close to 3mg/ml.

Mudge: Barde and Thoenen have argued that CNTF, a ciliary neuronotrophic factor (Barbin et al 1984), is unlikely to be an NGF-like molecule, because it is too abundant and its tissue distribution is too wide. If you can take a chick, grind it up, and actually dilute it and obtain effects, this is surprising. Could your factor be something like CNTF, which has a profound effect on survival but seems unlikely to be the rescuing factor for naturally occuring cell death?

Oppenheim: At present we cannot rule that out. However, by using an *in vivo* system and showing that the factor has a specificity for motoneurons and not for many other cells that are dying at the same time, it seems more likely that we are, in fact, dealing with a specific motoneuron survival factor.

Mudge: I could play Devil's Advocate here and argue that, since there are many small sensory (pain) fibres that project to muscle, you should be able to rescue those sensory neurons with a crude extract.

Oppenheim: I should stress that the experiments where we looked at the effects on dorsal root, sympathetic and parasympathetic ganglia were done only with the 25–75% ammonium sulphate fraction, and thus it is conceivable that there are factors in the other fractions that could affect a subpopulation of sensory neurons.

Vrbová: Do these chickens hatch that have so many surviving motoneurons? And what conditions are the embryos in when you inject them with extract?

Oppenheim: From their developmental stage, gross morphology, and the histology of the spinal cord, the treated embryos develop normally. We have not yet allowed embryos to survive beyond E10.

Vrbová: It would be interesting to see what motor units they have, with such a surplus of motoneurons.

Oppenheim: From our studies with neuromuscular blocking agents (such as curare or α-bungarotoxin), when we stop treating chick embryos with these agents their motor activity recovers (Oppenheim 1987). They then lose all the excess motoneurons that were rescued previously. So the only way to keep the extra motoneurons alive might be to treat them daily with muscle extract, right up to hatching. When we have a more purified factor, we may be able to do this.

Buller: Do you mean that they would eventually lose their (excess) motoneurons because of the wastage process that Gerta Vrbová has described, so it wouldn't be useful to let them hatch, because wastage would occur to a greater extent, in a later, secondary stage?

Oppenheim: The motoneurons that we have rescued by treating embryos with neuromuscular blockers regress and die when the block is removed during embryonic stages. However, if the neuromuscular blockade is maintained up to hatching, then the excess motoneurons survive for at least one week after hatching, despite the resumption of neuromuscular activity (Oppenheim 1984). Thus, there appears to be a critical period after which neuromuscular blockade is no longer required to maintain the excess neurons. It may be that the putative neurotrophic factor acts in the same manner.

Carlson: Saunders et al (1962) investigated cell death in the mesoderm of the posterior necrotic zone (PNZ) of the avian wing bud. They found that if PNZ mesoderm was grafted to other regions of the wing before a critical period of determination (HH stage 22), the cells were 'rescued' and remained alive instead of dying. There is not necessarily a target in this case, but I would not say that the cells are doomed to die. I gather that you feel that the motoneurons would disappear, once you eliminated that trophic factor?

Oppenheim: By extrapolation from the results of treating chick embryos with neuromuscular blocking agents (see above), rescuing the motoneurons with muscle extract would be expected to be transient during the embryonic period. It seems likely that to maintain the rescued cells one would have to continue to supply exogenous sources of the putative muscle-derived survival factor.

Van Essen: If the factor is disappearing by Day 15, in the chick, this suggests that the motoneurons have lost their susceptibility to it. By that measure perhaps they would be stable, once they had gone through this critical period.

Oppenheim: Or they may become dependent on a different neurotrophic factor that is present after that stage—the neurons might change their requirement for trophic factors during development. We simply don't know.

Lowrie: You have shown a clear dose–response relationship with the limb muscle extract in the embryonic chick. From what I remember of the early experiments in which the proportion of motor pool to target muscle was manipulated (Hollyday & Hamburger 1976, Lamb 1981, Denton et al 1985, Sohal et al 1986), the correlation between the amount of muscle available to be innervated and the survival of motoneurons was not so clear. Was this because of technical problems?

Oppenheim: Experiments in the chick in which the size of the targets has been manipulated have shown a proportional relationship between the size of the target (i.e., the number of primary myotubes present during the cell death period) and the number of surviving motoneurons (Tanaka & Landmesser 1986).

Lowrie: In these experiments, either muscles were made to maintain up to twice the normal number of motoneurons (Lamb 1981, Denton et al 1985), or the size of the neuron pool relative to its target was reduced but did not stop motoneuron death occurring (Sohal et al 1986). This would argue against a dose–response relationship for a trophic factor.

Nadal-Ginard: You said that you had ruled out the possibility that your factor is actually a growth factor for muscle, and that it was the target that had increased.

Oppenheim: Yes. It doesn't look as if the factor is affecting motoneuron survival indirectly by altering muscle development. Muscle development (size, myotube number, fast and slow properties) appears normal after treatment with muscle extract.

References

Barbin G, Manthorpe M, Varon F 1984 Purification of the chick eye ciliary neuronotrophic factor. J Neurochem 43:1468–1478

Denton CJ, Lamb AH, Wilson P, Mark RF 1985 Innervation pattern of muscles of one-legged *Xenopus laevis* supplied by motoneurons from both sides of the spinal cord. Dev Brain Res 17:85–94

Dohrmann U, Edgar D, Thoenen H 1987 Distinct neurotrophic factors from skeletal muscle and the central nervous system interact synergistically to support the survival of cultured embryonic spinal motor neurons. Dev Biol 124:145–152

Fahrbach SE, Truman JW 1987 Mechanisms for programmed cell death in the nervous system of a moth. In: Selective neuronal death. Wiley, Chichester (Ciba Found Symp 126) p 65–81

Henderson CE 1988 The role of muscle in the development and differentiation of spinal motoneurons: *in vitro* studies. In: Plasticity of the neuromuscular system. Wiley, Chichester (Ciba Found Symp 138) p 172–191

Hollyday M, Hamburger V 1976 Reduction of the naturally occurring motor neuron loss by enlargement of the periphery. J Comp Neurol 170:311–320

Lamb AH 1981 Selective bilateral motor innervation in *Xenopus laevis* tadpoles with one hind limb. J Embryol Exp Morphol 65:149–163

Manasek FJ 1969 Myocardial cell death in the embryonic chick ventricle. J Embryol Exp Morphol 21:271–284

Martin DP, Schmidt RE, DiStefano PS, Lowry OH, Johnson EM 1987 Inhibitors of protein synthesis and RNA synthesis prevent neuronal death caused by nerve growth factor deprivation. Abstr Soc Neurosci 13:922

Oppenheim RW 1984 Cell death of motoneurons in the chick embryo spinal cord. VIII. Motoneurons prevented from dying in the embryo persist after hatching. Dev Biol 101:35–39

Oppenheim RW 1987 Muscle activity and motor neuron death in the spinal cord of the chick embryo. In: Selective neuronal death. Wiley, Chichester (Ciba Found Symp 126) p 96–108

Saunders JW, Gasseling MT, Saunders LC 1962 Cellular death in morphogenesis of the avian wing. Dev Biol 5:147–178

Sohal GS, Stoney JR, Arumugam T, Yamashita T, Knox TS 1986 Influence of reduced neuron pool on the magnitude of naturally occurring motor neuron death. J Comp Neurol 247:516–528

Tanaka H 1987 Chronic application of curare does not increase the level of motoneuron survival-promoting activity in limb muscle extracts during the naturally occurring motoneuron cell death period. Dev Biol 124:347–357

Tanaka H, Landmesser L 1986 Cell death of lumbosacral motoneurons in chick, quail and chick–quail chimera embryos: a test of the quantitative matching hypothesis of neuronal cell death. J Neurosci 6:2889–2899

The role of muscle in the development and differentiation of spinal motoneurons: *in vitro* studies

Christopher E. Henderson*

Neurobiologie Moléculaire, Institut Pasteur, 25 rue du Dr Roux, 75724 Paris Cédex 15, France

Abstract. Results of *in vivo* experiments suggest that muscle cells, and probably other cell types, produce factors upon which motoneurons depend for survival and normal development. Most attempts to characterize such factors have used cultures of enriched or identified motoneurons, and have studied effects of muscle-derived substances on survival, neurite outgrowth and acetylcholine synthesis. Results from different laboratories vary widely, both in terms of the estimated abundance of motoneurons as a fraction of total dissociated spinal cord and in terms of the molecular weight estimates for factors tentatively proposed as candidate motoneuron growth factors. Nevertheless, there are several independent reports of 40–55 kDa species affecting each of the three parameters of spinal neuron development. We have begun to characterize one of these, partially purified from extracts of denervated muscle on the basis of its neurite-promoting activity for a subpopulation of 4.5-day embryonic chicken spinal neurons. Comparison between the factors under study in different systems, and confirmation of their importance *in vivo*, await the preparation of specific blocking antibodies.

1988 Plasticity of the neuromuscular system. Wiley, Chichester (Ciba Foundation Symposium 138) p 172–191

Introduction: motoneuron growth factors

The complex cellular environment of the spinal motoneuron is very likely to play an important role in determining its pattern of normal development (Fig. 1). Although both central and peripheral interactions have been shown to be vital for motoneuron survival in particular (Oppenheim & Haverkamp, this volume), most attention has been focused on the postsynaptic target of these neurons, skeletal muscle. Results from several *in vivo* experiments (reviewed in the preceding articles in this volume) have been interpreted in terms of hypothetical muscle-derived motoneuron growth factors (MNGFs) which still await characterization (Slack et al 1983). Such factors may be involved at

* *Present address*: Biochimie CNRS-INSERM, B.P. 5051, 34033 Montpellier Cedex, France.

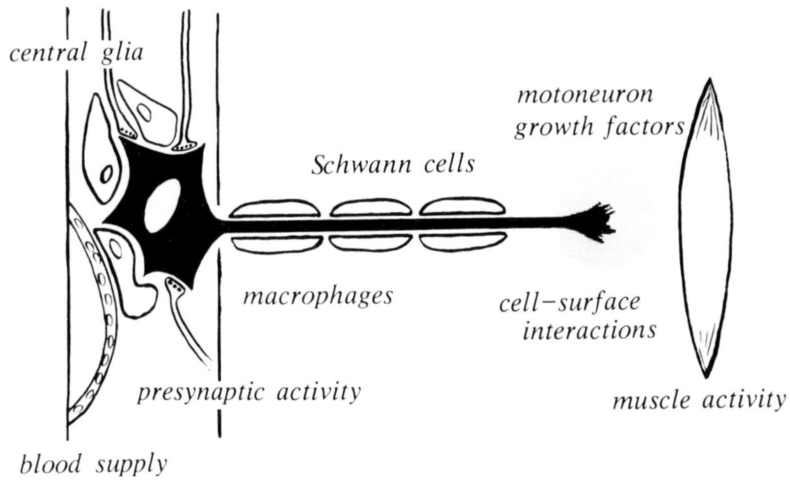

FIG. 1. Possible environmental influences on motoneuron survival, development and regeneration.

several developmental stages and in several different phenomena, such as cell survival, formation of muscle nerve branches, synapse stabilization and axonal sprouting. Given the diversity of the processes they are supposed to regulate, however, it is not clear how many different MNGFs are required in order for the motoneuron to reach successfully its adult configuration.

In the case of the only molecule whose role as a neuronal growth factor is rigorously proven, the nerve growth factor (NGF), a single polypeptide can enhance *in vitro* cell survival, neurite outgrowth and neurotransmitter synthesis of appropriate neurons. *In vivo*, however, its functions may be more restricted (Davies et al 1987). This illustrates the difficulty in drawing up an 'Identikit portrait' of the MNGF to be expected from the *in vivo* data. While it is probably true to say that most investigators seem to favour the concept of a pleiotropic neuronal growth factor, analogous to NGF, it is equally possible that different events at different stages of motoneuron development are regulated by quite independent factors, each of which could simultaneously play other roles in different regions of the developing embryo. This is not a banal problem, as it underlies the conception and the analysis of all *in vitro* experiments aimed at clarifying this aspect of the motoneuron's molecular interactions with its environment. Ideally, the significance of each molecule shown to be active *in vitro* should be tested by administering blocking antibodies *in vivo*, but the quantity of work this implies can prove dissuasive. Only in one case has this been possible for a candidate MNGF, neuroleukin, with results that on their own are inconclusive (see below). However, perhaps further experiments of this kind will allow the description of a series of molecules known and unknown, all of which are necessary for certain aspects of motoneuron development but none of which is alone sufficient.

In this article, I shall attempt briefly to review some of the different *in vitro* approaches being used to study candidate MNGFs, emphasizing one aspect (motoneuron abundance) for which divergent results have been obtained and another (molecular weight of MNGFs) where it is possible to imagine that a consensus might soon emerge. I shall then illustrate some of the problems and progress in the field by reference to our own work on muscle-derived neurite-promoting factors for embryonic spinal neurons.

Motoneurons in culture

In the mature spinal cord, the motoneurons make up an extremely small fraction of the total number of cells, identifiable mainly by their characteristic position in the anterior (ventral) horns. In the absence of specific markers for motoneurons, investigators have often resorted to using preparations in which they are simply enriched, either by density gradient centrifugation, or by taking advantage of the developmental advance of the motoneuron over most other spinal cord cells at early stages. These methods have the advantage of convenience and of requiring a minimum of experimental manipulations before cell culture. However, they can never provide positive identification of a given cell, even *a posteriori*.

The most frequently used technique for directly identifying motoneurons has been retrograde labelling by the injection into the limb bud of tracers that can subsequently be detected in populations of dissociated spinal neurons (Table 1). This approach has two major advantages: (a) providing that diffusion is prevented, the labelling of motoneurons is unambiguous; and (b) labelled motoneurons can be purified by flow cytometry. However, only motoneurons whose axon has reached the limb can be labelled, often with less than 100% efficiency, and tracer uptake and cell sorting can damage cells and alter their subsequent development *in vitro*. These problems may in part explain the ten-fold range in reported abundance of identified motoneurons in dissociated preparations of spinal neurons at quite similar developmental stages (Table 1). It is clear that less invasive methods, such as monoclonal antibodies to cell surface antigens (Tanaka & Obata 1984), would be of great interest, particularly for the study of early stages of motoneuron development.

Effects of identified molecules on spinal neurons *in vitro*

Factors affecting spinal neuron development *in vitro* have been classified according to their influence on the three major parameters regulated by NGF on its target neurons: cell survival, neurite outgrowth and neurotransmitter synthesis (Table 2). Only a few known growth-promoting molecules have

TABLE 1 Reported abundance of identified motoneurons in preparations of dissociated spinal neurons

Publication	*Embryonic age*	*Labelling method*	*Spinal cord region*	*Estimated abundance of motoneurons*
Berg & Fischbach, 1978	4 d chick	Synapse formation with myotubes *in vitro*	Total	50% of large neurons after 1 wk in culture
Calof & Reichardt, 1984	7 d chick	Retrograde WGA-LY	Total	9%
Dohrmann et al, 1987	6 d chick	Retrograde RhITC	Ventral	10%
Nurcombe et al, 1984	7 d chick	Retrograde HRP	Total	3%
O'Brien & Fischbach, 1986	6 d chick	Retrograde WGA-FITC	Total	2%
Schaffner et al, 1987	13 d mouse	Retrograde WGA-FITC	Ventral	14%
Schnaar & Schaffner, 1981	6–7 d chick	Metrizamide gradient fractionation	Total	17%
Smith et al, 1986	13–14 d rat	Retrograde WGA-LY	Ventral	4%
	13–14 d rat	CAT antibody	Total	5% (8% after 4 d in culture)
	13–14 d rat	CAT antibody	Ventral	11% (23% after 4 d in culture)
Tanaka, 1987	4 d chick	Retrograde DiI	Ventral	9%
	6 d chick	Retrograde DiI	Ventral	5%

Abbreviations: HRP, horseradish peroxidase; WGA, wheat germ agglutinin; LY, lucifer yellow; FITC, fluorescein isothiocyanate; CAT, choline acetyltransferase; DiI, dioctadecyl-tetramethyl-indocarbocyanine; RhITC, rhodamine isothiocyanate.
Unless otherwise indicated, estimated abundances are given as the percentage of dissociated spinal neurons labelled.

been reported to affect spinal neurons. Among these, basic fibroblast growth factor (bFGF) at 10 ng/ml enhances survival (Unsicker et al 1987) and choline acetyltransferase activity (T. Heuser & K. Unsicker, unpublished) of dissociated embryonic spinal neurons, while laminin (a basement membrane glycoprotein) promotes neurite outgrowth from identified motoneurons (Calof & Reichardt 1984).

Nanogram concentrations of neuroleukin, a novel candidate MNGF, permit the short-term survival of a sub-population of spinal neurons *in vitro* but have no effect on neurite outgrowth (Gurney et al 1986b). Monoclonal antibodies to neuroleukin, on the other hand, do not alter motoneuron numbers when adminstered *in vivo*, but do inhibit (by about 50%) the terminal sprouting in paralysed muscles of the adult mouse (Gurney et al 1986a). This apparent paradox is explained by the authors in terms of a model in which neuroleukin acts as a survival factor only during a critical period of development, and thereafter modulates neuronal growth but is not required for continued viability. However, these studies illustrate well the difficulty of defining a physiological role for a molecule identified *in vitro*, because another plausible explanation of these results might be that neuroleukin is a protein of growth cones or muscle that is merely permissive for axonal sprouting (thus explaining the inhibitory effect of antibodies) and that the survival-promoting activity of neuroleukin is expressed only in certain *in vitro* situations. The former hypothesis seems particularly relevant in the light of the observation (Gurney et al 1986b) that denervation caused no increase in the neuroleukin content of muscle, whereas the level of 'sprouting signal' was presumably greatly increased (Brown & Lunn, this volume).

The effects of FGF, laminin and neuroleukin are not limited to spinal neurons. Since in no case have antibodies to one of them been shown to block the same aspect of motoneuron development that they promote *in vitro*, it seems prudent at the present stage to suppose that each may play an important role in motoneuron development but that none of them seems to correspond to the NGF model for neuronal growth factors.

Partially characterized factors affecting spinal neuron development *in vitro*

The actions of muscle-derived substances on spinal neurons in culture have been studied in so many different systems that direct comparison is difficult. I have attempted to look for consistent features by classifying, according to the parameter of spinal neuron development that they affect, those factors for which tentative molecular weight data have been published (Table 2). It is at first sight striking that molecules in the molecular weight range 40–55 kDa support survival, neurite outgrowth and acetylcholine synthesis. This consensus, albeit partial, may reflect the pleiotropic effects of a single neurotrophic molecule. Unfortunately, however, since many proteins have molecular

TABLE 2 Molecular characterization of factors affecting three major parameters of spinal neuron development *in vitro*

	Enhancement of:		
Publication	*Survival*	*Neurite outgrowth*	*CAT activity*
Identified molecules			
Unsicker et al (1987), T. Heuser & K. Unsicker, unpublished	FGF (18 kDa)	—	FGF (18 kDa)
Calof & Reichardt (1984)	—	Laminin (400 kDa)	—
Gurney et al (1986b)	Neuroleukin (56 kDa)	—	—
Molecules not yet purified			
Dribin & Barrett (1982)	—	50 kDa 300 kDa	—
Giess & Weber (1984)	—	—	Cholinergic factor
Henderson et al (1981)	—	40 kDa 500 kDa 1000 kDa	—
Kaufman et al (1985)	—	—	40 kDa
Kaufman & Barrett (1983)	55 kDa	—	—
Smith et al (1985)	1.5 kDa	55 kDa 33 kDa	55 kDa 1.5 kDa
Tanaka & Obata (1982)	150 kDa 70 kDa 40 kDa	—	—

The values for apparent molecular weight (in kilodaltons) quoted are those cited in the articles indicated; no account is taken here of the different conditions and techniques used in estimating them.

weights in this range, this observation cannot yet be taken as strong support for such a hypothesis.

Neurite-promoting factors for embryonic chicken spinal neurons

The approach we have adopted for the study of candidate MNGFs is based on measurement of neurite outgrowth in cultures of total spinal neurons from 4.5-day chicken embryos (Fig. 2). The presence of muscle is apparently not required *in vivo* for the first stages of axonal growth outside the spinal cord (Oppenheim & Haverkamp, this volume). Nevertheless, we believe that this assay should be able to detect motoneuron growth factors, in the same way

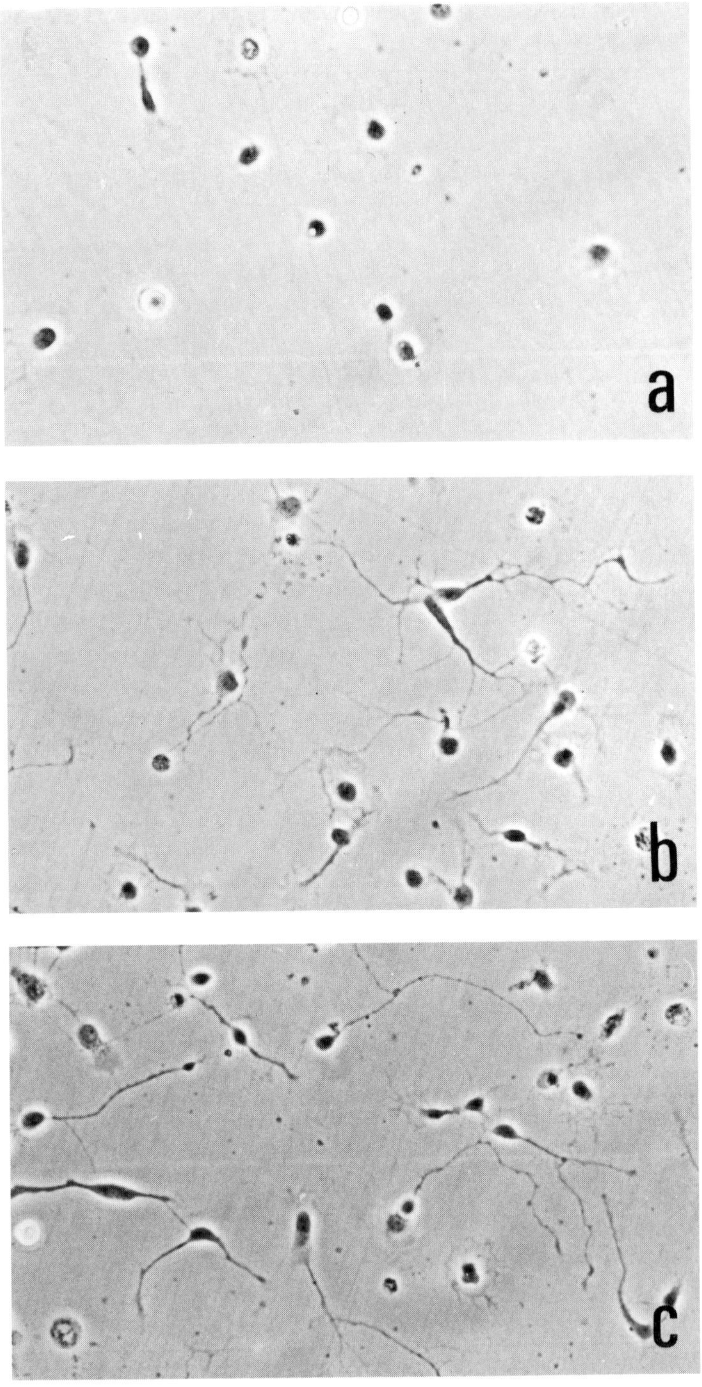
a
b
c

that NGF is quite reliably assayed by its effect on axonal regeneration *in vitro* even though it is probably not involved in axonal growth toward the target *in vivo* (Davies et al 1987). Rather than review our work on the regulation of muscle-derived neurite-promoting activities during development and by muscle activity (Henderson et al 1981, 1983, 1984), I wish here to emphasize those aspects which differentiate the factors under study from others reported in the literature and which suggest that our *in vitro* results may have some relevance for the *in vivo* situation.

We have identified two muscle-derived neurite-promoting activities for spinal neurons, apparently associated with different molecular species. The 'embryonic' activity, found in media conditioned by embryonic myotubes, is not detected at comparable levels in most other conditioned media tested (Henderson et al 1981). The 'neonatal' activity present in high-speed supernatants of leg-muscle from post-hatch chicks does not show such pronounced tissue specificity, as extracts of several other neonatal tissues also stimulate neurite outgrowth (Henderson et al 1984). However, levels of 'neonatal'

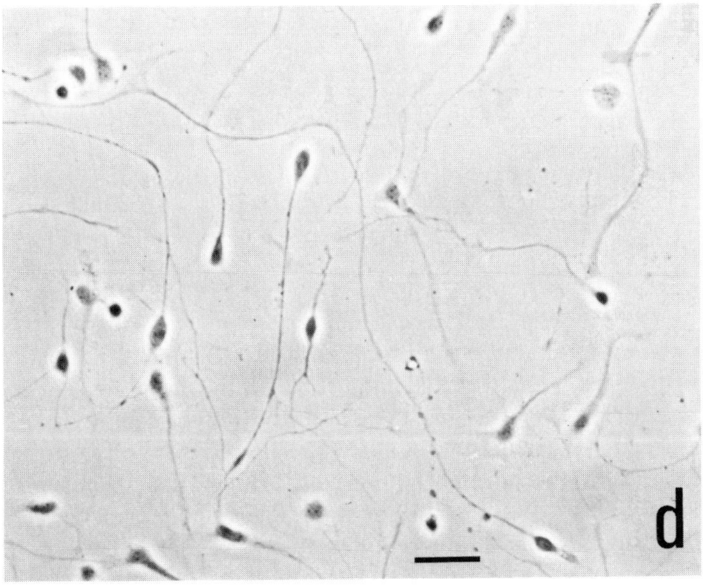

FIG. 2. Neurite outgrowth response of embryonic chicken spinal neurons to muscle-derived factors and to laminin. Neural tubes of 4.5-day chicken embryos were dissociated and cultured as described (Henderson et al 1984). (a) Spinal neurons cultured for 36 h on uncoated tissue culture plastic in serum-free F12 medium without supplements; (b, c) parallel cultures to which embryonic myotube-conditioned medium diluted 20-fold (b), or 0.5 μg/ml denervated muscle extract (Henderson et al 1983) (c), was added; (d) spinal neurons cultured for 36 h on a substratum precoated with polyornithine and laminin. Scale bar, 50 μm.

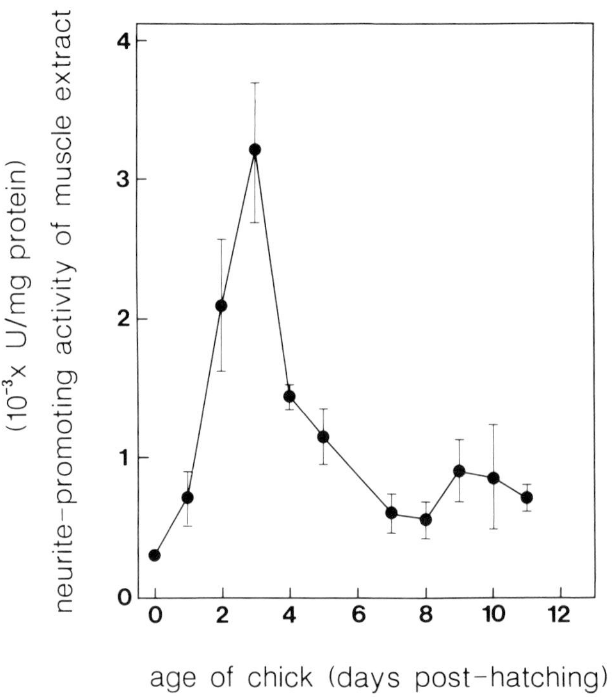

FIG. 3. Developmental regulation of levels of neurite-promoting activity for spinal neurons in extracts of chick leg muscle. Post-hatch chicks of the indicated ages were sacrificed, and soluble extracts of leg muscle prepared as described (Henderson et al 1984). Specific neurite-promoting activities are expressed in biological units per mg protein.

activity in muscle undergo a pronounced developmental regulation in the course of the first week after hatching (Fig. 3). Furthermore, denervation of neonatal muscle results in an increase of up to 15-fold in levels of neurite-promoting activity (Henderson et al 1983). This increase follows a time course comparable to that required for observation of the first sprouts after partial denervation (Brown & Lunn, this volume). Preliminary evidence (Henderson et al 1984) suggests that the type of neurite-promoting activity produced is a function of the age of the muscle, because embryonic myotubes that have been allowed to age *in vitro* produce a neurite-promoting activity that shares properties with the 'neonatal' factor. The activities of such muscle-derived preparations are probably high enough to be biologically relevant: total protein concentrations on the order of 1 μg/ml are sufficient to give half-maximal effects on neurite outgrowth, although the factor itself is probably present with low abundance (see below).

Both the 'embryonic' and 'neonatal' neurite-promoting activities thus show biological properties consistent with those that might be predicted for an

MNGF. Because we have used cultures of total spinal cord, however, it is not possible to state with certitude that all the observed effects are on motoneurons. However, in cultures from 4.5-day chick embryos, motoneurons should be enriched as compared to later stages (Berg & Fischbach 1978). About 25–30% of the neurons in our cultures respond to the neurite-promoting factors described above; comparison with the data in Table 1 for spinal neurons labelled mostly at later stages makes it seem not impossible that this fraction of motoneurons be present in our cultures. In addition, subdissection of the 4.5-day spinal cord has revealed that all the responsive neurons are situated in the ventral regions (C.E. Henderson, unpublished work 1987). In preliminary studies, we have also shown that many neurons in these cultures are recognized by a monoclonal antibody specific for motoneurons and ventral epithelial cells (Tanaka & Obata 1984, H. Tanaka & C.E. Henderson, unpublished work 1987).

Purification of the active factor from extracts of denervated muscle is currently being undertaken using a combination of conventional and high-pressure chromatography. Although the procedure developed is not perfectly reproducible, the most highly purified preparations obtained, after isoelectric precipitation, anion-exchange chromatography, chromatofocusing and gel filtration, were apparently enriched approximately 10^4-fold with respect to total protein, and had specific activities in the neurite outgrowth assay on the order of 10^6 to 10^7 biological units/mg protein (comparable to that, in a different system, of highly purified NGF) (T. Taguchi & C.E. Henderson, unpublished work 1986). At this degree of purity, and at low protein concentrations, the activity migrates on gel filtration columns with an apparent molecular weight of 50 kDa. However, preliminary experiments for detection of neurite-promoting activity after separation on SDS-polyacrylamide gels suggest that activity is associated with a 25 kDa polypeptide. Current work is directed toward more detailed biochemical and functional characterization of these molecules.

The MNGF classically envisaged should promote not only neurite outgrowth but also cell survival and acetylcholine synthesis. In our culture system, effects on cell survival are rarely observed, owing probably to the extreme simplicity of the culture medium and substratum. Two observations, however, raise the possibility that the neurite-promoting factors under study might also have survival-promoting properties. Firstly, partially purified preparations from denervated muscle have very pronounced and selective effects (Fig. 4) on both neurite outgrowth and survival of a responsive subpopulation of telencephalic neurons (Taguchi et al 1987). Secondly, the same preparations enhance the survival of about 20% of total spinal neurons cultured in serum-free medium on laminin-coated dishes (C.E. Henderson, unpublished work 1986). We have not so far studied effects on total choline acetyltransferase activity.

In so far as possible, we wished to exclude the possibility that we were

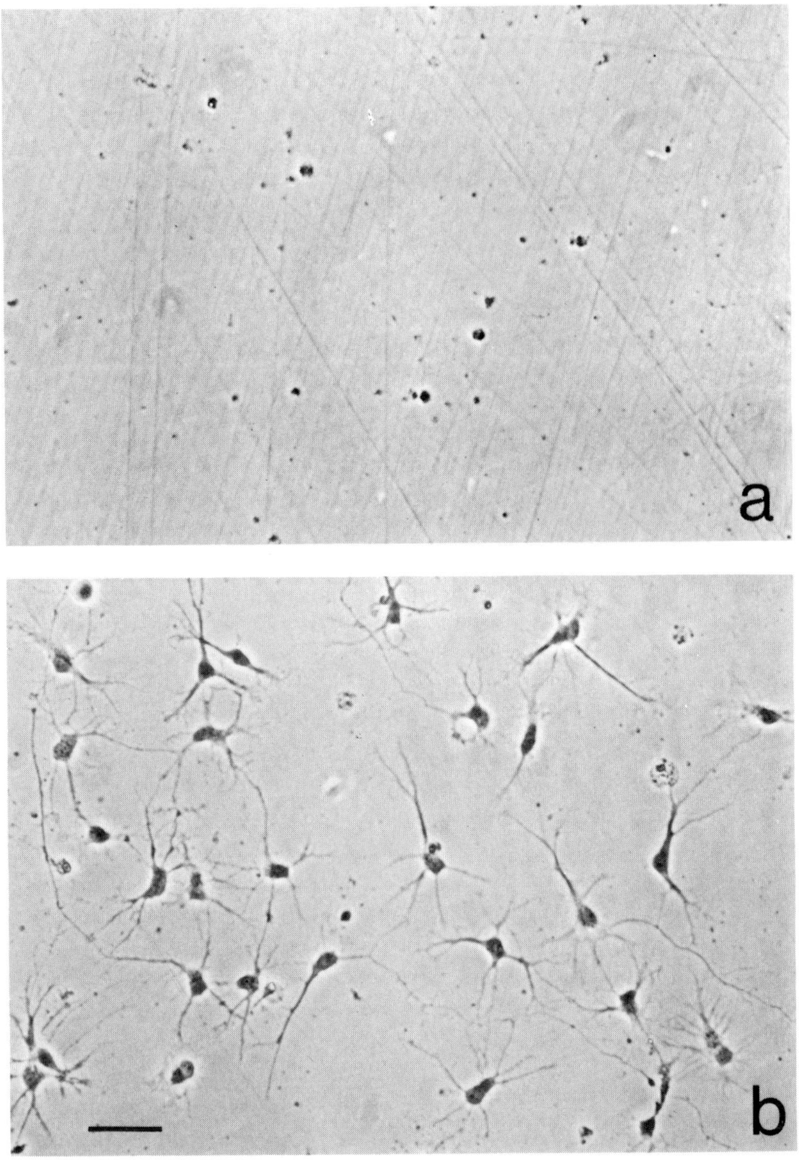

FIG. 4. A DEAE-Sepharose fraction containing neurite-promoting activity for spinal neurons selectively enhances the survival of a subpopulation of neurons from the embryonic telencephalon. Fractions of denervated muscle extract eluted from a DEAE-Sepharose column at 0.25 M NaCl were applied at 1 μg/ml protein to cultures of dissociated telencephalic neurons from 5-day chicken embryos (Taguchi et al 1987). After 3 days, almost no cells survived without the DEAE fraction (a), whereas a subpopulation survived well in its presence (b). Scale bar, 50 μm.

TABLE 3 Comparison between partially purified neurite-promoting factor and laminin

Property	*Neurite-promoting factor*	*Laminin*
Maximal percentage of total spinal neurons responding	30%	90%
Selective effect on ventral spinal neurons	Yes	No
Neurite-promoting activity when cells cultured on uncoated plastic	Optimal	Zero
Neurite-promoting effect on embryonic mesencephalic neurons (Taguchi et al 1987)	Very low	High
Apparent molecular weight in non-denaturing conditions	50 kDa	500 kDa

observing the effects of a molecule already characterized (C.E. Henderson, unpublished work 1987). Having shown that insulin, transferrin, NGF and several other proteins had no effect on neurite outgrowth in this system, we turned our attention to the molecules already mentioned, FGF, laminin and neuroleukin, known to affect spinal neurons. In our hands, bFGF (a generous gift of Dr M. Sensenbrenner) had no effect on neurite outgrowth; in addition, the neurite-promoting activity in denervated muscle extracts showed little affinity for heparin–Sepharose, which is known to bind FGF very tightly. Neuroleukin, which is in any case reported to be without effect on neurite outgrowth, has affinities for columns of immobilized Procion Red and hydroxylapatite that are radically different from those we have observed for our neurite-promoting factors.

It seemed particularly important to show that the muscle-derived activities were not due to laminin or one of its proteolytic fragments, since laminin had been claimed to be responsible for the neurite-promoting activity of many conditioned media. The results of a series of experiments (M. Roa & C.E. Henderson, unpublished work 1987) are shown in Table 3. They clearly exclude any biochemical or functional similarity between laminin and our neurite-promoting factors. In addition, dose–response curves obtained using the two in combination have shown that all spinal cells responsive to the muscle factors are also responsive to laminin but that, on the other hand, many laminin-responsive neurons do not put out neurites in the presence of the muscle-derived factors alone.

Conclusions and perspectives

Work from several different laboratories provides evidence that is consistent with the existence of MNGFs affecting one or more parameters of spinal neuron or motoneuron development *in vitro*. However, only when purified molecules and blocking antibodies are available will it be possible to determine on one hand, whether these *in vitro* activities faithfully reflect *in vivo* roles and, on the other, to what extent the resulting MNGFs (if any) correspond to the 'Identikit portrait' established for them from interpretation of the results of *in vivo* manipulations. Although molecular characterization of these factors has progressed only slowly, the relative accessibility of the neuromuscular junction at all developmental stages should make it a favourable system for subsequent analysis of the physiological role of candidate MNGFs.

Acknowledgements

I wish to thank wholeheartedly my collaborators in the laboratory of Jean-Pierre Changeux — Monique Huchet, Takahisa Taguchi and Jorge Kirilovsky — for many stimulating discussions. I gratefully acknowledge funding from the Association Française contre les Myopathies.

References

Berg DK, Fischbach GD 1978 Enrichment of spinal cord cell cultures with motoneurons. J Cell Biol 77:83–98

Brown MC, Lunn ER 1988 Mechanism of interaction between motoneurons and muscles. In: Plasticity of the neuromuscular system. Wiley, Chichester (Ciba Foundation Symposium 138) p 78–96

Calof AL, Reichardt LF 1984 Motoneurons purified by cell sorting respond to two distinct activities in myotube-conditioned medium. Dev Biol 106:194–210

Davies AM, Bandtlow C, Heumann R, Korsching S, Rohrer H, Thoenen H 1987 Timing and site of nerve growth factor synthesis in developing skin in relation to innervation and expression of the receptor. Nature (Lond) 326:353–358

Dohrmann U, Edgar D, Thoenen H 1987 Distinct neurotrophic factors from skeletal muscle and the central nervous system interact synergistically to support the survival of cultured embryonic spinal motor neurons. Dev Biol 124:145–152

Dribin LB, Barrett JN 1982 Two components of conditioned medium increase neuritic outgrowth from rat spinal cord explants. J Neurosci Res 8:271–280

Giess MC, Weber MJ 1984 Acetylcholine metabolism in rat spinal cord cultures: regulation by a factor involved in the determination of the neurotransmitter phenotype of sympathetic neurons. J Neurosci 4:1442–1452

Gurney ME, Apatoff BR, Heinrich SP 1986a Suppression of terminal axonal sprouting at the neuromuscular junction by monoclonal antibodies against a muscle-derived antigen of 56 000 daltons. J Cell Biol 102:2264–2272

Gurney ME, Heinrich SP, Lee ML, Yin H-S 1986b Molecular cloning and expression of neuroleukin, a neurotrophic factor for spinal and sensory neurons. Science (Wash DC) 234:566–574

Henderson CE, Huchet M, Changeux JP 1981 Neurite outgrowth from embryonic chicken spinal neurons is promoted by media conditioned by muscle cells. Proc Natl Acad Sci USA 78:2625–2629

Henderson CE, Huchet M, Changeux JP 1983 Denervation increases a neurite-promoting activity in extracts of skeletal muscle. Nature (Lond) 302:609–611

Henderson CE, Huchet M, Changeux JP 1984 Neurite-promoting activities for embryonic spinal neurons and their developmental changes in the chick. Dev Biol 104:336–347

Kaufman LM, Barrett JN 1983 Serum factor supporting long-term survival of rat central neurons in culture. Science (Wash DC) 220:1394–1396

Kaufman LM, Barry SR, Barrett JN 1985 Characterization of tissue-derived macromolecules affecting transmitter synthesis in rat spinal cord neurons. J Neurosci 5:160–166

Nurcombe V, Hill MA, Eagleson KL, Bennett MR 1984 Motor neuron survival and neuritic extension from spinal cord explants induced by factors released from denervated muscle. Brain Res 291:19–28

O'Brien RJ, Fischbach GD 1986 Isolation of embryonic chick motoneurons and their survival *in vitro*. J Neurosci 6:3265–3274

Oppenheim RW, Haverkamp LJ 1988 Neurotrophic interactions in the development of spinal cord motoneurons. In: Plasticity of the neuromuscular system. Wiley, Chichester (Ciba Foundation Symposium 138) p 152–171

Schaffner AE, St John PA, Barker JL 1987 Fluorescence-activated cell sorting of embryonic mouse and rat motoneurons and their long-term survival *in vitro*. J Neurosci 7:3088–3104

Schnaar RL, Schaffner AE 1981 Separation of cell types from embryonic chicken and rat spinal cord; characterization of motoneuron-enriched fractions. J Neurosci 1:204–217

Slack JR, Hopkins WG, Pockett S 1983 Evidence for a motor nerve growth factor. Muscle & Nerve 6:243–252

Smith RG, McManaman J, Appel SH 1985 Trophic effects of skeletal muscle extracts on ventral spinal cord neurons *in vitro*: separation of a protein with a morphologic activity from proteins with cholinergic activity. J Cell Biol 101:1608–1621

Smith RG, Vaca K, McManaman J, Appel SH 1986 Selective effects of skeletal muscle extract fractions on motoneuron development *in vitro*. J Neurosci 6:439–447

Taguchi T, Huchet M, Roa M, Changeux JP, Henderson CE 1987 A subpopulation of embryonic telencephalic neurons survive and develop *in vitro* in response to factors derived from the periphery. Dev Brain Res 37:125–132

Tanaka H 1987 Chronic application of curare does not increase the level of motoneuron survival-promoting activity in limb muscles during the naturally occurring motoneuron cell death period. Dev Biol 124:347–357

Tanaka H, Obata K 1982 Survival and neurite growth of chick embryo spinal cord cells in serum-free culture. Dev Brain Res 4:313–321

Tanaka H, Obata K 1984 Developmental changes in unique cell surface antigens of chick embryo spinal motoneurons and ganglion cells. Dev Biol 106:26–37

Unsicker K, Reichert-Preibsch H, Schmidt R, Pettmann B, Labourdette G, Sensenbrenner M 1987 Astroglial and fibroblast growth factor have neurotrophic functions for cultured peripheral and central nervous system neurons. Proc Natl Acad Sci USA 84:5459–5463

DISCUSSION

Pette: Can you comment on the identification of neuroleukin as glucose-6-phosphate isomerase? How could this enzyme act as a trophic factor?

Henderson: Chaput et al (1988) suggest that it may act *in vitro* by binding to sugars on the surface of the cultured cells. (See also Faik et al 1988 and reply by Gurney 1988).

Pette: It may not be the enzymic activity which is functional in this trophic effect.

Henderson: That would be very interesting!

Gordon: If a factor is present in conditioned medium, does that mean that it's a soluble factor?

Henderson: If a factor is in a conditioned medium, it must exist, at least in part, in a water-soluble form. By that definition, NGF is a soluble factor. The fact that such factors are soluble has led some to suggest the possibility of diffusion in the embryo over relatively long distances. However, we know that many growth factors have a strong affinity for cationic culture substrata such as polylysine and polyornithine, so there might not be great diffusion of even the soluble factors *in vivo*. They may be released from, and then directly taken up by, the nerve terminal, or they might be immobilized on the extracellular matrix and used by the incoming nerve.

Ribchester: How species-specific is the monoclonal antibody which binds to motoneurons?

Henderson: This is Dr Tanaka's antibody (Gunma University, Japan). It is very species-specific and doesn't react with motoneurons from rat or mouse.

Ribchester: Have you found evidence of factors which cause the regression of neurites that have been previously stimulated, using any of your bioassay systems? And have you had the opportunity to study the interaction in culture between, say, two growth cones of different motoneurons, and seen any mutual contact inhibition or regression of the neurons as described by Kapfhammer & Raper (1987)?

Henderson: Only recently have we been able to say that we are looking at motoneurons, so we have not looked at motoneuron–motoneuron interactions for the moment.

It is hard to study factors that might cause regression in culture until we know exactly what substance we are dealing with. It would be difficult to distinguish between factors that are simply killing the cells, and a factor causing regression. We looked at extracts of muscle from patients with spinal muscular atrophies, where we and others postulate that something might be wrong with the trophic support system for the motoneurons (Henderson et al 1987). We found that muscle extracts from such patients inhibit our 'neonatal' trophic activity, but not the 'embryonic' activity. But, again, this is not the same as causing regression once growth has occurred.

Walsh: Are standard protein chemical purification methods likely to be adequate to give one enough trophic factor for N-terminal sequencing?

Henderson: I think so. If we need to purify 10000-fold and we want 100 μg of protein, then at, say, 1% yield we will need 100g of protein, which is possible using denervated chicken muscles. However, we haven't succeeded in doing it, so I may be being optimistic!

Walsh: Are you discounting methods such as the *Xenopus* oocyte system used for studying other trophic factors, where it is possible to go directly to cloning the molecule? In the immune system, a number of growth factors have been cloned in this way.

Henderson: For this approach to work, one would need to have a bioassay that was not only extremely sensitive, but also highly selective. I personally don't believe it is likely to work in our system, given the number of different factors that probably affect motoneuron development *in vitro*.

Hoffman: With an antibody that blocked the trophic effect, you would be in a good position to use that antibody for clonal selection, to generate a cDNA probe for your trophic factor.

Henderson: If I had an antibody that blocked the effects, I would be in a good position for many experiments!

Walsh: What effects do you see with fibroblast growth factor (FGF) in your culture model?

Henderson: We have not detected any effect of FGF on neurite outgrowth in our system, although Unsicker et al (1987) found enhanced neuronal survival and CAT activity.

Walsh: Could the factor identified by Appel and colleagues (Smith et al 1985), which induces choline acetyl transferase, or the factor which induces choline uptake, be FGF?

Henderson: Given that FGF has been reported in forms ranging from 18000 to 25000Da subunit molecular mass, it is important to test possible identity with FGF in every case. Our factor, for instance, has only low affinity for heparin–Sepharose, and FGF doesn't work in our system; so we think it unlikely that our factor is FGF.

Lowrie: You are measuring the initial outgrowth of a neurite from a motoneuron and the effect of your muscle extract on that. We have heard in Cynthia Lance-Jones' paper that muscle probably isn't necessary *in vivo* for the initial outgrowth of the axons, but may be important for the branching of axons, near or inside the muscle. In some of your micrographs the motoneurons seem to have branched neurites. Have you looked at this aspect?

Henderson: Your first point has indeed worried me, because it is clear now that the first stages of motoneuron outgrowth don't require the muscle; and in fact it would be surprising if they did, because the distance is too great. However, in the sensory system we now know that NGF production only begins once the growth cones arrive at the target. Nevertheless, neurite outgrowth *in vitro* is an adequate test of NGF activity for most purposes. Therefore, while I would not wish to say that muscle-derived factors must be involved in the initial neurite outgrowth *in vivo*, an *in vitro* system such as this seems to me to be a valid test with which to look for and purify candidate trophic factors. Once that stage has been reached, one can then ask what, if anything, they are doing *in vivo*.

In answer to your second question, we have not done quantitative studies on neurite branching, but it is clear that it is highly dependent on the type of culture substratum used.

Lowrie: Is it possible that extract prepared from denervated muscle is more potent not because it contains a larger amount of a trophic factor but because it is easier to extract this substance from a denervated muscle?

Henderson: We don't know why the denervated muscle extract has more neurite-promoting activity. It might be because the muscle is inactive, but it could equally be one of many other things, like products of nerve degeneration, Schwann cell proliferation, or interruption of retrograde transport. I don't think it is a difference in extraction, because we have made these comparisons in many different extraction protocols, with pH values ranging from 2.0 to 10.0. The difference, although it may vary quantitatively, is always maintained.

Sanes: Is there any preferential effect of the factor on a subset of neuronal processes in culture? In other words, are some of the neurites potentially dendrites and others axons, as in hippocampal cultures (Cáceras et al 1986)?

Henderson: That's an interesting idea in relation to the motoneuron, whose dendrites develop in the centre and whose axon develops, at least at later stages, in the periphery. Unfortunately, the tools hardly exist for doing this. We haven't done MAP-2 (microtubule-associated protein 2) staining, but I believe it wouldn't be useful for motoneurons, because MAP-2 is in both central and peripheral processes.

Thesleff: Botulinum toxin type A induces marked sprouting in motor nerves. This might be a good way to test for the presence of such a factor in muscle.

Henderson: We would like to do that when a pure factor is available.

Vrbová: I would like to relate Dr Henderson's and Dr Oppenheim's work to the recent finding on sympathetic neurons, where dendrites develop only when the neurons are in contact with the target (Vojvodic 1987). This might be important for the development of the motoneuron, in that if they don't develop dendrites, those synapses that are on the cell body and on the initial segment may persist there longer. Does survival depend on the development of dendrites?

Oppenheim: We examined the differentiation of chick motor neurons after early (E2) limb bud removal. Dendrite development was normal up to the stage (E6) when the neurons begin to die *en masse* (Oppenheim et al 1978). We have shown previously that survival is dependent on afferent input (Okado & Oppenheim 1984). This effect may require normal dendrite development.

Lance-Jones: *In vivo*, muscle cells or muscle cell precursors might not be required for early motoneuron outgrowth but other specific mesenchymal cell types might be. In attempting to relate *in vivo* with *in vitro* results. I was wondering if your muscle extract might contain components derived from connective tissue cells as well as muscle cells?

Henderson: Conditioned media from pure muscle cell lines enhance neurite outgrowth in our system, but we cannot exclude that other cell types contribute to the effects we observe using muscle extracts.

Lance-Jones: The factors that control axon outgrowth to a specific target and those that control ramification within a muscle are likely to be different. Studies of early axon outgrowth certainly don't rule out the possibility that once a motoneuron reaches its target, it is dependent on muscle-derived factors for further development.

Henderson: Concerning the stage at which muscle-derived factors are involved, I agree entirely, and I tried to make the point that there could be different factors for different developmental stages. In particular, I specifically excluded the suggestion that the first stages of neurite outgrowth might be regulated by muscle. I agree that nerve branching within the growing muscle is quantitatively important in terms of axon outgrowth, and it is necessary to think in terms of something that regulates that.

Buller: It would certainly be remarkable if the same determinant satisfied both functions, the initial outgrowth without much large sprouting (there are branches in axons half way down the leg which pass to the same distal muscle, as Sherrington showed) and then most of the sprouting occurs once the axon has reached the target area, in a much more constrained way.

Rieger: Have you tried your factor on PC12 cells, which are able to make contacts with muscle cells?

Henderson: We have tried muscle extracts on PC12 cells (R. Defez & C.E. Henderson, unpublished work). Using very large concentrations of neonatal muscle extracts we could produce NGF-like responses (we haven't tried the partially purified factors). However, I put little reliance in results obtained using such high protein concentrations *in vitro*.

Walsh: Was this experiment carried out also with control cultures containing anti-NGF antibodies?

Henderson: Yes. The effect proved to be NGF independent. We have tested our 'neonatal' and 'embryonic' activities on sensory neurons and found no effect and, as I said, the vast majority of brain neurons don't respond either.

Vrbová: I would like to turn the whole topic of the target dependence of motoneurons upside down and suggest that it would be useful to look at what factor in the spinal cord could lead to the death of motoneurons. What kills them so suddenly? Perhaps if one knew this one could think of how to prevent it, or what is needed to provide the cell with, so that it is not killed.

Henderson: I don't think we know what kills these cells in the spinal cord. The problem in studying this would be that if the choice between cell survival and death were made early and 'killing factors' existed simply to clear away those cells whose elimination had previously been determined, it would be difficult to get at the underlying mechanisms.

Buller: Is it right to suggest that cell death in the spinal cord is entirely determined *within* the cord? Surely it is because there is no activity in the periphery?

Vrbová: I agree that cell death is determined by something that the target imparts to the motoneuron, but maybe one could discover what it is by investigating why these cells die. Perhaps it is because they don't have enough glutamate receptors, or perhaps the target induces glutamate receptors, or dendrite growth? If one knew that motoneurons die because they have too much or too little of one or the other, one could then see what would stimulate the production of whatever they have too much or too little of.

Buller: I have always thought that it was a question of redundancy—that axons grow out, make contact, then find themselves in excess to the need for motor innervation, have nothing to do, and the excess motoneurons die as a result. This is why it is so surprising that there has to be a substance produced to keep some of them alive.

Vrbová: Yes, but what is it, in the phenotype of the cell, that is needed for its survival? This survival factor may be target derived, and it may depend on the interaction with the target. But for discovering the cause of motoneuron death it is important to know the phenotypic expression of this induction. I am only suggesting that one could look at this from the opposite direction.

Mudge: I think that what Gerta is suggesting implies some interaction between motoneurons and other spinal cord cells. But surely the action of NGF on sympathetic and sensory neurons is a good model for target-dependent motoneuron survival? Sympathetic neurons grown in isolation from other cell types, but in the presence of NGF, live; remove the NGF, and the neurons die. A reasonable explanation is that NGF switches genes that are vitally important for neuronal survival 'on' or 'off'. In fact, it has been shown recently that one of the earliest genes that NGF induces in PC12 cells encodes a protein which looks like a transcriptional regulatory factor (Milbrandt 1987). If you think of the action of survival factors in these terms, it doesn't seem too mysterious.

Oppenheim: The results obtained using protein synthesis inhibitors, both *in vivo* and *in vitro* (see Discussion, p 167), suggest that perhaps all neurons have an intrinsically determined death programme which is overridden only when the cell takes up a trophic factor that changes the programme.

Buller: That seems to me remarkable, but I understand your description.

References

Cáceras A, Banker GA, Binder L 1986 Immunocytochemical localization of tubulin and microtubule associated protein 2 during the development of hippocampal neurons in culture. J Neurosci 6:714–722

Chaput M, Claes V, Portetelle D et al 1988 The neurotrophic factor neuroleukin is 90% homologous with phosphohexose isomerase. Nature (Lond) 332:454–455

Faik P, Walker JIH, Redmill AAM, Morgan MJ 1988 Mouse glucose-6-phosphate isomerase and neuroleukin have identical 3′ sequences. Nature (Lond) 332:455–456

Gurney ME 1988 Reply. Nature (Lond) 332:456–457

Henderson CE, Hauser SL, Huchet M et al 1987 Extracts of muscle biopsies from patients with spinal muscular atrophies inhibit neurite outgrowth from spinal neurons. Neurology 37:1361–1364

Kapfhammer J, Raper J 1987 Collapse of growth cone structure on contact with specific neurites in culture. J Neurosci 7:201–212

Milbrandt J 1987 A nerve growth factor-induced gene encodes a possible transcriptional regulatory factor. Science (Wash DC) 128:797–799

Okado N, Oppenheim RW 1984 Cell death of motoneurons in the chick embryo spinal cord. IX. The loss of motoneurons following removal of afferent inputs. J Neurosci 4:1639-1652

Oppenheim RW, Chu-Wang IW, Maderdrut JL 1978 Cell death of motoneurons in the chick embryo spinal cord. III. The differentiation of motoneurons prior to their induced degeneration following limb-bud removal. J Comp Neurol 177:87–112

Smith RG, McManaman J, Appel SH 1985 Trophic effects of skeletal muscle extracts on ventral spinal cord neurons *in vitro*: separations of a protein with morphologic activity from proteins with cholinergic activity. J Cell Biol 101:1608–1621

Unsicker K, Reichert-Preibsch H, Schmidt R, Pettmann B, Labourdette G, Sensenbrenner M 1987 Astroglial and fibroblast growth factors have neurotrophic functions for cultured peripheral and central nervous system neurons. Proc Natl Acad Sci USA 84:5459–5463

Vojvodic JT 1987 Development and regulation of dendrites in the rat superior cervical ganglion. J Neurosci 7:904–913

Distinct roles of neurofilament and tubulin gene expression in axonal growth

Paul N. Hoffman

Departments of Ophthalmology and Neurology, The Johns Hopkins School of Medicine, Baltimore, Maryland 21205, USA

Abstract. Tubulin and the neurofilament (NF) proteins, which are major constituents of the axonal cytoskeleton, play distinct roles in longitudinal and radial growth of axons. The predominantly longitudinal growth of axons in developing neurons correlates with relatively high levels of expression for a particular tubulin gene, encoding the class II beta tubulin isotype, whereas levels of gene expression for two other beta tubulin isotypes (classes I and IV) are comparable in developing and maturing neurons. Gene expression for the 68 kDa NF protein (NF68) is low during longitudinal growth. Conversely, relatively high levels of gene expression for NF68 and low levels for class II beta tubulin in maturing neurons correlate with the radial growth of axons. The developmental pattern of gene expression is recapitulated during axonal regeneration. Expression of the class II beta tubulin isotype correlates with the outgrowth (elongation) of regenerating sprouts.

1988 Plasticity of the neuromuscular system. Wiley, Chichester (Ciba Foundation Symposium 138) p 192–204

The potential for axonal growth is expressed in mature neurons during axonal regeneration. When an axon in the mammalian peripheral nervous system is interrupted by cutting or crushing a nerve (axotomy), the distal stump degenerates, and is replaced by the outgrowth (elongation) of regenerating sprouts arising from the proximal stump (Ramón y Cajal 1928). It is generally assumed that developmental mechanisms for axonal growth are recapitulated during axonal regeneration. The principal cytoskeletal proteins of the axon, tubulin and the neurofilament (NF) proteins, are directly involved in the process of axonal growth (Lasek & Hoffman 1976). In this review I shall consider the distinct roles of these cytoskeletal proteins in longitudinal and radial growth of axons. In addition I shall examine recent evidence indicating that the developmental pattern of gene expression for these proteins is recapitulated during axonal regeneration.

Axonal transport of the cytoskeleton

The axonal cytoskeleton is a three-dimensional network consisting of microtubules (MT), NF, and cross-linking structures (Hirokawa 1982). NF are heteropolymers composed of three distinct protein subunits (with molecular weights of 68, 145 and 200 kDa, respectively) (Hoffman & Lasek 1975). Tubulin, the major subunit protein of MT, is a dimer comprising one alpha and one beta tubulin subunit. Since the axon is devoid of ribosomes, axonal proteins are synthesized in the neuron cell body (soma) before undergoing somatofugal translocation within the axon via the axonal transport system. Tubulin and the NF proteins are transported at velocities of 1–2 mm/day in the slow component of axonal transport (Hoffman & Lasek 1975). NF proteins appear to be assembled in the cell body shortly after synthesis and undergo axonal transport in the form of NF polymers (Black & Smith 1988); axons do not contain a significant pool of unassembled NF protein (Morris & Lasek 1982). In contrast, it is likely that both assembled MT and unassembled tubulin subunits undergo axonal transport (Brady et al 1984, D.F. Watson, J.W. Griffin & P.N. Hoffman, unpublished observations). The transport of unassembled tubulin provides the potential for the assembly of MT in the distal axon.

Cytoskeletal gene expression

The mRNAs encoding tubulin and the NF proteins are transcribed from genes located in the cell nucleus and translated into proteins on ribosomes located in the neuron cell body. Each of the NF proteins is encoded by a separate gene (Lewis & Cowan 1985, Julien et al 1986). The situation is more complex for tubulin; alpha and beta tubulin are each encoded by multigene families (Sullivan & Cleveland 1986). Within these families distinct tubulin isotypes can be distinguished on the basis of their unique C-terminal amino acid sequences. There are five beta tubulin isotypes; each is encoded by a separate gene (Sullivan & Cleveland 1986). Genes encoding three of these isotypes (classes I, II and IV) have been cloned from rat (Bond et al 1984, Farmer et al 1984, Sullivan & Cleveland 1986). *In situ* hybridization studies with specific cDNA probes demonstrate that each of these beta tubulin isotypes is expressed in rat sensory neurons (Hoffman & Cleveland 1988).

Recently, we examined the differential expression of genes encoding these beta tubulin isotypes (classes I, II and IV) and the 68 kDa NF protein (NF68) in rat sensory neurons during development, maturation, and axonal regeneration (Hoffman & Cleveland 1988). In developing neurons of five-day-old animals axonal growth is primarily longitudinal; significant radial growth has not yet begun (N.A. Muma, E.H. Koo & P.N. Hoffman, unpublished observations). Conversely, in maturing neurons of 10-week-old animals longitudinal growth is largely completed (Martinez & Friede 1970) and radial growth predominates (Hoffman et al 1985a). Expression of the specific tubulin gene encoding the class II isotype was induced to high levels during both development and axonal regeneration, whereas the expression of genes encoding the two other isotypes (classes I and IV) remained comparable to mature levels. Conversely, expression of NF68 was relatively low during both development and regeneration. In maturing neurons, levels of gene expression were relatively high for NF68 and low for class II beta tubulin. Thus, inverse patterns of gene expression for NF68 and the class II tubulin isotype were found in developing and maturing neurons. The development pattern was recapitulated in maturing neurons during axonal regeneration.

Tubulin expression and axonal growth

The expression of a specific tubulin gene, encoding a distinct beta tubulin isotype (class II), is selectively induced during neuronal development and axonal regeneration. It is not surprising that tubulin is expressed at high levels during development and regeneration because MT are the principal cytoskeletal elements in both developing axons and regenerating sprouts (Berthold 1978). It has been suggested that the ability of MT to undergo reversible assembly and disassembly contributes to the plasticity of developing axons (Lasek 1981).

Tubulin participates directly in the mechanism of axonal elongation (Yamada et al 1971) and the delivery of tubulin by slow axonal transport may limit the rate at which axons elongate during regeneration (Lasek & Hoffman 1976, Wujek & Lasek 1983). Increased expression of the class II isotype after axotomy correlates with increases in regenerating nerve fibres in both the total amount of pulse-labelled tubulin undergoing axonal transport and the relative amount of this tubulin conveyed in the faster phase of slow axonal transport (SCb) (Hoffman & Lasek 1980, Hoffman et al 1985b). Thus, increased amounts of class II beta tubulin may be transported in SCb in regenerating nerve fibres, thereby increasing the delivery of this isotype to regenerating sprouts. This raises the possibility that the class II beta tubulin isotype plays a special role in the outgrowth of regenerating sprouts.

Morphological correlates of NF expression

NF are major intrinsic determinants of axonal calibre in myelinated nerve fibres, where NF number correlates directly with the axonal cross-sectional area (Friede & Samorajski 1970, Weiss & Mayr 1971, Berthold 1978, Hoffman et al 1984). The low level of NF expression is consistent with the observed paucity of axonal NF in developing neurons (Peters & Vaughn

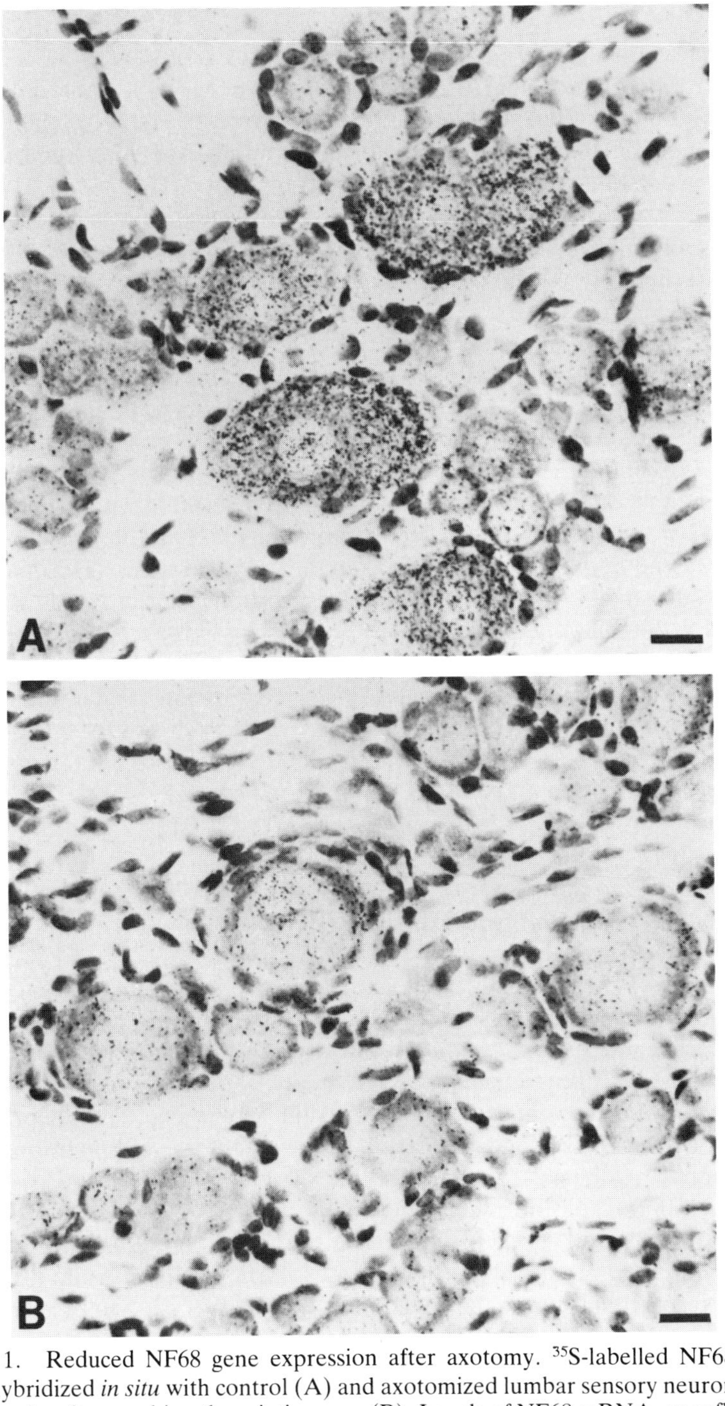

FIG. 1. Reduced NF68 gene expression after axotomy. ^{35}S-labelled NF68 cDNA was hybridized *in situ* with control (A) and axotomized lumbar sensory neurons of rat, two weeks after crushing the sciatic nerve (B). Levels of NF68 mRNA, as reflected by the density of silver grains in these autoradiograms, are normally greater in large than in small sensory neurons (A). After axotomy, mRNA levels are reduced in both large and small neurons (B). Bars, 10 μm. (Reproduced from Hoffman et al 1987).

1967). Before undergoing radial growth, developing axons are small in calibre (less than 1mm in diameter), are highly enriched in MT, and contain few NF (Peters & Vaughn 1967, Berthold 1978). In mammals, radial growth is restricted to myelinated axons. Axons arising from neurons with low levels of NF gene expression remain small throughout ontogeny (Hoffman et al 1987), and include normal unmyelinated fibres which are morphologically indistinguishable from embryonic axons (Berthold 1978). The induction of NF gene expression during postnatal development correlates with the radial growth of myelinated nerve fibres (N.A. Muma, E.H. Koo & P.N. Hoffman, unpublished observations), a process which begins at approximately 10 days of age (N.A. Muma, E.H. Koo & P.N. Hoffman, unpublished observations) and continues until at least six months of age in rat (Hoffman et al 1985a).

The decrease in axonal calibre which occurs in the proximal stump after axotomy (Greenman 1913) correlates with reduced NF gene expression (Fig. 1) (Hoffman et al 1987). This axonal atrophy is associated with proportional reductions in axonal NF content (Hoffman et al 1984) and in the amount of NF protein undergoing axonal transport in the proximal stump (Hoffman et al 1985b). Since this atrophy begins near the cell body and proceeds somatofugally along the fibre at a rate equal to the velocity of NF transport, we have termed this process somatofugal axonal atrophy Hoffman et al 1987). Moreover, since axonal calibre is the single most important determinant of conduction velocity in myelinated nerve fibres (Hursh 1939), somatofugal axonal atrophy is associated with reduced conduction velocity in the proximal stumps of regenerating nerve fibres (Cragg & Thomas 1961, Gillespi & Stein 1983).

Regulation of cytoskeletal gene expression

Maintenance of the pattern of gene expression found in maturing neurons (i.e., high for NF and low for class II beta tubulin) may be regulated by neuron–target interactions. Somatofugal atrophy, the morphological correlate of reduced NF gene expression (Hoffman et al 1987) is initiated when a sensory neuron is disconnected from its peripheral targets (Czéh et al 1977). If the sciatic nerve is cut and regeneration prevented, the proximal stumps of axons remain atrophic throughout the life of the animal (Weiss et al 1945, Dyck et al 1984). On the other hand, if the nerve is crushed and effective regeneration allowed, calibre recovers in the proximal stump (Kuno et al 1974, Hoffman et al 1984). The recovery of normal calibre in the proximal axon, which correlates with the restoration of NF expression to pre-axotomy levels, coincides with the reinnervation of target tissues (Kuno et al 1974). Thus, specific interactions between neurons and their targets may play important roles in both the induction of NF gene expression and the down-regulation of class II beta tubulin expression. The transition from the developmental to the maturational pattern of gene expression at approximately 10 days of age in rat sensory neurons (N.A. Muma, E.H. Koo & P.N. Hoffman, unpublished observations) coincides in a general manner with the time when sensory axons establish their mature patterns of peripheral innervation (Diamond 1981).

An unresolved issue is the extent to which neuron–target interactions specify the level of NF gene expression. Each class of target cells is innervated by axons of a specific calibre. For example, extrafusal muscle fibres and encapsulated sensory receptors are innervated by large myelinated axons, intrafusal muscle fibres are innervated by small myelinated axons, and free nerve endings, which lack specific receptors, arise from small unmyelinated sensory fibres (Brodal 1981). This relationship between the physiological function of targets and axonal calibre raises the possibility that target cells help to specify the level of NF gene expression, and thereby determine axonal calibre. According to this hypothesis, the quantity or quality of target-induced signal would determine the rate of NF gene expression, and each class of target cells would specify characteristic levels of signal.

The nature of the signal mediating these neuron–target interactions is unknown. One possibility is that the signal, or signals, generated by these interactions travels, by retrograde transport, from the axon terminals to its site of action in neuronal perikarya. This signal could be either a molecule produced by target cells and transferred transsynaptically to axon terminals or a molecule produced in neurons which is modified upon entering the axon terminals and returns to the cell body as a signal.

Distinct patterns of gene expression correlate with longitudinal and radial growth of axons

Rapid longitudinal growth of axons correlates with the expression of class II beta tubulin. Radial growth is associated with NF gene expression. The developmental pattern of gene expression (relatively high for class II beta tubulin and low for NF) correlates with the phase of predominantly longitudinal growth of developing axons. Similarly, recapitulation of the developmental pattern during axonal regeneration correlates with the outgrowth of regenerating sprouts. In contrast, the inverse pattern of gene expression (high for NF and low for class II beta tubulin) in maturing neurons correlates with the radial growth of axons. Recapitulation of the developmental pattern after axotomy (i.e., reduced NF expression) correlates with the cessation of radial growth and, instead, with axonal atrophy. Thus, the developmental pattern of gene expression for NF and class II beta tubulin correlates with the longitudinal growth of axons, while the maturational pattern is associated with radial growth.

Acknowledgements

Special thanks to Drs John Griffin, Daniel Watson and Thomas Crawford for their helpful suggestions for the manuscript. Aspects of this work were supported by grants from the National Institutes of Health (NS-22849 and NS-20164). The author is the recipient of a Research Career Development Award from the National Institutes of Health (NS-00896).

References

Berthold CH 1978 Morphology of normal peripheral axons. In: Waxman SG (ed) Physiology and pathobiology of axons. Raven Press, New York, p 3–63

Black MM, Smith W 1988 Regional differentiation of the neuronal cytoskeleton with an appendix: diffusion of proteins in the neuron cell body–mathematical approximations and computer simulations. In: Lasek RJ, Black MM (eds) Intrinsic determinants of neuronal form and function. Alan R. Liss, New York, p 463–486

Bond JF, Robinson GS, Farmer SR 1984 Differential expression of two neural cell-specific beta-tubulin mRNAs during rat brain development. Mol Cell Biol 4:1313–1319

Brady ST, Tytell M, Lasek RJ 1984 Axonal tubulin and axonal microtubules: biochemical evidence for cold stability. J Cell Biol 99:1716–1724

Brodal A 1981 Neurological anatomy in relation to clinical medicine, 3rd edn. Oxford University Press, New York

Cragg BG, Thomas PK 1961 Changes in conduction velocity and fibre size proximal to peripheral nerve lesions. J Physiol (Lond) 157:315–327

Czéh G, Kudo N, Kuno M 1977 Membrane properties and conduction velocity in sensory neurones following central and peripheral axotomy. J Physiol (Lond) 270:165–180

Diamond J 1981 The recovery of sensory function in skin after peripheral nerve lesions. In: Gorio A et al (eds) Posttraumatic peripheral nerve regeneration. Raven Press, New York, p 533–548

Dyck PJ, Nukada H, Lais AC, Karnes JL 1984 Permanent axotomy: a model of chronic neuronal degeneration preceded by axonal atrophy, myelin remodeling, and degeneration. In: Dyck PJ et al (eds) Peripheral neuropathy. Saunders, Philadelphia, p 666–690

Farmer SR, Bond JF, Robinson GS, Mbangkollo D, Fenton MJ, Berkowitz EM 1984 Differential expression of the rat beta-tubulin multigene family. In: Borisy GG et al (eds) Molecular biology of the cytoskeleton. Cold Spring Harbor Laboratory, Cold Spring Harbor, NY, p 333–342

Friede RL, Samorajski T 1970 Axon caliber related to neurofilaments and microtubules in sciatic nerve fibers of rats and mice. Anat Rec 167:379–387

Gillespi MJ, Stein RB 1983 The relationship between axon diameter, myelin thickness and conduction velocity during atrophy of mammalian peripheral nerves. Brain Res 259:41–56

Greenman MJ 1913 Studies on the regeneration of the peroneal nerve of the albino rat: number and sectional areas of fibers: area relation of axis to sheath. J Comp Neurol 23:479–513

Hirokawa N 1982 Cross-linker system between neurofilaments, microtubules, and membranous organelles in frog axons revealed by the quick-freeze, deep-etching method. J Cell Biol 94:129–142

Hoffman PN, Cleveland DW 1988 Neurofilament and tubulin expression recapitulates the developmental program during axonal regeneration: induction of a specific beta tubulin isotype. Proc Natl Acad Sci USA 85:4530–4533

Hoffman PN, Lasek RJ 1975 The slow component of axonal transport: identification of the major structural polypeptides of the axon and their generality among mammalian neurons. J Cell Biol 66:351–366

Hoffman PN, Lasek RJ 1980 Axonal transport of the cytoskeleton in regenerating motor fibers: constancy and change. Brain Res 202:317–333

Hoffman PN, Griffin JW, Price DL 1984 Control of axonal caliber by neurofilament transport. J Cell Biol 99:705–714

Hoffman PN, Griffin JW, Gold BG, Price DL 1985a Slowing of neurofilament transport and the radial growth of developing nerve fibers. J Neurosci 5:2920–2929

Hoffman PN, Thompson GW, Griffin JW, Price DL 1985b Changes in neurofilament transport coincide temporally with alterations in the caliber of axons in regenerating motor fibers. J Cell Biol 101:1332–1340

Hoffman PN, Cleveland DW, Griffin JW, Landes PW, Cowan NJ, Price DL 1987 Neurofilament gene expression: a major determinant of axonal caliber. Proc Natl Acad Sci USA 84:3472–3476

Hursh JB 1939 Conduction velocity and diameter of nerve fibers. Am J Physiol 127:131–139

Julien J-P, Meyer D, Flavell D, Hurst J, Grosveld F 1986 Cloning and developmental expression of the murine neurofilament gene family. Mol Brain Res 1:243–250

Kuno M, Miyata Y, Munoz-Martinez EJ 1974 Properties of fast and slow alpha motoneurones following motor reinnervation. J Physiol (Lond) 242:273–288

Lasek RJ 1981 The dynamic ordering of neuronal cytoskeletons. Neurosci Res Program Bull 19:7–32

Lasek RJ, Hoffman PN 1976 The neuronal cytoskeleton, axonal transport and axonal growth. Cold Spring Harbor Conf Cell Proliferation 3:1021–1049

Lewis SA, Cowan NJ 1985 Genetics, evolution and expression of the 68,00-mol-wt neurofilament protein: isolation of a cloned cDNA probe. J Cell Biol 100:843–850

Martinez AJ, Friede RL 1970 Changes in nerve cell bodies during the myelination of their axons. J Comp Neurol 138:329–338

Morris JR, Lasek RJ 1982 Stable polymers of the axonal cytoskeleton: the axoplasmic ghost. J Cell Biol 92:192–198

Peters A, Vaughn JE 1967 Microtubules and filaments in the axons and astrocytes of early postnatal optic nerves. J Cell Biol 32:113–119

Ramón y Cajal S 1928 Degeneration and regeneration of the nervous system. Hafner, New York

Sullivan KF, Cleveland DW 1986 Identification of conserved isotype-defining variable region sequences for four vertebrate beta tubulin polypeptide classes. Proc Natl Acad Sci USA 83:4327–4331

Weiss P, Edds MV Jr, Cavanaugh M 1945 The effect of terminal connections on the caliber of nerve fibers. Anat Rec 92:215–233

Weiss PA, Mayr R 1971 Organelles of neuroplasmic ('axonal') flow: neurofilaments. Proc Natl Acad Sci USA 68:846–850

Wujek JR, Lasek RJ 1983 Correlation of axonal regeneration and slow component *b* in two branches of a single axon. J Neurosci 3:243–251

Yamada KM, Spooner BS, Wessels NK 1971 Ultrastructure and function of growth cones and axons of cultured nerve cells. J Cell Biol 49:614–635.

DISCUSSION

Kernell: You found a strong correlation between axon diameter and the number of neurofilaments and you observed no very good correlation between axon diameter and the number of microtubules. Does that mean that the amount of microtubules per cross-sectional area might vary with the functional state of the neuron? This would imply that the capacity for fast axonal transport might be changeable.

Hoffman: I agree. It is quite possible that different populations of axons with comparable cross-sectional areas may differ in the density of microtubules.

Kernell: Do you see any systematic difference between smaller and larger alpha axons, or between alpha and gamma axons, in the density of microtubules per cross-sectional area?

Hoffman: We haven't looked specifically at larger and smaller alpha axons. If we compare large and small motor fibres in the ventral root, which are presumably alphas and gammas, the density is higher for the smaller fibres (Friede & Samorajski 1970). Using our relatively crude scheme I am not sure how we would differentiate large and small alphas. Would you just use cross-sectional area as the criterion? We could do that.

Kernell: It would be interesting to do.

Buller: However, there are variations in diameter along the length of alpha fibres.

Kernell: Yes, but for these comparisons between large and smaller alpha axons it would primarily be of interest to rank the nerve fibres according to size (e.g., mean diameter or cross-sectional area). Absolute sizes would presumably be less important.

Lowrie: You described the atrophy of the axon after axotomy followed by recovery and regeneration. We have been looking at the effect of crushing the sciatic nerve in young rats in the first week after birth (Lowrie et al 1982, 1987). After crushing at five days of age we see regeneration of apparently all the motoneurons back to the muscle, and we get some recovery of the muscle, but we have been struck by the fact that the sciatic nerve, both proximal and distal to the point of crush, remains rather thin. We haven't measured the axons, but my guess is that axonal diameters are reduced. Have you any comment on that, in relation to your findings on interaction with the target?

Hoffman: The induction of neurofilament expression in rat sensory neurons between five and ten days of age correlates with the onset of radial growth in myelinated axons (N.A. Muma & P.N. Hoffman, unpublished work). It is conceivable that axonal injury prevents this induction, thereby impairing radial growth.

Lowrie: We find differences both in the muscle and in the motoneurons. There is a change in the size distribution of the motoneurons, and, in the fast muscles at least, a reduction in the size of the motor units. Both could be related to reduction in the size of the axon.

Nadal-Ginard: Are the changes in neurofilament mRNA and tubulin mRNA transcriptionally regulated, or is it a half-life regulator?

Hoffman: In this *in vivo* system we are limited in the number of cells that we obtain, so we cannot look at transcriptional levels. We can only guess as to whether we are looking at increased turnover of neurofilament message or reduced transcription.

Nadal-Ginard: You emphasized the kinetics of the changes of different steady-state levels for these two mRNAs. Do you know whether the neurofilament mRNA has a longer half-life than tubulin mRNA? Would a difference in half-life explain the asynchronous behaviour of the two?

Hoffman: We have no information on the half-life or the transcriptional rates of either mRNA in this system.

Nadal-Ginard: With an *in vivo* system, however, you could look at half-life by inhibition of RNA synthesis.

Zak: There is a potential hazard of the assay for measuring the quantity of messenger RNA. If the data are expressed per unit of RNA and you find for example that the abundance of mRNA is half, and the regenerating tissue has twice as great a concentration of RNA, the total amount of mRNA will not change. I would interpret some of your *in situ* hybridizations as showing less change than the blotting assay.

Hoffman: It is our impression that the RNA blots are more sensitive than *in situ* hybridization for detecting changes in mRNA levels. A 10-fold reduction in neurofilament mRNA levels measured on RNA blots corresponds to a threefold reduction in grain density over neurons examined by *in situ* hybridization (Hoffman & Cleveland 1988).

Zak: From the sequence of your probe, wouldn't it be advisable to design a generic probe which would tell you the total amount of mRNA and use it as a reference for the expression of specific mRNA?

Hoffman: Yes, when we examined the expression of beta amyloid in regenerating sensory neurons we found essentially no change in mRNA levels (E.H. Koo, N.A. Muma, P.N. Hoffman & D.L. Price, unpublished work). This suggests that there is little, if any, change in the total RNA content of these neurons.

Oppenheim: Have experiments been done to see whether the target-derived factor is NGF? This would indicate whether you are dealing with a sensory neuron.

Hoffman: We don't know, and we should investigate this, using NGF or an antibody to NGF, or both.

Van Essen: Much of your measured response may be the active response to the obligation to regenerate that these nerves have, but part of it may be the actual disconnection. Has anyone looked at the effects of blocking neuromuscular transmission, to see what changes in either axonal diameter or cytoskeletal composition occur?

Hoffman: Blocking neuromuscular transmission with botulinum toxin produces somatofugal axonal atrophy in motor neurons (B.G. Gold, J.W. Griffin, A. Pestronk, P.N. Hoffman, E.F. Stanley & D.L. Price, unpublished work). We are currently examining cytoskeletal gene expression in this system (T.O. Crawford, J.W. Griffin & P.N. Hoffman, unpublished work).

Henderson: I am interested in what is really happening to the neurofilaments when, for instance, a nerve terminal dies back or is removed during the regression of polyneuronal innervation. Initially, there is a stack of neurofilaments which give a certain diameter to the axon. Are they maintained as a stack while being degraded by the calcium-dependent protease from the end?

Hoffman: This is not known.

Kernell: How does this system work at branchings? Do you preserve the same total number of neurofilaments throughout the whole branching tree of an axon?

Hoffman: We have not examined branch points, but it is clear that neurofilament number is not conserved along unbranched portions of myelinated nerve fibres. In large-calibre fibres the axonal cross-sectional area may be as much as 10-fold greater at internodes than at nodes. Since neurofilament density is comparable at nodes and internodes, neurofilament number is substantially greater along internodal segments (Berthold 1978). This suggests that there must be a reorganization of the cytoskeleton as neurofilaments are transported through nodes.

There also appears to be reorganization of axonal neurofilaments during radial growth of myelinated nerve fibres. Examination of transport kinetics reveals that there is a continuous decline in the velocity of pulse-labelled neurofilament protein with increasing distance along the nerve (Hoffman et al 1985a). Therefore, neurofilaments enter each region of an axon (e.g., an internodal segment) slightly faster than they leave, leading to a net accumulation of neurofilaments over time. We believe that this contributes directly to the radial growth of myelinated axons, a process which occurs continuously during the first six months of postnatal development in peripheral nerve fibres of the rat (Hoffman et al 1985a).

Kernell: Can one see neurofilament 'bricks' lying around in the axon?

Hoffman: Our information is quite limited here. We do not know whether neurofilaments are relatively short, or whether they are infinitely long and folded back on themselves within each internodal segment (see Hoffman et al 1984).

Van Essen: When you post-label the neurofilaments, how sharp is the region of labelling as it migrates down?

Hoffman: The peak is initially sharp and spreads as it moves along the nerve (Hoffman et al 1983).

Sanes: Is there any manipulation that keeps the production of neurofilaments up? If so, how is axon calibre regulated? Is there a limit?

Hoffman: Although the available evidence indicates that the level of neurofilament gene expression is a major determinant of axonal calibre in large myelinated nerve fibres, an unresolved issue is the role of target cells in regulating neurofilament gene expression. One possibility is that the level of neurofilament expression is specified by trophic interactions between neurons and their targets. Alternatively, both the level of neurofilament expression and target preference could be intrinsic properties of neurons. In the latter case, the target would play a permissive (rather than deterministic) role in regulating neurofilament gene expression; that is, target cell interactions would be required for expression to occur, but the level of expression would be determined by other factors.

Walsh: Can you detect temporal changes in the expression of the three neurofilament protein subunits in your lesion model?

Hoffman: No. In axonal transport experiments we found that levels of radioactivity in each of the three neurofilament proteins are reduced proportionately during regeneration (Hoffman et al 1985b). More recently, using cDNA probes, we found comparable reductions in mRNA levels for all three neurofilament proteins. Thus, it appears that the expression of these proteins is coordinated.

Kernell: Are dendrites regulated in the same way as axons?

Hoffman: We haven't looked. In sensory neurons there is however a correlation between neurofilament expression and cell body size, as well as axonal diameter. In the axotomy model, when the axons shrink, the cell bodies are also smaller. That was shown by Cavanaugh (1951) and we found the same. Of course, there are more things in the cell body than neurofilaments, so I don't want to say that neurofilaments control cell body size, but they contribute.

Kernell: There is also a larger total sum of dendritic stem diameters going out from a large than from a smaller motoneuronal cell body (e.g. Zwaagstra & Kernell 1981).

Hoffman: Yes, and there is evidence that the large dendrites also have neurofilaments, and that the neurofilament density is comparable in large dendrites and in large axons, at about $100/\mu m^2$.

Zak: During branching, the diameter of the axon gets smaller. Is there any increased vulnerability of the axons of small diameter?

Hoffman: We have no data on this.

References

Berthold CH 1978 Morphology of normal peripheral axons. In: Waxman SG (ed) Physiology and pathobiology of axons. Raven Press, New York, p 3-63

Cavanaugh MW 1951 Quantitative effects of the peripheral innervational area on nerves and spinal ganglion cells. J Comp Neurol 94:181–219

Friede RL, Samorajski T 1970 Axon caliber related to neurofilaments and microtubules in sciatic nerve fibers of rats and mice. Anat Rec 167:379–387

Hoffman PN, Cleveland DW 1988 Neurofilament and tubulin expression recapitulates the developmental program during axonal regeneration: induction of a specific beta tubulin isotype. Proc Natl Acad Sci USA 85:4530–4533

Hoffman PN, Lasek RJ, Griffin JW, Price DL 1983 Slowing of the axonal transport of neurofilament proteins during development. J Neurosci 3:1694–1700

Hoffman PN, Griffin JW, Price DL 1984 Neurofilament transport in axonal regeneration: implications for the control of axonal caliber. In: Elam JS, Cancalon P (eds) Axonal transport in neuronal growth and regeneration. Plenum Press, New York, p 243–260

Hoffman PN, Griffin JW, Gold BG, Price DL 1985a Slowing of neurofilament transport and the radial growth of developing nerve fibers. J Neurosci 5:2920–2929

Hoffman PN, Lasek RJ, Griffin JW, Price DL 1985b Changes in neurofilament transport coincide temporally with alterations in the caliber of axons in regenerating motor fibers. J Cell Biol 101:1332–1340

Hoffman PN, Koo EH, Muma NA, Griffin JW, Price DL 1988 Role of neurofilaments in the control of axonal caliber in myelinated nerve fibers. In: Lasek RJ, Black MM (eds) Intrinsic determinants of neuronal form. Alan R Liss, New York, p 389–402

Lowrie MB, Krishnan S, Vrbová G 1982 Recovery of slow and fast muscles following nerve injury during early post-natal development in the rat. J Physiol (Lond) 331:51–66

Lowrie MB, Krishnan S, Vrbová G 1987 Permanent changes in muscle and motoneurons induced by nerve injury during a critical period of development of the rat. Dev Brain Res 31:91–101

Zwaagstra B, Kernell D 1981 Sizes of soma and stem dendrites in intracellularly labelled alpha-motoneurons of the cat. Brain Res 204:295–309

General discussion II

Plasticity, specialization and adaptability

Buller: It still remains unknown whether neurons are predetermined and will specifically innervate only particular types of muscle, or whether motor axons, having reached the target, are informed in some way by the target cell how their cell bodies ought to develop. I don't know the opinion of those currently in the field, but it seems to me that since the Konstanz symposium (Pette 1980) there has been a move away from 'plasticity' (the idea that motoneurons are not predetermined) back to the earlier views stressing the specificity of motoneurons. Would anyone like to comment on that? At the beginning of the symposium, I felt that Dirk Pette was constraining muscle fibres to work only within fixed limits.

Pette: In fact, the more we study populations of muscle fibres, the more we detect diversity *and* specialization. For example, certain troponinT subunits display muscle-specific distribution patterns in hindlimb, masticatory or tongue muscles (Moore et al 1987). In addition, within the same muscle, fibres may be histochemically classified as identical, yet microbiochemical analyses indicate discrete differences in the expression of certain proteins (Pette & Spamer 1986, Staron & Pette 1987a,b). These variations may reflect specialization brought about by the functional demand, but they also reflect the dynamic state of muscle fibres—that is to say, the plasticity of muscle.

Vrbová: We should perhaps try to define plasticity; as a term it is rather diffuse. What I understand by plasticity is the ability of a structure, whether a neuron or nervous system or muscle, to adapt to functional demands. And because we are dealing with systems made up of several components, we can talk about plasticity only in relation to particular components and a particular function. This is why the myosin molecule and the expression of its various isoforms will probably be related to its function—that is, to the mechanical conditions under which it works and to the amount of work that it does. It is important to emphasize that plasticity is not just an overall response of the cell. It can be coordinated, but it need not be.

Buller: That is what I was suggesting, that we are now constraining plasticity, a concept that went through a period where it was believed that everything can do almost anything.

Kernell: We need a term that implies neither limitless plasticity, nor that everything is fixed: something like 'controlled adaptability'.

Rubinstein: Some of the difference in emphasis between the Konstanz symposium and this one is that then we talked about what we could do to cells, to make them different, with papers on regeneration, cross-innervation, and the effects of chronic stimulation. We are now asking what normally occurs during development and during disease processes. We have discovered that this isn't quite what we thought. At the time of the earlier symposium, we saw muscle cells as blank slates and we thought that exogenous factor(s) would come in and determine what each fibre became. We know now that this is not true.

Zak: It may be useful to think about these problems in terms of differentiation programmes. Perhaps we should start from the fully undifferentiated cell (the premyoblast). These cells go through several stages of differentiation, but in each case there is a programme which is 'possible' and another which is prohibited. It is a question of having certain 'windows' for phenotypic decisions, and within each window there is a certain adaptability that cannot be exceeded.

Gordon: I like that way of looking at it, and my paper will be concerned with what one could consider to be pathology but I think is plasticity, after reinnervation. To some extent, regeneration involves dedifferentiation or reversion, and is similar to development. The window concept may be important here. Much of the response to injury is a limited and also an adaptive dedifferentiation. It may not go back to the earliest stages of development. Many of the specific cues are lost in the plasticity of the adult. We have evidence (Gordon et al, this volume) that in regeneration the process of 'getting back' is a non-specific one, but the residual plasticity of the neuromuscular system allows for recovery of the original properties.

Buller: In other words, some doors or windows that are gone through during development are closed, and you cannot go back by the same route.

Gordon: Exactly!

Van Essen: The other point to remember is that these windows seem narrowed or constrained only in relation to the infinite initial possibilities. The number of decisions that can still be taken within a window is much larger than we might have realized earlier, and to choose among 10^9 or so myosin isoform possibilities, or to choose pathways with as much specificity as we now know exists, in some respects signifies that a vast amount of plasticity is present.

Nadal-Ginard: As we learn more about the neuromuscular system, it is logical that it seems more complex and more specialized. One issue that is confusing is the fact that we are detecting more specialization, but that doesn't necessarily mean that we are constraining the plasticity. Dr Van Essen just said that the neuromuscular system may be more plastic than we expected, because there are many more decisions to be made than we thought. The question is what these ideas about 'windows' and 'programmes' mean in biochemical terms. For neuromuscular gene expression, we need to know which phenotypes are allowed and which are not allowed, and why.

One way of addressing these issues is to look for interspecific differences in gene expression. Closely related animal species may have different developmental programmes, especially for the contractile proteins. What does this mean biochemically? Does it mean that these developmental differences are encoded in the sequence of the structural genes, or depend on other regulatory genes? In other words, are these differences *cis* or *trans* regulated? For some of the myosin genes we know that certain differences among mammalian species are *trans* regulated.

Buller: Are there examples of genes that can be turned on in development but cannot be turned on again in the adult?

Nadal-Ginard: There are examples of developmentally dependent genes, but the contractile protein genes that have been studied so far are not of this type, but are reversibly induced. The number of examples of genes that cannot be turned on again is more limited than the literature would suggest. In some instances the cell type that was expressing the gene early in development has disappeared (e.g. by programmed cell death), and this is why the gene appears not to be re-expressed.

Pette: Chronic stimulation experiments disclose pronounced species-specific differences in responsiveness. We have compared the effect of 10 hours per day of indirect (10Hz) stimulation of the tibialis anterior (TA) muscle in mouse, guinea-pig, rat and rabbit (Simoneau & Pette 1988). To our surprise, the same stimulus pattern evoked very different responses in these small mammals. Stimulation brought about only slight increases in enzyme activities of aerobic-oxidative metabolism in mouse TA, but produced several-fold increases in rabbit TA. Rat and guinea-pig were intermediate (Simoneau & Pette 1988). The question arises whether the mouse, being a small and active animal, already has maximally expressed these enzymes, whereas the sedentary rabbit normally operates at a submaximal level of expression.

Vrbová: It is difficult to know whether a cell has lost the ability to express a new gene,or whether we do not know how to switch it on. For example, if you look at the cable properties of membranes of slow-tonic and fast-twitch muscle fibres, once they are established you can't alter them and they remain recognizably different. This doesn't mean that if one can find the right stimulus, the potential is not there.

Gordon: One of the best examples of that is the growth-associated proteins (GAPs). Growing neurons in the central nervous system all express these proteins. Central neurons don't regenerate; this is because they lack the right environment in which to grow. If they are induced to grow into the Schwann cell tubes of the peripheral nervous system they can re-express GAPs. So it is the induction of the protein, or the inductive stimulus, which may be lost, or appear to be lost.

Vrbová: It could be our limitations that prevent us from finding the appropriate stimulus.

Zak: An example of a 'one-way' gene is the embryonic haemoglobin, which is never re-expressed in the adult.

Nadal-Ginard: This is because the cells that make it, in the liver, have disappeared in the adult. As an example of the complexity of the regulatory system, Thy-1 antigen expression in humans and mice is very different. This difference is encoded in the sequence of the human and mouse genes. When the human gene is introduced into mice in transgenic animals it retains the human pattern of expression. This is not the case for myosin. The human has a cardiac myosin pattern of expression that is very different from rat or mouse. This different programme is not encoded in the myosin genes. Human genes in mouse cells behave like mouse genes and vice versa. So here are two examples of species-specific phenotypic differences that are regulated differently. Thy-1 seems to be *cis* regulated. The Thy-1 gene dictates the pattern of expression independently of the species, whereas myosin is *trans* regulated. The gene does not have the information for the species-specific difference, but the cell does. A human cell imposes a pattern of expression on a human as well as a mouse myosin gene introduced into it, and vice versa.

Eisenberg: The question is whether muscle is able to go through all the gene programmes again, or whether we are unable to trigger the correct response. When we study an adult cell (as we do with our invasive techniques) we are not producing all the events that happened originally in development. A physiologist looks at properties that we cannot recapitulate—the kind of work that a muscle fibre does, the load against which it is now working. In the adult, the connective tissue is in place already and the tendon ends are connected up. In some pathological situations, for example where overloading of chicken ALD is extreme, you can get a population of migrating satellite cells that form new myotubes which express a myosin characteristically found only during embryonic development (Kennedy et al 1988).

Zak: On the potential effect of the microenvironment, it must be an extremely complicated situation. Each gene family of muscle proteins has a completely different expression programme and a different regulatory programme (Swynghedauw 1986). One kind of stimulus affects one gene family and may cause opposite changes in another gene family.

Oppenheim: With regard to the issue of species differences, we usually study highly domesticated species (as in Dr Pette's studies of rat, rabbit, and other species) where selection pressures may have loosened. Consequently, the differences that we are seeing may not have much biological significance. The kinds of species differences that would be more biologically interesting are those expressed in wild-type or non-domesticated animals. Until this study is done it seems premature to argue that species differences in muscle properties of highly domesticated animals are biologically meaningful.

References

Gordon T, Bambrick L, Orozco R 1988 Comparison of injury and development in the neuromuscular system. In: Plasticity of the neuromuscular system. Wiley, Chichester (Ciba Found Symp 138) p 210–226

Kennedy JM, Eisenberg BR, Reid SK, Sweeney LJ, Zak R 1988 Nascent muscle fiber appearance in overloaded chicken slow-tonic muscle. Am J Anat 181:203–215

Moore GE, Briggs MM, Schachat FH 1987 Patterns of troponin T expression in mammalian fast, slow and promiscuous muscle fibres. J Muscle Res Cell Motil 8:13–22

Pette D (ed) 1980 Plasticity of muscle. Walter de Gruyter, Berlin/New York

Pette D, Spamer C 1986 Metabolic properties of muscle fibers. Fed Proc 45:2910–2914

Simoneau JA, Pette D 1988 Species-specific effects of chronic nerve stimulation upon tibialis anterior muscle in mouse, rat, guinea pig and rabbit. Pfluegers Arch Eur J Physiol 412:86–93

Staron RS, Pette D 1987a The multiplicity of myosin light and heavy chain combinations in histochemically typed single fibres of rabbit soleus muscle. Biochem J 243:687–693

Staron RS, Pette D 1987b The multiplicity of myosin light and heavy chain combinations in histochemically typed single fibres of rabbit tibialis anterior muscle. Biochem J 243:695–699

Swynghedauw B 1986 Development and functional adaptation of contractile proteins in cardiac and skeletal muscles. Physiol Rev 66:710–770

Comparison of injury and development in the neuromuscular system

Tessa Gordon, Linda Bambrick and Roberto Orozco

Department of Pharmacology, University of Alberta, Edmonton, Canada T6G 2H7

Abstract. Comparisons of development and regeneration have suggested that axotomized motoneurons and denervated muscles undergo dedifferentiation to an embryonic state with recovery of adult properties after reinnervation. Using electrophysiological and radioligand-binding techniques to monitor axonal size and numbers of extrajunctional acetylcholine receptors in axotomized motoneurons and denervated muscles respectively, we have demonstrated that this dedifferentiation is limited. We suggest that this limited dedifferentiation may be adaptive for survival, regeneration and reinnervation.

Correlative physiological and histochemical studies of reinnervated motor units in cat and rat hindlimb muscles show that the processes of regeneration and reinnervation differ in a number of fundamental ways from developmental processes of axonal growth and muscle innervation. Enlargement of motor units after partial nerve injuries does not appear to be limited to the size of the neonatal motor unit as originally suggested but may be influenced by factors operating at the level of axonal branching. Regeneration after complete and partial nerve injuries is a random process in contrast to the specific nature of the innervation of targets during development. Regenerating axons frequently fail to make connections with their original muscles and newly reinnervated motor units contain muscle fibres which formerly belonged to several different motor units. Despite this misdirection of regenerating nerve fibres, neuromuscular plasticity restores neuromuscular properties to the extent that these are appropriate at the single motor unit level for the gradation of force by the orderly recruitment of units during movement.

1988 Plasticity of the neuromuscular system. Wiley, Chichester (Ciba Foundation Symposium 138) p 210–226

Neurons and muscle fibres share many features with other cells. Even voltage-dependent ion channels and contractile proteins are not unique to excitable cells. It is the high degree of morphological specialization, large diversity of shape and the particular location that distinguishes neurons and enables them to communicate signals to precise targets. In muscle tissue, contractile proteins are highly organized in unusually long multinucleated cells which allows for processing in the nervous system to be translated ultimately into movement.

Neurons also differ from other tissues in their inability to replicate themselves in the adult. While many other cells derived from the ectoderm undergo a natural cell turnover in the adult animal, neuroblasts which withdraw from the cell cycle and differentiate fully are not replaced in the lifetime of the animal, even when the mature neurons are damaged. Similarly, the adult muscle fibres which differentiate from myoblasts do not normally undergo cellular replacement. They differ from neurons in that damaged muscle fibres can be repaired by the proliferation and differentiation of satellite cells in the basal lamina, processes which recapitulate myogenesis (Allbrook 1981).

The neuromuscular system is also distinct in that the development and maintenance of the adult phenotype depend on trophic cell-to-cell interaction (Vrbová et al 1978). Nerve injuries causing disruption of this interaction lead to regressive changes in axotomized motoneurons and denervated muscles. For example, axotomized neurons re-express growth-associated proteins (GAPs) (Bisby et al 1988) and reduce the slow axonal transport of neurofilaments relative to microtubules and actin filaments (Hoffman et al 1985, Hoffman, this volume). Denervated muscles incorporate acetylcholine (ACh) receptors in the extrajunctional membrane (Fambrough 1979) and express glycoproteins such as neural cell adhesion molecules (N-CAMs) which are normally expressed only in immature muscle (Sanes et al 1986).

Even undamaged motoneurons appear to revert to a more immature form after partial nerve injuries. They undergo some of the electrophysiological changes of axotomized motoneurons (Huizar et al 1977) and emit growth cones. These growth cones, which emerge from nodes of Ranvier or from unmyelinated terminals within the muscle to give rise to nodal and terminal sprouts respectively, reinnervate denervated muscle fibres (Brown et al 1981). The large number of terminal branches which result from sprouting appear to be a re-expression in the adult motoneuron of the neonatal motoneuron's large number of nerve terminals (Brown et al 1976, Brown & Ironton 1978). Accordingly, Brown recently referred to sprouting as 'a recapitulation of ontogeny' (Brown 1984).

Recovery of neuromuscular function after nerve injuries must depend on these regressive changes which are permissive for axonal regeneration and muscle reinnervation, because the damaged motoneurons are not replaced. If nerve regeneration and muscle reinnervation were able to recapitulate development, these processes in the adult would re-establish former nerve and muscle properties as well as the precise neuromuscular connections. In the past few years, my colleagues and I have used electrophysiological, histochemical and radioligand-binding techniques to study the neuromuscular response to nerve injuries to determine, first, to what extent axotomized motoneurons and denervated muscles actually dedifferentiate and, second, to what extent nerve regeneration and muscle reinnervation recapitulate development. The results of these studies will be summarized in the next few sections which are headed by a series of questions. Since much of this work has been published, the reader is referred to the original articles for details of methodology.

To what extent do denervated muscles and axotomized motoneurons dedifferentiate to an embryonic state?

Denervated muscle

Probably one of the most dramatic and well-known consequences of muscle denervation is that the extrajunctional muscle membrane becomes sensitive to acetylcholine (ACh), the so-called denervation supersensitivity (reviewed by Fambrough 1979). Because embryonic muscle is also sensitive to ACh all over its surface membrane, as first shown by Ginetzinsky and Shamarina in 1942 and later by Diamond and Miledi in 1962, it has often been assumed that muscle denervation leads to a dedifferentiation of the muscle to an embryonic state (see for example Vrbová et al 1978).

However, this dedifferentiation appears to be limited, as we have shown recently using radioligand-binding assays designed to make a quantitative comparison of the number of ACh receptors in developing and denervated muscles. The increase in numbers of ^{125}I-bungarotoxin-binding sites with time after denervation of adult rat muscle (shown by the + symbol in Fig. 1A) is compared with the developmental decline in the numbers of sites during neonatal life. ACh receptors are normally confined in high density to the endplate regions, which make up a very small percentage of the muscle membrane and are not detected by the assay. As the extrajunctional areas of the membrane incorporate ACh receptors (Merlie et al 1984), the number of binding sites increases (Bambrick & Gordon 1987). This increase follows a simple exponential to reach a plateau by two weeks after nerve section. The time course of the incorporation of ACh receptors into denervated muscle membranes is remarkably similar to the exponential fall in receptors during the maturation of neonatal muscle (compare time constants of eight and six days in Fig. 1A). In this sense, denervation reverses the developmental maturation. However, the final level of ACh receptors in the extrajunctional membrane falls short of the maximum numbers in the developing muscles.

In terms of the number of extrajunctional ACh receptors, denervation causes adult muscle to revert to a partially dedifferentiated state which corresponds to the immature muscle of a one-week-old neonatal rat (Fig. 1) and may be maintained for long periods in the absence of innervation (Steinbach 1981). Interestingly, if neonatal muscles are denervated at this early stage of development (one week after birth), the same relative increase in ^{125}I-bungarotoxin-binding site numbers (Fig. 1B) returns the muscle to a stage of development which corresponds to 18 days *in utero*. This is the develop-

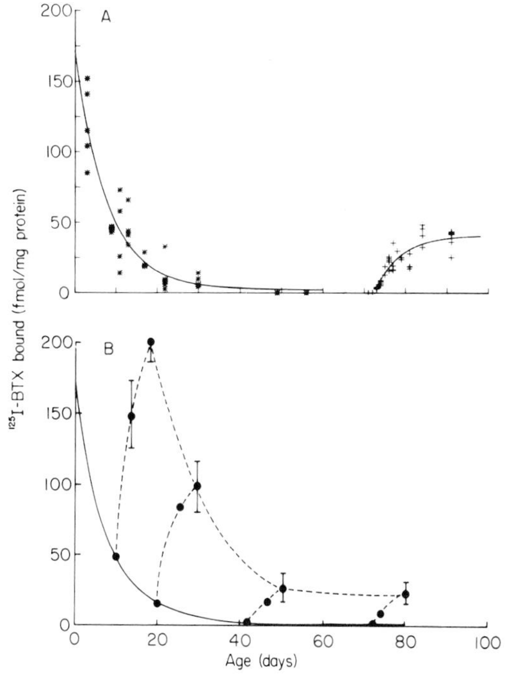

FIG. 1. (A) Comparison of the exponential decline in ^{125}I-bungaratoxin (BTX)-binding sites in rat hindlimb triceps surae muscles during postnatal development (*) with the exponential rise in sites after denervation in the adult (+). The time constants of the falling and rising exponential curves (fitted as described by Bambrick & Gordon 1987) were eight and six days, respectively. (B) Deviation from the developmental curve of denervated muscles with respect to the mean (± SE) of the number of binding sites is shown by dotted lines which rise from the solid curve. The exponential line is shown for the developmental curve (● —— ●). A second line is drawn (by eye) through the mean values in one-week denervated muscles at four stages of development (● - - - - ●).

mental stage just after synaptogenesis when ACh receptor levels are highest (Bevan & Steinbach 1977). When the motor nerve is cut at progressively later times of postnatal development and the extent of dedifferentiation is compared, an interesting trend emerges (Fig. 1B). Denervation increases the number of ACh receptors by a similar proportion and the dedifferentiation is to progressively later stages of postnatal development. By adult life, the muscle is dedifferentiated to a relatively late stage of development with regard to its ACh receptors. It is significant that it is at this stage of development, namely a week after birth, that muscles in the rat lose their critical dependence on the nerve and will survive without innervation (Vrbová 1952).

Adult rat muscles therefore appear to respond to nerve injury by dedifferentiation to an immature state which can be maintained without innervation and may therefore be conducive to reinnervation by regenerating nerves. Re-expression of glycoproteins in the extracellular matrix of the denervated muscles, particularly at the endplate region of the denervated muscle, guides growing axons and favours reinnervation at the endplate site (Sanes et al 1986).

Axotomized nerves

Using chronic electrophysiological recording techniques, we demonstrated that axotomized neurons also show a limited dedifferentiation with respect to the size of their axons. By recording the compound action potential (CAP) on a branch of the sciatic nerve in cat and rabbit hindlimbs before and after cutting the nerve below the recording cuff electrodes, we monitored the time course of atrophy of the axotomized nerve (Davis et al 1978, Orozco et al 1985). Consistent with the morphological evidence for a decline in fibre diameter, the amplitude of the CAP, which reflects the diameter of the constituent nerve fibres, declined after the nerve was cut. Yet, as shown in Fig. 2, the CAP reached a stable although lower level which could be maintained for long periods of time. Nerve fibres that remake connections will fully recover (Gordon & Stein 1982a) but, even in the absence of connections,

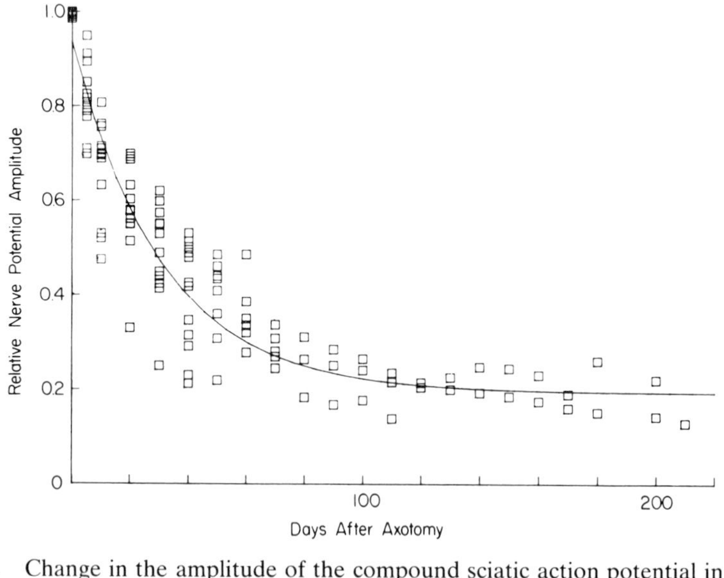

FIG. 2. Change in the amplitude of the compound sciatic action potential in rabbit hindlimbs evoked in response to stimulation of common peroneal nerves proximal to the level of nerve section and ligation, measured relative to potentials recorded before nerve section. Nerve amplitude declines to a lower stable level of 20% of initial values with a time constant of 29 days. The method of chronic recording of compound action potentials is described in detail by Davis et al (1978).

nerve fibres survive and maintain their smaller calibre indefinitely (Fig. 2).

Reduced axonal transport of neurofilaments accounts for the reduced fibre diameter and corresponding reduced CAP amplitude (Dyck et al 1985, Hoffman et al 1985). Although this change has been interpreted as a reversion of the axons toward the embryonic state (Hoffman, this volume), axotomized neurons differ from embryonic neurons in that a small component of neurofilaments appears to be able to maintain the axons at calibres which are considerably larger than those of the fine axons in the early embryo. Furthermore, the spectrum of size of adult myelinated fibres which evolves some time after birth in the rat (Vejsada et al 1985) remains after axotomy, despite atrophy (Gordon & Stein 1982a, Gillespie & Stein 1983). The myelin sheath is fully retained with respect to both the number of myelin turns and the internodal length (Gillespie & Stein 1983). These observations indicate that nerve fibres do not dedifferentiate to very early stages of development. Rather, the nerve fibres survive and may readily recover their former dimensions and properties once functional contacts are remade in the periphery (reviewed by Gordon 1983).

Is the enlarged motor unit in partially denervated muscles a re-expression of the neonatal motor unit size?

The initial formation of nerve–muscle synapses by embryonic motoneurons establishes large motor units with 3–5 times the adult complement of muscle fibres as a result of the polyneuronal innervation of immature muscle fibres. Loss of polyneuronal innervation establishes the smaller adult motor unit and single innervation of each muscle fibre (Brown et al 1976, O'Brien et al 1978). Thompson & Jansen (1977) and Brown & Ironton (1978) made the interesting observation that the size of motor units in partially denervated muscles in adult rat and mouse hindlimbs increased to a maximum of 3–5 times the normal adult size, which corresponds to the larger neonatal motor unit size. An intriguing explanation is that the intact motoneurons after partial nerve injuries re-express the enlarged motor unit size of the neonatal muscle through sprouting. Implicit in the explanation is that the response of motoneurons to partial nerve injury involves a dedifferentiation to an earlier developmental state in which the motoneuron expressed the maximal number of nerve terminal branches and therefore the maximal number of muscle fibres innervated.

If this hypothesis holds true, the extent of sprouting should be limited to the establishment of neonatal unit size in a number of muscles from different species. Instead, recent studies of a number of hindlimb muscles in the cat indicate that the expansion of adult motor units by sprouting is not limited to the size of the neonatal unit (Hatcher et al 1985, Gordon & Orozco 1987). Removal of 7–94% of the innervation of the gastrocnemii and soleus muscles

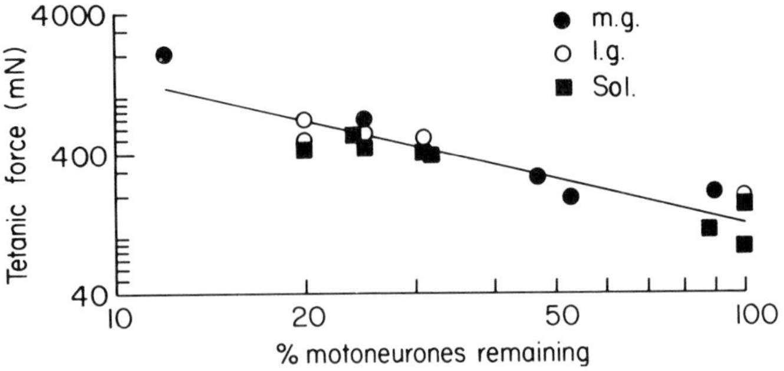

FIG. 3. Mean tetanic force of motor units in medial gastrocnemius (m.g.), lateral gastrocnemius (l.g.) and soleus (Sol.) muscles of the cat hindlimb after section of the L7 or S1 root, 4–14 months previously, plotted as a function of the percentage of motoneurons remaining in the uncut root. A slope of unity on these double-logarithmic plots shows that force output of motor units in partially denervated muscles varies inversely with the number of motoneurons which remain, as would be predicted for complete compensation for partial denervation.

(Gordon & Orozco 1987) or the flexor digitorum longus muscle (Hatcher et al 1985) by section of one of two contributing ventral roots was fully compensated by sprouting of terminals from motoneurons exiting the cord in the remaining root, which enlarged motor units beyond the 3–5-fold limit previously described in murine muscles. Force in the enlarged motor units varied as an inverse function of the numbers of motoneurons supplying the muscle, as shown by a slope of unity in the double logarithmic plots in Fig. 3 and as would be predicted for the reinnervation of all denervated muscle fibres by the remaining motoneurons.

Therefore, reinnervation of partially denervated muscles does not recapitulate development, in the sense that the size of the adult motor unit may greatly exceed that of neonatal units in some muscles. It is also unlikely that limited motor unit size in murine muscles is simply due to an intrinsic limit to the number of axonal branches. A more likely explanation may lie in consideration of the branching pattern of nerves reinnervating muscles. Growing axons must be guided to denervated muscles. For partially denervated muscles, the axonal sprouts from undamaged motoneurons grow within the limited microenvironment of the empty Schwann cell sheaths (Brown et al 1981). It is conceivable that sprouts which emanate from nodal branches at a distance from denervated muscles would branch more extensively to supply more muscle fibres than terminal sprouts. Consequently, the relative proportion of nodal and terminal sprouts may be a more important factor determining the size of motor units in partially denervated muscles than the intrinsic growth capacity of the motoneurons. The fact that terminal sprouts predominate in soleus muscles in which sprouting does not compensate for extensive partial denervation (Brown et al 1980) is consistent with this explanation. Whether nodal sprouts predominate in the cat muscles in which compensation is impressive is experimentally testable.

Does nerve regeneration recapitulate the specificity of nerve–muscle connections established during development?

Type grouping, in which muscle fibres with similar histochemical reactivity are grouped together in muscle cross-sections, is so characteristic of reinnervated muscles that it is used routinely as a diagnostic indication. This fibre distribution differs dramatically from the normal mosaic distribution of muscle fibre types in adult muscle, where similar staining fibres are intermingled with fibres of different reactivity (Dubowitz & Brooke 1963). The latter is established during synaptogenesis and by removal of polyneuronal innervation in the neonatal animal (Thompson 1986, Balice-Gordon & Thompson 1988). The fact that the normal distribution is altered to type grouping in reinnervated muscles is the simplest demonstration of the inability of regenerating motor nerves to find their original muscle fibres (eg. Gillespie et al 1987).

When peripheral nerves are cut and the Schwann cell sheaths disrupted, regenerating axons enter the vacated sheaths in the distal stump to navigate toward their denervated targets. However, in contrast to the growing axons in the developing animals which make highly specific connections with the appropriate target muscles (Landmesser 1986), regenerating axons frequently fail to make appropriate connections. The Schwann cells, which may provide highly specific guidance cues in the immature animal, appear to lose this potential and regenerating axons frequently enter inappropriate Schwann cell tubes (Hafteck & Thomas 1968). It is nevertheless quite clear that they play a pivotal supportive role and function to guide regenerating axons to their target cells (Keynes 1987).

The misdirection of regenerating axons has severe functional consequences. For example, voluntary activation of muscle controlling the first finger leads to contraction of the little finger as a result of the misdirection of regenerating ulnar nerve fibres (Thomas et al 1987). Thus several claims for some degree of specificity of regenerating axons for their former targets should be assessed carefully. As discussed elsewhere in detail (Gillespie et al 1986), most of the experiments test the efficacy of reinnervation of a target by a foreign and original nerve in choice paradigms which are complicated by the surgical manipulations of the two nerves. The complications introduced experimentally are considerably greater than the more common clinical situation in which the proximal and distal end of a cut nerve are approximated. Even for a nerve branch containing nerves to only two muscles, such as the

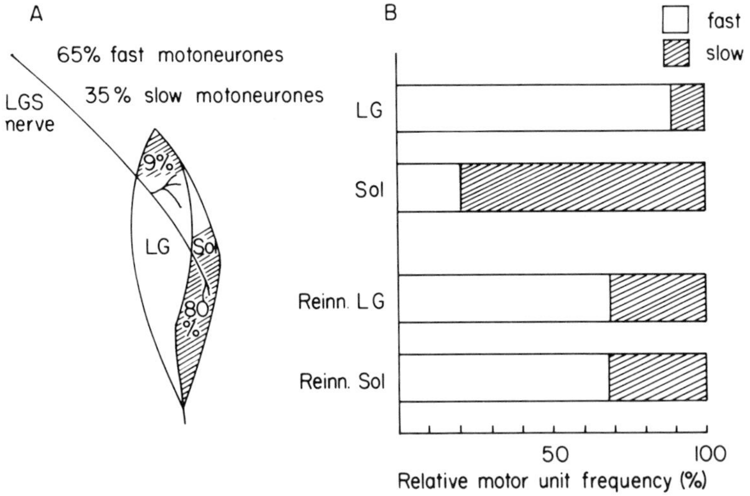

FIG. 4. (A) Diagrammatic representation of the relative proportions of motor axons in the common LGS nerve which supplies the fast and slow motor units in fast-twitch lateral gastrocnemius (LG: 9% slow) and slow-twitch soleus (Sol:80% slow) muscles in the rat hindlimb. (B) The relative frequencies of slow motor units in LG and Sol muscles are dramatically different in the normally innervated muscles but are very similar in the reinnervated muscles. The latter corresponds to the relative proportion of slow motor axons in the common LGS nerve that reinnervates the muscle, which is the result predicted by random reinnervation of denervated muscle fibres by regenerating axons. (Modified from Gillespie et al 1986.)

common lateral gastrocnemius soleus (LGS) nerve to the fast-twitch gastrocnemius and the slow-twitch soleus muscles, we recently demonstrated, as shown in Fig. 4, that regenerating nerves in the cut and sutured nerves failed to show any specificity for their appropriate muscles. The majority of the slow motoneurons contained in the common LGS nerve normally supply the predominantly slow soleus muscle. Similarly, most of the fast motoneurons in the nerve supply the lateral gastrocnemius muscle. If the regenerating nerves preferentially grew toward their former muscles, the normal distribution of fast and slow motor units in the two muscles should be evident in the reinnervated muscle. This was not so. Each reinnervated muscle contained the same proportion of fast and slow motor units, and this proportion was in good agreement with the proportion of the fast and slow motoneurons in the regenerating LGS nerve, showing that muscle reinnervation is a random process.

Is recovery of muscle properties a recapitulation of the differentiation of immature muscle?

Whereas nerve regeneration and synaptogenesis fail to restore the original

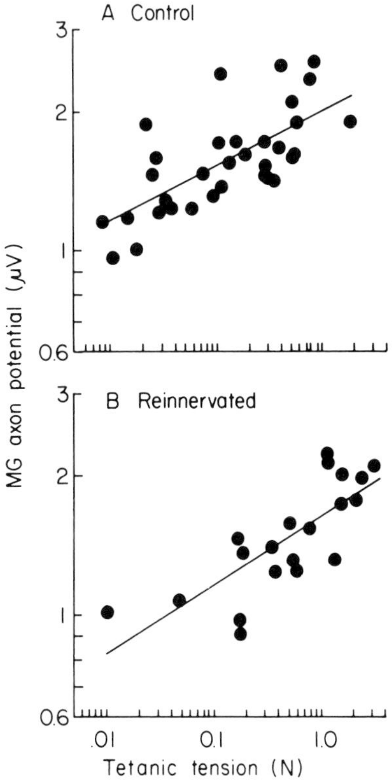

FIG. 5. Relationships between tetanic force of motor units and axon potential amplitude in (A) a normal medial gastrocnemius (MG) muscle in the cat hindlimb muscle and (B) a MG muscle reinnervated by regenerating axons 500 days after section of the MG nerve and suture of the distal stump to the denervated MG muscle. Reinnervation restores the normal range of axonal size, as measured by potential amplitude, and unit force and the normal relationships which are appropriate for the orderly recruitment of motor units. (Modified from Gordon & Stein 1982b.)

nerve–muscle connections, the ability of the regenerating nerve to determine the properties of the muscle fibres that it innervates is strikingly similar to the ability of the immature motoneuron to specify the properties of immature muscles (reviewed by Vrbová et al 1978). Since regenerating motor nerves do not return to their former muscle fibres but supply muscle fibres that formerly belonged to several different motor units, the muscle fibre composition of the units is initially heterogeneous. This heterogeneity is reflected in the metabolic heterogeneity of the muscle fibres and the randomness of the force output of the motor units (Warszawski et al 1975, Gordon & Stein 1982b). With time, the histochemical properties become homogeneous under the influence of the single motoneuron supplying them (Kugelberg et al 1970, Totosy de Zepetnek et al 1987), and the relationships between motor unit contractile force and axonal size (measured electrophysiologically) returns (Fig. 5; Gordon & Stein 1982a,b, Gordon 1988). The innervating motoneurons therefore specify how much force the innervated muscle units will develop, the speed of contraction and their metabolic characteristics, as they do during normal development. To restore force, the number of muscle fibres that are reinnervated is controlled in addition to the size of the innervated muscle fibres (Stein et al 1988). Re-specification of the contractile speed and metabolic properties of the reinnervated fibres involves the control of gene expression by the motoneuron by as yet unknown mechanisms, although it is clear that the pattern and/or amount of neuromuscular activity is important (Pette & Vrbová 1985).

Thus, although the processes of nerve regeneration and muscle reinnervation do not recapitulate developmental processes, the considerable plasticity of the neuromuscular system permits a high degree of recovery of function. Muscle fibres are re-specified with respect to their histochemical properties, their contractile speed and their force output. The net result of neuromuscular plasticity is the re-establishment of motor units whose force output varies over a 100-fold range which is appropriate for the gradation of force by orderly motor unit recruitment during movement. Nevertheless, because the specificity of nerve growth and innervation of target cells is lost in the adult, the recovery of function after nerve injuries is often disappointing.

Acknowledgements

The financial support of the Muscular Dystrophy Association of Canada and the Medical Research Council of Canada is gratefully acknowledged. T.G. is presently a Research Scholar supported by the Alberta Heritage Foundation for Medical Research.

References

Allbrook D 1981 Skeletal muscle regeneration. Muscle & Nerve 4:234–245

Balice-Gordon RJ, Thompson WJ 1988 Synaptic rearrangements and alterations in motor unit properties in neonatal rat extensor digitorum longus muscle. J Physiol (Lond) in press

Bambrick L, Gordon T 1987 Acetylcholine receptors and sodium channels in denervated and botulinum-toxin-treated adult rat muscle. J Physiol (Lond) 382:69–86

Bevan S, Steinbach JH 1977 The distribution of α-bungarotoxin binding sites on mammalian skeletal muscle developing *in vivo*. J Physiol (Lond) 267:195–213

Bisby MA, Redshaw JD, Carlsen RC, Reh TA, Zwiers H 1988 Growth associated proteins (GAPS) and axonal regeneration. In: Gordon T et al (eds) The current status of peripheral nerve regeneration. Alan R Liss, New York, p 35–52

Brown MC 1984 Sprouting of motor nerves in adult muscles: a recapitulation of ontogeny. Trends Neurosci 7:10–14

Brown MC, Ironton R 1978 Sprouting and regression of neuromuscular synapses in partially denervated mammalian muscles. J Physiol (Lond) 278:325–348

Brown MC, Jansen JKS, Van Essen D 1976 Polyneuronal innervation of skeletal muscle in newborn rats and its elimination during maturation. J Physiol (Lond) 261:387–422

Brown MC, Holland RL, Ironton R 1980 Nodal and terminal sprouting from motor nerves in fast and slow muscles of the mouse. J Physiol (Lond) 306:493–510

Brown MC, Holland RL, Hopkins WG 1981 Motor nerve sprouting. Annu Rev Neurosci 4:17–42

Davis LA, Gordon T, Hoffer JA, Jhamandas J, Stein RB 1978 Compound action potentials recorded from mammalian peripheral nerves following ligation and resuturing. J Physiol (Lond) 285:543–559

Diamond J, Miledi R 1962 A study of foetal and new-born rat muscle fibres. J Physiol (Lond) 162:393–408

Dubowitz V, Brooke MH 1963 Muscle biopsy: a modern approach. Saunders, London

Dyck PJ, Lais A, Karnes J, Sparks M, Dyck PJB 1985 Peripheral axotomy induces neurofilament decrease, atrophy, demyelination and degeneration of root and fasciculus gracilis fibers. Brain Res 340:19–36

Fambrough DM 1979 Control of acetylcholine receptors in skeletal muscle. Physiol Rev 59:165–226

Gillespie MJ, Stein RB 1983 The relationship between axon diameter, myelin thickness and conduction velocity during atrophy of mammalian peripheral nerves. Brain Res 259:41–56

Gillespie MJ, Gordon T, Murphy PR 1986 Reinnervation of the lateral gastrocnemius and soleus muscles in the rat by their common nerve. J. Physiol (Lond) 372:485–500

Gillespie MJ, Gordon T, Murphy PR 1987 Motor units and histochemistry in rat lateral gastrocnemius and soleus muscles: evidence for dissociation of physiological and histochemical properties after reinnervation. J Neurophysiol 57:921–937

Ginetzinsky AG, Shamarina NM 1942 The tonomotor phenomenon in denervated muscle. Uspekhi Sovremennoi Biologii 15:283–294

Gordon T 1983 Dependence of peripheral nerves on their target organs. In: Burnstock G et al (eds) Somatic and autonomic nerve–muscle interactions. Elsevier, New York, p 289–325

Gordon T 1988 To what extent can the normal organization of motor unit properties be re-established after muscle reinnervation? In: Gordon T et al (eds) The current status of peripheral nerve regeneration. Alan R Liss, New York, p 275–285

Gordon T, Orozco R 1987 Complete compensation for partial nerve injuries in cat muscles. Neuroscience 22 Suppl:5649

Gordon T, Stein RB 1982a Time course and extent of recovery in reinnervated motor units of cat triceps surae muscles. J Physiol (Lond) 323:307–323

Gordon T, Stein RB 1982b Reorganization of motor-unit properties in reinnervated muscles of the cat. J Neurophysiol 48:1175–1190

Hafteck J, Thomas PK 1968 Electron-microscope observations on the effects of localized crush injuries on the connective tissues of peripheral nerve. J Anat 103:233–243

Hatcher DD, Luff AR, Westerman RA, Finkelstein DI 1985 Contractile properties of cat motor units enlarged by motoneurone sprouting. Exp Brain Res 60:590–593

Hoffman PN 1988 Distinct roles of neurofilament and tubulin gene expression in axonal growth. In: Plasticity of the neuromuscular system. Wiley, Chichester (Ciba Found Symp 138) p 192–204

Hoffman PN, Thompson GW, Griffin JW, Price DL 1985 Changes in neurofilament transport coincide temporally with alterations in the caliber of axons in regenerating motor fibers. J Cell Biol 101:1332–1340

Huizar P, Kuno M, Kudo N, Miyata Y 1977 Reaction of intact spinal motoneurones to partial denervation of the muscle. J Physiol (Lond) 265:175–191
Keynes RJ 1987 Schwann cells during neural development and regeneration: leaders or followers? Trends Neurosci 10:137–139
Kugelberg E, Edstrom L, Abbruzzese M 1970 Mapping of motor units in experimentally innervated rat muscle. J Neurol Neurosurg Psychiatry 33:319–329
Landmesser L 1986 Axonal guidance and the formation of neuronal circuits. Trends Neurosci 9:489–492
Merlie JP, Isenberg KE, Russell SD, Sanes JR 1984 Denervation supersensitivity in skeletal muscle: analysis with a cloned cDNA probe. J Cell Biol 99:332–335
O'Brien R, Ostberg A, Vrbová G 1978 Observations on the elimination of polyneuronal innervation in developing muscle. J Physiol (Lond) 282:571–582
Orozco R, Gordon T, Davis LA, Goldsand G 1985 Chronic electrical stimulation of axotomized neurons in the rabbit. Soc Neurosci Abstr 11:1146
Pette D, Vrbová G 1985 Neural control of phenotypic expression in mammalian muscle fibers. Muscle & Nerve 8:676–689
Sanes JR, Schachner M, Covault J 1986 Expression of several adhesive macromolecules (N-CAM, L1, J1, NILE, uvomorulin, laminin, fibronectin, and a heparan sulfate proteoglycan) in embryonic, adult, and denervated adult skeletal muscle. J Cell Biol 78:176–186
Stein RB, Gordon T, Totosy de Zepetnek J 1988 Mechanisms for respecifying muscle properties following reinnervation. In: Binder M, Mendell LM (eds) The segmental motor system. Oxford University Press, Oxford, in press
Steinbach JH 1981 Neuromuscular junctions and α-bungarotoxin-binding sites in denervated and contralateral cat skeletal muscles. J Physiol (Lond) 313:513–528
Thomas CK, Stein RB, Gordon T, Lee RG, Elleker MG 1987 Patterns of reinnervation and motor unit recruitment in human muscles after complete ulnar and median nerve section and resuture. J Neurol Neurosurg Psychiatry 50:259–268
Thompson WJ 1986 Changes in the innervation of mammalian skeletal muscle fibers during postnatal development. Trends Neurosci 9:25–28
Thompson W, Jansen JKS 1977 The extent of sprouting of remaining motor units in partly denervated immature and adult rat soleus muscle. Neuroscience 2:523–535
Totosy de Zepetnek J, Zung HV, Gordon T 1987 Motor unit force, innervation ratio and muscle fiber size in normal and reinnervated muscle. Soc Neurosci Abstr 13:874
Vejsada R, Palecek J, Hnik P, Soukup T 1985 Postnatal development of conduction velocity and fibre size in the rat tibial nerve. Int J Dev Neurosci 3:583–595
Vrbová G 1952 The dependence of the rate of atrophy of skeletal muscle on age. (In Russian) Physiol Bohemoslov 1:22–25
Vrbová G, Gordon T, Jones R 1978 Nerve–muscle interaction. Chapman & Hall, London
Warszawski M, Telerman-Toppet N, Durdu J, Graff GLA, Coers C 1975 The early stages of neuromuscular regeneration after crushing sciatic nerve in the rat. Electrophysiological and histological study. J Neurol Sci 24:21–32

DISCUSSION

Thesleff: In relation to the numbers of ACh receptors and sodium channels in muscle, the physiologically important variable is the number of such structures per unit of membrane area; but you have measured the total number in the minced muscle, so you include in your measurement the number of receptors and channels inside the fibre which are synthesized but not yet expressed in the surface. Secondly, you express your numbers in relation to mg of protein and not in relation to membrane area. I wonder how well your data represent physiologically relevant numbers?

Gordon: In trying to measure the total numbers of ACh receptors one has to consider their sites, as you suggest. We chose to prepare muscle homogenates rather than to do a membrane purification because one has a variable yield of membrane in the latter. So we express the numbers of ^{125}I-BTX-binding sites per mg homogenate protein, which provides a reasonable estimate of total numbers of ACh receptors. But we are aware that it is the surface membrane which counts functionally. We measure muscle fibre diameters to normalize the number of ^{125}I-BTX-binding sites/mg protein by calculated muscle fibre surface area and thereby obtain estimates of binding site densities in the surface. This does make the assumption that the binding sites are primarily located in the sarcolemma. Clearly, you are right that we are measuring the total rather than the incorporated number of surface receptors. In the case of ^{3}H-labelled saxitoxin (^{3}H-STX)-binding sites, where up to 50% of the sites are located in T-tubules, estimates of sodium channel density in the sarcolemma are significantly changed by these normalization procedures. Thus only when we took into account fibre volume and the distribution of sodium channels in the fibre did the total number of ^{3}H-STX-binding sites measured in muscle homogenates become meaningful (for further details, see Bambrick & Gordon 1987).

Thesleff: There are two types of sodium channels in denervated mammalian muscles, one type with high affinity for saxitoxin and tetrodotoxin, which are both blockers of sodium channels, and one type with low affinity. Have you taken that into account when you estimate the total number of sodium channels?

Gordon: We measure only the high affinity site in these experiments. In other experiments we were asking whether there is also dedifferentiation with regard to sodium channels. If we measure the number of saxitoxin-binding sites, not normalized, there is no change per mg of protein after denervation. If you take into account the change in fibre diameter and the large proportion of sodium channels in the T-tubules, the decline in ^{3}H-STX-binding site density in the sarcolemma is consistent with a decline of about 30% in sodium channel density in the sarcolemma. This decline is consistent with the data of Pappone (1980) on the proportion of sodium current which becomes TTX resistant. Rogart & Regan (1985) have demonstrated the reappearance of the low affinity sodium channel in denervated mammalian muscle. So the high affinity sodium channel is declining in adult innervated muscle to about 70% levels, concurrently with the reappearance of the TTX-insensitive channel, which is a sodium channel with a low affinity for TTX after denervation.

Crowder: It seems that the concept of dedifferentiation for neurons is useful, but I am not so sure for muscle, at least in terms of acetylcholine receptors. We don't need to be so mysterious about it there, because nerve impulses and muscle activity are known to be the environmental cue that affects the type and concentration of receptors and contractile isoforms.

Gordon: Part of the developmental process is a repression of the expression of extrajunctional ACh receptors and is mediated by activity. The fact that there is dedifferentiation in muscles reflects the fact that the denervated muscles aren't active any more. The fact that adult muscle expresses only the level of ACh receptor synthesis seen in the 10-day neonate goes together with the fact that other properties of the muscle which are also dedifferentiating are doing so to a stage in development in which these muscle fibres can survive; they are not regressing to the very early stages of development where nerve-muscle interaction is critical for survival.

Crowder: Could it be that at this developmental stage the myotubes are multiply innervated, and the extra ACh receptors are just from the clusters at those additional synapses and not from an extraordinary increase in extrasynaptic receptors, upon denervation of the embryonic muscle?

Gordon: After denervation, the endplate ACh receptors are quite stable and decline with a slower time course and to slightly lower levels than the normally innervated endplates; so there isn't an increase of receptors in that area. On the point about multiple sites, when the muscles are reinnervated the nerve fibres come back to the original endplate, so it really is an extrajunctional event.

Zak: In your dedifferentiation studies you count the number of ACh receptors, but the subunit composition also changes during the development and after denervation. Perhaps you have reached a complete biochemical dedifferentiation, yet the number of receptors does not tell the entire story. Can you measure some pharmacological property of the receptors in denervated muscle? I think it is possible in this way to distinguish between types of receptors.

Gordon: I haven't looked at the change in open and closing rates of acetylcholine channels, but others have. With respect to the number of ACh receptors, one is looking at an expression of the protein, and if one is assaying the number of ACh receptors in a homogenate all the way through denervation (the expression of these receptors in the whole muscle, which is really what we are considering), it is not a total dedifferentiation. Whether or not their ACh receptors are incorporated into the sarcolemmal membrane—and I think you and Professor Thesleff are right that this is unclear from our measurement—it is nevertheless clear that the total number of receptors is not increasing to the high levels seen in the muscle during fetal development.

Nadal-Ginard: This concept of dedifferentiation is misleading. It presupposes that one understands the mechanism responsible for it. It also implies that the increase or decrease of a given phenotype is produced by the same mechanism as is at work during normal development. In fact, I don't think we know. Also, are we going to talk about dedifferentiation when all the traits of the cell are changing in a very well-defined direction that confers a new phenotype? If we use this type of nomenclature we shall find that on the one hand the cell is dedifferentiated, but on the other hand it continues to have a well-defined and well-differentiated phenotype that happens to be different from the original one.

Second, the concept of dedifferentiation has the teleological implication that we know what is best, and we then find ourselves with the contradiction of saying that dedifferentiation is adaptive. This is another way of saying that dedifferentiation is another type of differentiation anyway. A more useful concept would be to talk of the 're-expression' of a fetal or earlier phenotype. The term 're-expression' only describes the phenomena and avoids implications about mechanisms.

Vrbová: Let me add one example here. After denervation, a tonic muscle expresses a more differentiated form of myosin (S. Kamel & R. Zak, personal communication). I think we are still at a descriptive stage, and I agree with Dr Nadal-Ginard that when one has to describe what happens after denervation one should not call it dedifferentiation. It's another state of the cell.

Gordon: I agree that 'dedifferentiation' is not the best term, and I take this last point too, but I was trying to consider the concept in an historical way. Regressive changes in nerve and muscle have long been regarded as processes of dedifferentiation; in other words, there is a recapitulation of development. The point that I wanted to make is that the nervous system differs from other systems in that it doesn't undergo normal turnover in which there is the potential for expression throughout life. The way the nervous system repairs itself is adaptive in that, although precursor cells cannot replace damaged cells, the damaged neurons undergo regressive changes to resemble, in a qualitative way, embryonic neurons. These changes are permissive for neuronal growth and the formation of synapses. I wanted to suggest that the injured mature neuron does not dedifferentiate to the earliest stage of development where nerves and muscles are critically dependent on their interaction for survival. They appear to regress to a relatively late state where, although interaction is important for the maintenance of their properties, nerves and muscles can survive without that interaction in a state which is adaptive for recovery. This long-term survival and adaptation for recovery is clear from the experiments of Aitken et al (1947). They cut a peripheral nerve and prevented it from regenerating for a few months by ligating the nerve. If the nerve was then permitted to grow, the nerve made connections with denervated targets and recovered, despite the delay. So one is keeping these nerve cells and muscle cells in a state which is not the mature phenotype, but where they can survive and express relevant genes for recovery. 'Dedifferentiation' may be a poor term. It is really an adaptive maintenance.

Hoffman: It was very instructive that axonal calibre does not increase in neurons forced to increase their area of innervation. This suggests that interactions between neurons and targets are permissive rather than deterministic with regard to levels of neurofilament expression. If the amount of a target-derived trophic factor reaching a neuron determined the level of neurofilament expression, we would expect increased expression (and axonal calibre) in neurons innervating a larger number of targets. Instead, it appears that interactions with targets permit neurons to express neurofilaments at predetermined levels.

Gordon: Indeed this result suggests that adult motoneurons show a more limited plasticity than the muscle fibres that they supply. The retrograde influence of muscle fibres appears to allow recovery of a differentiated state. Others have shown, including Kuno et al (1974), that motoneurons recover their properties irrespective of their target muscles of reinnervation (but see Foehring et al 1988).

References

Aitken JT, Sharman MP, Young JZ 1947 Maturation of regenerating nerve fibres with various peripheral connections. J Anat 81:1–22

Bambrick L, Gordon T 1987 Acetylcholine receptors and sodium channels in denervated and botulinum-toxin-treated adult rat muscle. J Physiol (Lond) 382:69–86

Foehring RC, Sypert GW, Munson JB 1988 Interaction between motoneurons and the muscle fibers they innervate. In: Gordon T et al (eds) The current status of peripheral nerve regeneration. Alan R Liss, New York, p 287–296

Kuno M, Miyato Y, Muñoz-Martinez EJ 1974 Properties of fast and slow alpha motoneurones following motor reinnervation. J Physiol (Lond) 240:725–739

Pappone PA 1980 Voltage-clamp experiments in normal and denervated mammalian skeletal muscle fibres. J Physiol (Lond) 154:190–205

Rogart RB, Regan LJ 1985 Two types of sodium channel with tetrodotoxin sensitivity and insensitivity detected in denervated mammalian skeletal muscle. Brain Res 329:314–318

Model for the study of plasticity of the human nervous system: features of residual spinal cord motor activity resulting from established post-traumatic injury

Milan R. Dimitrijevic

Division of Restorative Neurology and Human Neurobiology, Baylor College of Medicine, 7000 Fannin Suite 2140, Houston, Texas 77030, USA

Abstract. Established post-traumatic spinal cord injuries can serve as an 'experimental model' in which trauma has partially separated the 'spinal neuronal pool' from supraspinal influence. Our findings show that: (1) when the muscle is deprived of upper motor neuron activity, fatigue resistance is diminished and external, electrically induced daily contractions will restore the level of fatigue resistance close to that of muscles in healthy, active subjects; (2) the spinal interneuron network, when completely deprived of brain influence, is a 'spinal reflex centre' with a relatively restricted and low excitability level; and (3) the 'discomplete spinal cord injury' model illustrates that spasticity is of suprasegmental origin and that there are two basic features of brain motor control of the spinal interneuron system: the command to restrict interneuronal pool activity and the command to activate the interneuronal network. Moreover, I have described the modifiability of fatigue resistance, locomotor patterns and different alternatives in the neurocontrol of motor activity, depending on the kind and degree of residual brain influence. Such significant modifiability can be thought of as plasticity of the neuromuscular system and impaired control of the nervous system.

1988 Plasticity of the neuromuscular system. Wiley, Chichester (Ciba Foundation Symposium 138) p 227—239

In this chapter I shall discuss the fatigue resistance properties of muscles in chronically paralysed spinal cord-injured individuals and the features of motor activity which result from different degrees of impairment of brain and spinal cord communication. The results presented here support the view that motor neuron activity is an important factor in determining the properties of skeletal muscle fibres, such as their ability to withstand fatigue, and that increased fatigability can be reversed by chronic muscle electrical simulation.

Furthermore, it is proposed that in addition to the model of the 'transected spinal cord injury' (see Fig. 2A, p 230) and partially injured spinal cord (Fig. 2C and D), there is a third model with a 'discomplete' lesion (Fig. 2B). This finding indicates that it is necessary to revise the conventional concept of 'released segmental functions' in patients with apparently transected spinal cords. Studies of neurocontrol in subjects with incomplete recovery of motor functions show that upper motor neuron paresis is an expression of weaker control resulting from the loss of functional axons, as well as an *alternative in neurocontrol* ensuing from suprasegmental and segmental mechanisms within residual structures and their functional performance.

Observations

These observations are based on a study of approximately 750 chronic spinal cord-injured patients examined since 1971. All patients were evaluated clinically by experienced professionals in the field of spinal cord injury — neurologists, physiatrists (physicians specializing in physical medicine and rehabilitation), orthopaedic surgeons, neurosurgeons, and physical therapists.

The results of these observations are presented under four subheadings: (1) alteration and modification of muscle fatigue resistance; (2) spinal cord motor activity and subclinical brain influence; (3) locomotor patterns and features of neurocontrol; and (4) short- and long-term effects of peripheral nerve electrical stimulation on neurocontrol.

1. Alteration and modification of muscle fatigue resistance

The fatigue index (the ratio of the tension after three minutes of stimulation to the initial tension, expressed as a percentage; Fig. 1A) was measured in the dorsal ankle flexors of spinal cord-injured patients. The average value was 38%, in comparison with 82% in a control group of 10 healthy subjects (Fig. 1B). Thus, continued disuse due to upper motor neuron dysfunction leads to increased fatigability of the skeletal muscles (Vrbová et al 1987).

The common peroneal nerve of the paralysed tibialis anterior of the spinal cord-injured patients, who had no volitional control of the ankle dorsal flexors, was stimulated daily with a train of 50 Hz (preferably twice a day for half an hour each time) for three months. At the end of this time the fatigue index increased from 20–30% to 50–60% (Fig. 1C). This observation, repeated in three different subjects, illustrates that resistance to fatigue of muscles with longstanding upper motor neuron paralysis can be prolonged by daily electrical stimulation over a period of several months (Vrbová et al 1987).

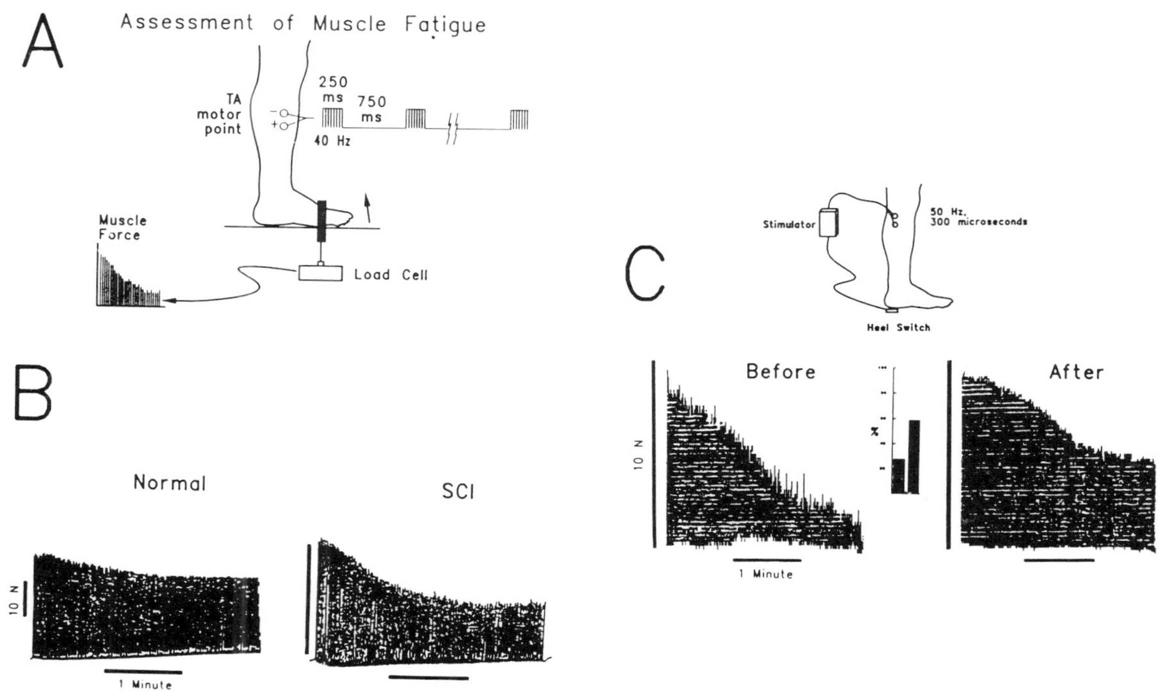

FIG. 1. The alteration of muscle fatigue by electrical stimulation. (A) A 5 cm^2 carbon rubber cathode electrode was placed over the motor point of the ankle dorsiflexors with a distal anode. 250 ms trains of stimuli at 40 Hz were repeated at one-second intervals for three minutes, while the force of contraction was plotted in response to each burst measured by a load cell attached to a strap over the foot. (B) The force of contraction for each burst of stimulation is shown for a healthy subject and for a paralysed spinal cord-injured (SCI) patient. (C) After six months of functional electrical stimulation, the resistance to fatigue increased markedly in this second SCI patient.

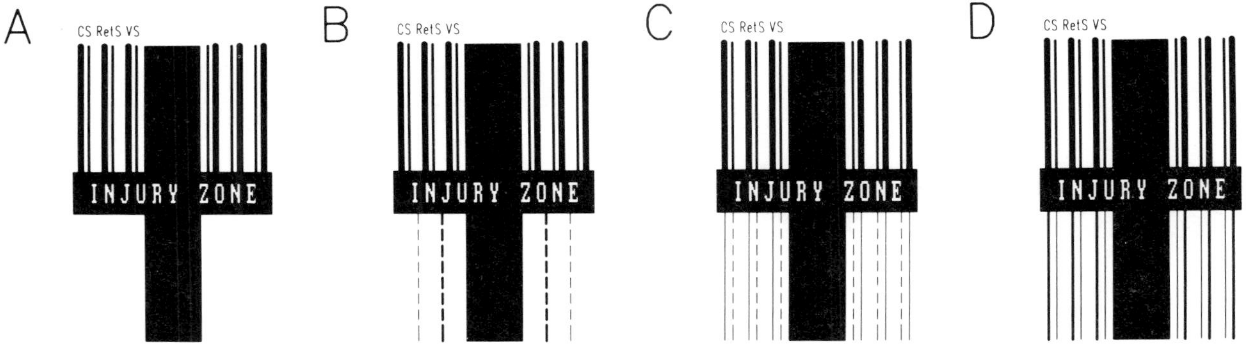

FIG. 2. The varied degree of residual descending tract function. The central vertical bar represents the spinal cord gray matter. The three pairs of lines on each side represent crossed and uncrossed corticospinal (CS), reticulospinal (RetS) and vestibulospinal (VS) pathways interrupted to varying degrees by the lesion. (A) A clinically and neurophysiologically complete lesion is schematized here, with no neurally mediated suprasegmental influence possible below the lesion. (B) The diffuse, sparsely preserved fibres shown here serve only to trigger segmentally organized patterns of activity. (C) With some preserved organization, but relatively poor connections, some suprasegmentally organized patterns (e.g. gross flexion) may be present. (D) Relatively well-preserved axons in discrete tracts mark the presence of organized but weak control of a variety of movements.

2. *Spinal cord motor activity and subclinical brain influence*

When the spinal cord is fully isolated from the influence of the brain, motor activity can be elicited by stretch and exteroceptive reflexes. In other words, 'spinal interneurons' become a 'spinal reflex centre' which governs spinal reflex activities. Reflex responses are isolated only to one limb or side and are shortlasting and phasic. However, spasticity, even in the absence of any volitionally induced movement, reveals the presence of residual, subclinical brain influence resulting in the modification of the behaviour of spinal reflexes (Dimitrijevic 1984, Dimitrijevic et al 1983a).

Experimental animal studies on the recovery of spinal cord functions have suggested the presence of four underlying mechanisms involved in the restitution of functions after spinal cord injury: sprouting, synaptogenesis, restoration of function to uninjured and uncrossed fibres, and remyelination of demyelinated injured fibres, as well as restoration of the ability to conduct impulses through the injured fibres in the absence of myelin. However, our studies on paraplegic patients strongly suggest that plasticity after spinal cord injury is expressed by neurocontrol resulting from suprasegmental and segmental mechanisms within residual structures and their functional performance (Dimitrijevic 1987).

In spastic paraplegic patients, reinforcement manoeuvres can activate a few spinal motor neurons of the 'spinal reflex centre' which, although deprived of multi-pathway descending communication with the brain, has preserved minimal residual brain influence (Fig. 2B and 3B). The segmental interneuronal pool will respond with restricted activation to the arriving suprasegmental input (R1) followed by generalized R2 responses (Dimitrijevic et al 1984). Furthermore, in a similar population of spastic paraplegics we have demonstrated the presence of the brain's residual, subclinical ability to suppress peripherally induced segmental plantar reflex responses (Cioni et al 1986).

In summary, when comparing the plasticity of the 'spinal brain' that has been completely deprived both clinically and subclinically of any brain influence, with the plasticity of the 'spinal brain' that has partial subclinical brain influence, the latter exhibits more pronounced vertical and horizontal radiation of excitation within the segments below the injured portion. Moreover, residual brain influence will generate tonic patterns in reflex activity.

3. *Locomotor patterns and features of neurocontrol*

The observations based on this group of patients represent an 'experimental model' in which trauma instead of the surgical knife has partially separated the 'spinal neuronal pool' from the supraspinal locomotor centres in subthalamic, mesencephalic and pontine nuclei. A portion of the population of axons of the vestibulo-, reticulo-, rubro- and cerulospinal pathways, as well as the crossed and uncrossed corticospinal tracts, will still be operational, along with part of the population of axons that constitute the long ascending tracts (Fig. 2C, D and Fig. 3C, D). In these patients we have studied patterns of volitional muscle activation during four different locomotor acts: flexion and extension of the whole lower limb, and flexion and extension only of the ankle while in the supine position (Dimitrijevic et al 1983b, Dimitrijevic & Lenman 1980). Clinical observations of movement were combined with photographic and video records. Surface electrodes recorded the motor unit activity from lumbar paraspinal and abdominal muscles, both quadriceps, adductors, hamstrings and plantar and dorsal flexors of the ankle. Through the analysis of the surface electromyographic record thus obtained, it became apparent that residual volitional control can generate two different locomotor patterns: one in which the activation of motor units is restricted to selected muscle groups, and one in which the activation of motor units loses all selectivity and is not restricted to those muscle groups involved in flexion or extension. These observations suggest that separate programmes for the implementation of different volitional motor acts are generated in the suprasegmental locomotor centres, from where they are conducted to the injured zone in the spinal cord. Subsequently, these programmes are mediated through the injured portion of the cord. The neurocontrol that is finally generated will selectively or less selectively activate the flexor and extensor muscle groups, depending upon the number and kinds of surviving axons.

If this suggestion is taken one step further, it can be proposed that when the residual descending control is expressed through a large population of spared axons (Fig. 3D), the locomotor programmes will selectively arrive and activate their specific spinal motor nuclei to generate flexion and extension. When a smaller population of axons is functional (Fig. 3C), the resulting volitional neurocontrol will generate non-selective extensor and flexor locomotor patterns; the motor patterns for multi- and single-joint motor acts will not exhibit distinct differences. It appears that the primary role of this 'spinal interneuronal network', partially separated from the brain, in addition to being the 'reflex centre', also becomes in part a 'spinal pattern generator' responding to suprasegmental command. A much reduced population of axons is capable of activating segmental interneuronal networks; however, the greater the reduction in the functioning axon population the less the activation of selective patterns will be possible (Malezic et al 1984).

Observations in this group of patients led to the conclusion that locomotor patterns during ambulation largely depend upon suprasegmental programmes, while 'pattern generators' at the segmental level contribute to the interaction between different portions of committed and uncommitted interneurons in response to suprasegmental commands to reinforce and integrate the programmes coming from the higher levels.

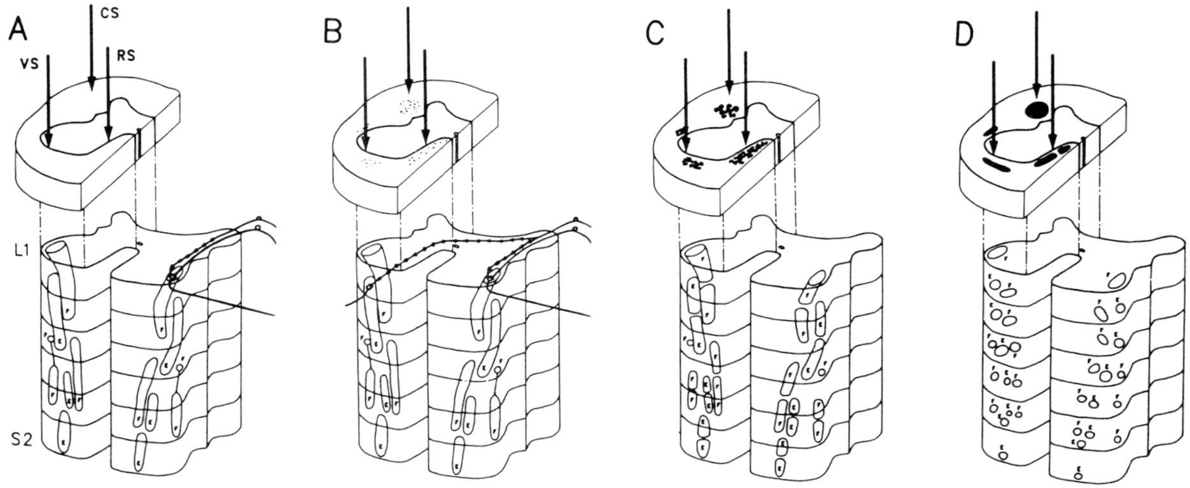

FIG. 3. The proposed functional organization of observed residual functions. Shown in this figure are the gray matter segments L1 to S2, together with a representation of the white matter descending organization. The three arrows represent the three tracts as described in Fig. 2. The functional organization of the motor nuclei is shown schematically within the gray matter, to illustrate (with a drawing of large spinal motor nuclei) the reflexly and suprasegmentally induced 'patterned' motor unit activity, and, by drawing smaller nuclei, to point out that by improving brain influence, more discrete activation of motor units becomes possible. (A) In the isolated spinal cord, only non-sustained unilateral responses are evident, and are grossly organized, with a loss of specificity of response to a limited stimulus. (B) With the presence of some descending fibres, unilateral stimulation can evoke bilateral responses, but still grossly organized. (C) When descending pathways are better preserved, the motor nuclei in the spinal gray matter will react with more discrete motor responses. (D) Well-preserved descending tracts convey facilitation, inhibition, and organized motor nuclei responses in multi-joint, single-joint patterned and skilful movements.

4. Short- and longterm effects of peripheral nerve electrical stimulation on neurocontrol

Let us discuss some as yet unreported observations on what will happen to the just described locomotor patterns of ambulatory spinal cord-injured patients if, in addition to a reduced but present suprasegmentally induced locomotor pattern, we add an externally controlled train of electrical stimuli over the motor points, mixed nerve trunks and cutaneous nerves. The extensor pattern will be enhanced by the stimulation of the gluteus maximus, vastus lateralis and medialis of the quadriceps group and triceps surae. Facilitation of the flexors of the locomotor pattern can be achieved by stimulation of the rectus femoris, dorsal flexors and part of the hamstring muscles. These effects are the result of direct muscle contraction through stimulation of motor nerve fibres or the conditioned responsiveness of particular 'spinal interneurons' achieved through depolarization of large primary and secondary nerve fibres (Dimitrijevic 1983).

What is the longterm effect of the prolonged use of multi-site functional electrical stimulation? We have found in a study of six ambulatory spinal cord-injured patients that daily multi-site functional electrical stimulation during walking for less than a year does not change the features and neurocontrol of locomotor patterns. A definite change in the neurocontrol of locomotor patterns was shown in two patients with a walking ability of only a few steps, in whom stimulation was applied daily, for a minimum of two hours and for a period of two years. In both cases, the improved locomotor pattern persisted and they continued to be able to walk unsupported for a year after stimulation was discontinued. In a control non-functional ambulatory spinal cord-injured patient, who walked with support daily for six years, no changes were induced in the neurocontrol of locomotor patterns.

These observations suggest that the time factor is important if one is to modify the neurocontrol of locomotor patterns in ambulatory spinal cord-injured patients. Why is this so and where is the site of action? Follow-up motor studies after a prolonged stimulation period reveal a definite improvement in volitionally induced neurocontrol. Is this the result of structural reorganization of the 'segmental interneuronal pool', or rather the adaptation of the suprasegmental organization to the restricted population of conducting axons? In order to answer these questions it is necessary to have an experimental animal model.

Summary and conclusion

When the muscle is deprived of activity, fatigue resistance is diminished and external, electrically induced daily contractions will restore the level of fatigue resistance close to that of muscles in healthy and active subjects. The spinal interneuronal network, when completely deprived of brain influence, is a 'spinal reflex centre' with a relatively restricted and low excitability level. Such a 'spinal reflex center' generates unsustained and phasic reflex responses as opposed to tonic spinal reflex activity in patients with 'discomplete' spinal cord injury, when spasticity results from recovery in conduction of the descending axons through the injured portion of the spinal cord. The activity of a small number of functional axons can exert an excitatory as well as an inhibitory effect on the spinal interneuron system in the absence of residual volitional movement in spastic paraplegics. This particular model illustrates that there are two basic features of brain motor control of the spinal interneuronal system in humans: the command to restrict interneuronal pool activity and the command to activate the interneuronal network.

Thus, primary, secondary and tertiary highly connected spinal interneurons are continuously shared by long descending tracts of different brain structures during the preparation and execution of suprasegmentally induced programmes. Connectivity between interneurons allows brain programmes to activate and condition this 'spinal pattern generator' at a certain stage of activity established by central and peripheral mechanisms of previous motor tasks. With the progressive reduction of functioning axons, this connectivity of spinal interneurons starts to contribute to the development of new features of the neurocontrol of locomotor activity. Finally, by augmenting locomotor patterns in ambulatory spinal cord-injured patients by means of functional electrical stimulation, it is possible, after prolonged daily use of this clinical system, to enhance gait performance and alter the previously existing neurocontrol which will later be retained, even in the absence of electrical stimulation.

Overall, it has been possible to demonstrate plasticity as the significant adaptability and modifiability of fatigue resistance of paralysed muscles and the neurocontrol of locomotor patterns. At the same time, the intact spinal interneuron system, completely or partially deprived of brain influence, can develop different alternatives in the neurocontrol of motor activity with features depending upon the size and kind of residual brain influence. We must determine whether to consider these examples of new alternatives as released phenomena, as compensatory mechanisms, or as the outcome of newly established anatomical connections and their expression of plasticity.

Acknowledgements

This work was generously supported by the Vivian L. Smith Foundation for Restorative Neurology, Houston, Texas.

References

Cioni B, Dimitrijevic MR, McKay WB, Sherwood AM 1986 Voluntary supraspinal suppression of spinal reflex activity in paralyzed muscles of spinal cord injury patients. Exp Neurol 93:574–583

Dimitrijevic MR 1983 Physiological procedures for the modification of abnormal motor control: functional electrical stimulation and spinal cord stimulation. In: Pedersen E et al (eds) Actual problems in multiple sclerosis research. Fadl's forlag, Copenhagen (Proc of the Jubilee Conference on Multiple Sclerosis, Copenhagen, Denmark, 1982) p 114–116

Dimitrijevic MR 1984 Neurocontrol of chronic upper motor neurone syndromes. In: Shahani B (ed) Electromyography in CNS disorders: central EMG. Butterworth, London, p 111–127

Dimitrijevic MR 1987 Residual motor functions in spinal cord injury. Adv Neurol 47:139–155

Dimitrijevic MR, Dimitrijevic M, Faganel J, Sherwood AM 1984 Suprasegmentally induced motor unit activity in paralyzed muscles in patients with established spinal cord injury. Ann Neurol 2:216–221

Dimitrijevic MR, Faganel J, Lehmkuhl LD, Sherwood AM 1983a Motor control in man after partial or complete spinal cord injury. Adv Neurol 39:915–926

Dimitrijevic MR, Faganel J, McKay WB, Sherwood AM 1983b Studies of spinal cord upper motor neurone dysfunction during volitional, non-volitional motor activity. Abstract (7th International Congress of Electromyography, Munich, Germany)

Dimitrijevic MR, Lenman JAR 1980 Neural control of gait in patients with upper motor neurone lesions. In: Feldman R et al (eds) Spasticity: disordered motor control. Symposia Specialists, Miami, p 101–114

Malezic M, Dimitrijevic MM, Dimitrijevic MR, Sherwood AM 1984 Activation of ankle flexor and extensor muscles during stretch reflex, volitional activity and posture in ambulatory spinal cord injury patients. Proc Eighth International Symposium on ECHE, Dubrovnik, Yugoslavia, p 421–428

Vrbová G, Dimitrijevic MM, Partridge M, Halter J, Verhagen Metman L 1987 Reversal of increased muscle fatigue in paraplegic patients by electrical stimulation. Soc Neurosci Abstr 13:1698 (470.2)

DISCUSSION

Crowder: Do the two neurological functions (those you have identified with the preparatory functions of the motor act, making the circuitry ready for the generation of movement, and the execution of the actual movement) have correlates in the corticospinal tracts?

Dimitrijevic: There is at present no direct evidence from human studies indicating which population of descending axons contributes to the brain's neurocontrol of movement. In our studies on the neurocontrol of paralysis in patients with established spinal cord injury we have found brain influence on spinal interneurons (facilitation and suppression) even in the absence of volitional movements (Dimitrijevic et al 1984, Cioni et al 1986). We have also made some preliminary observations which suggest that reticulospinal pathways are primarily involved in motor set mechanisms and that corticospinal control influences the variables of volitional motor activity (Dimitrijevic et al 1988).

Buller: You mentioned magnetic cortical stimulation. Have you much experience of that? Have you used it clinically?

Dimitrijevic: Yes, we use daily magnetic as well as electrical transcranial stimulation for our physiological, pathophysiological and clinical studies. Magnetic stimulation has the advantage that it is painless but the stimulated area of the cortex is rather large when compared to that in transcranial high voltage electrical stimulation.

Holder: Have you gone back to look at the lesions in the patients after spinal cord injury, to see if there is any eventual alteration in the damage site after treatment?

Dimitrijevic: Morphological studies of the human spinal cord are not routine for several reasons. It takes some skill accompanied by the necessary tools to remove the spinal cord from the spine. This difficulty can be overcome when neuropathologists are actively involved in studies of acute and chronic spinal cord injury patients, which is what has happened at the Department of Neuropathology of the Perth School of Medicine in Australia, where Professor Byron Kakulas has for many years systematically studied the morphological features of spinal cord injury in acute, subacute and chronic patients and has correlated these features with neurological findings (Kakulas 1985).

Another drawback is that restorative procedures are applied at a time when patients are in very specialized centres for restorative neurology. These patients usually come from all over the world and when they die they are far away from centres where morphological studies of the spinal cord are made. We must also bear in mind that today's life expectancy of spinal cord injury patients is practially identical to that of a person who has never suffered from spinal cord injury. We are, therefore, restricted to well-designed, meticulous and comprehensive studies of the physiological and functional characteristics of the human spinal cord or to pathological specimens from patients suffering from cancer. Some restorative procedures have been applied in the course of treatment for neurological pain or paralysis: it is thus that morphology became available for comparison at a later date. At present, there are studies on dorsal root entry zone (DREZ) lesions for the modification of intractable pain in patients with cancer. I anticipate that in the near future we shall succeed in obtaining more morphological information from the human spinal cord in similar circumstances.

Gordon: I am interested in the patients with very little residual function, many of whom have severe spasticity. You alluded to the fact that you get rhythmical flexion and extension in such patients with functional electrical stimulation. Do you manage to do that in a large proportion of your patients, to eliminate the spasticity, and to get rhythmic activation, and how long does it take? Is the effect generalized in some sense, because you use stimulation of the spinal cord as well as of peripheral nerves?

Dimitrijevic: Yes, we are very successful in controlling spasticity by cutaneous mixed nerve and spinal cord electrical stimulation (M.M. Dimitrijevic et al 1986a,b, M.R. Dimitrijevic et al 1986). We have recently been able to repeat reported results on the successful control of spasticity by means of the intrathecal application of morphine and baclofen (Erickson et al 1985, Penn & Kroin 1984).

We constantly bear in mind that spasticity is the expression of altered sensory-motor integration (Dimitrijevic 1973). As restorative neurologists we first examine and describe altered sensory-motor integration and then influence it by means of physical, physiological, structural and pharmacological restorative procedures. This practice makes it possible to recognize immediately that closed spinal cord injuries are not restricted to well-defined segments and portions of the spinal cord. They can be minute and severe with asymmetries at different levels of the spinal cord; therefore, it is essential to trace these new post-traumatic conditions before going on to apply restorative procedures.

Oppenheim: Has there been any success in getting ambulation through electrical stimulation in patients with complete spinal cord transection?

Dimitrijevic: Yes, there have been successful demonstrations of the restoration of ambulation in patients with complete spinal cord lesions in laboratory conditions (Marsolais & Kobetic 1983, Bajd et al 1983, Petrofsky et al 1984). However, I should add that morphologically complete transected spinal cord injuries are very rare (Dimitrijevic 1987, 1988).

Henderson: One of the problems, paradoxically, of working on spinal cord regeneration in animal models is to know exactly what lesions to put in the spinal cord, because an apparently small lesion can have a big effect and an apparently total lesion can leave the animal almost untouched. You were talking about a group of patients that you could identify from, presumably, their clinical symptoms as being ones that would be receptive to this type of treatment. Have enough of these patients died, for different reasons, to say whether there is a morphological correlate to that in the type of lesion within the spinal cord?

Dimitrijevic: Fortunately for spinal cord injury patients, there are specialized programmes for their treatment and rehabilitation. Unlike the conditions prevalent after World Wars I and II, these patients live long and, especially in Western countries which offer developed medical care, they are frequently productive and independent citizens. However, in the tragic cases of patients with central nervous system neoplasms and intractable pain it is necessary to intervene surgically to disconnect the spinal cord ascending and descending tracts. This model has been used to illustrate the basic rule you mentioned that there is no direct relationship between the size of the lesion and the degree of dysfunction. I have specifically in mind the work of Peter Nathan and Marian Smith (Nathan & Smith 1973).

References

Bajd T, Kralj A, Turk R, Benko H, Sega J 1983 The use of a four channel electrical stimulator as an ambulatory aid for paraplegic patients. Phys Ther 63:1116-1120

Cioni B, Dimitrijevic MR, McKay WB, Sherwood AM 1986 Voluntary supraspinal suppression of spinal reflex activity in paralyzed muscles of spinal cord injury patients. Exp Neurol 93:574-583

Dimitrijevic MM, Dimitrijevic MR, Verhagen-Metman L, Partridge M 1986a Modification of muscle tone in patients with upper motor neuron dysfunctions by electrical stimulation of the sural nerve. Am Acad Clin Neurophysiol Abstr 2:9

Dimitrijevic MM, Dimitrijevic MR, Illis LS, Nakajima K, Sharkey PC, Sherwood AM 1986b Spinal cord stimulation for the control of spasticity in patients wth chronic spinal cord injury. I. Clinical observations. Cent Nerv Syst Trauma 3(2):129–144

Dimitrijevic MR 1973 Spasticity—a clinical and neurophysiological entity or an alteration of sensory-motor behaviour. In: Fields WS (ed) Neural organization and its relevance to prosthetics. Symposia Specialists, Miami

Dimitrijevic MR 1987 Neurophysiology in spinal cord injury. Paraplegia 25:205–208

Dimitrijevic MR 1988 Residual motor functions in spinal cord injury. Adv Neurol 47:139–155

Dimitrijevic MR, Dimitrijevic M, Faganel J, Sherwood AM 1984 Suprasegmentally induced motor unit activity in paralyzed muscles in patients with established spinal cord injury. Ann Neurol 16:216-221

Dimitrijevic MR, Illis LS, Nakajima K, Sharkey PC, Sherwood AM 1986 Spinal cord stimulation for the control of spasticity in patients with chronic spinal cord injury. II. Neurophysiological observations. Cent Nerv Syst Trauma 3(2):145-152

Dimitrijevic MR, Eaton WJ, Sherwood AM, Van Der Linden C 1988 Assessment of corticospinal tract integrity in human chronic spinal cord injury. In: Rossini PM, Marsden CD (eds) Non-invasive stimulation of brain and spinal cord: fundamentals and clinical applications. Alan R Liss, New York, in press

Erickson DL, Blalock JB, Michaelson M, Sperling KB, Lo JN 1985 Control of spasticity by implantable continuous flow morphine pump. Neurosurgery 16:215–217

Kakulas BA 1985 Pathomorphological evidence for residual spinal cord functions. In: Eccles J, Dimitrijevic MR (eds) Recent achievements in restorative neurology: upper motor neuron functions and dysfunctions. Karger, Basel

Marsolais E, Kobetic R 1983 Functional walking in paralyzed patients by means of electrical stimulation. Clin Orthop Rel Res 175:30–36

Nathan PW, Smith MC 1973 Effects of two unilateral cordotomies on the motility of the lower limbs. Brain 96:471–494

Penn RD, Kroin JS 1984 Intrathecal baclofen alleviates spinal cord spasticity. Lancet 1:1078

Petrofsky JS, Phillips CA, Heaton HH 1984 Feedback control system for walking in man. Comp Biol Med 14:135–149

Responses of diseased muscle to electrical and mechanical intervention

V. Dubowitz

Department of Paediatrics and Neonatal Medicine, Royal Postgraduate Medical School, Hammersmith Hospital, Du Cane Road, London W12 OHS, UK

Abstract. It is well established that the properties of muscle fibres are influenced by their neurons and that this is at least in part mediated by the pattern of activity. Application of this knowledge has led to the experimental trial of electrical stimulation in diseased muscle, both in the dystrophic mouse and in children with Duchenne muscular dystrophy. This has shown a beneficial effect of slow frequency stimulation. Another route through which muscle properties can be influenced is by changing the load by procedures such as tenotomy. This has been studied by complete tenotomy in normal animals and recently by selective partial procedures in human disease. Y. Rideau has shown that release of early shortening (contractures) of several muscles, a consistent feature in Duchenne muscular dystrophy, has a beneficial effect on muscle function. From personal observations on a number of Rideau's patients who have undergone this procedure the improvement in function seems disproportionate to what could be explained on simple biomechanical grounds alone and suggests some more fundamental change in the contractile properties of the muscle.

1988 Plasticity of the neuromuscular system. Wiley, Chichester (Ciba Foundation Symposium 138) p 240–255

Studies in the 1960s on crossing the nerves between fast and slow muscles in cats or rabbits showed that the physiological characteristics of the muscle were dependent on their innervation (Buller 1960) and that the histochemical pattern, as well as the biochemical composition of the fibres, could be changed by changing the innervation (Dubowitz 1967), and that presumably the gene expression for fast or slow myosins and other proteins such as the troponins might also be modified in this way.

Although Buller et al (1960) originally postulated that the influence of the nerve on muscle might be chemically mediated, Vrbová and her colleagues established by a number of direct stimulation studies in animals that the activity pattern transmitted to the muscle was the factor determining the twitch characteristics of the muscles (Salmons & Vrbová 1969). These observations formed the basis for subsequent experimental studies of chronic nerve stimulation in normal and diseased human muscle. In cats and rabbits, chronic low frequency stimulation of the fast tibialis anterior (TA) and extensor digitorum longus (EDL) muscles at a frequency resembling that of slow motoneurons (10 Hz) produced an increased resistance to fatigue (Brown et al 1976). A similar study in human volunteers showed that comparable changes could be induced in human muscles such as TA and EDL and that after chronic slow frequency stimulation they became more fatigue resistant (Scott et al 1985).

The techniques consisted of measuring the tension of maximum voluntary contraction (MVC) and maximal electrically evoked contraction of TA and EDL. Contractions elicited at 40 Hz for 250 ms produced forces up to 44% of MVC. Tetanus was produced after prolonged stimulation at 40 Hz for 250 ms every second for five minutes, and a fatigue index was calculated from the ratio of tetanic tension at three minutes to initial tension. In our initial study, four adult volunteers stimulated their TA and EDL muscles with a small battery-operated stimulator at a frequency of 8–10 Hz for an hour, three times daily for six weeks. The contralateral leg acted as control. They were able to continue with their usual daily activities. Initial baseline data were established before stimulation and recorded again after three and six weeks. After six weeks of stimulation there was an increase in the fatigue index (that is, increased resistance to fatigue) in all subjects (with a statistically significant increase in the mean from 66.5% to 79.4%). There was no change in MVC.

In a study of the effects of chronic low frequency stimulation in dystrophic mice, Vrbová and her colleagues (Luthert et al 1980) showed an improvement in function and a reduced loss of muscle fibres together with raised levels of oxidative enzymes. In a comparable study of six young ambulant boys with Duchenne muscular dystrophy we found that intermittent chronic low frequency stimulation of the tibialis anterior (cycle 1.5 s on, 1.5 s off, for one hour three times a day, at a frequency of 5–10 Hz, for periods from seven to 11 weeks) produced a significant increase in the mean of the maximum voluntary contraction compared with the control contralateral leg (Scott et al 1986).

In a further study, the effect of chronic high frequency stimulation (30 Hz) was studied in a separate group of six boys with Duchenne muscular dystrophy (DMD), aged four to seven years, who stimulated their TA for one hour three times daily for at least seven weeks. Recordings were made of maximum voluntary contraction and electrically evoked contraction plus an assessment of the boys' physical abilities. After initial baseline data had been established, further testing was done after four and eight weeks of stimulation. There was a small decrease in MVC of both the stimulated and the control legs during eight weeks of stimulation (Fig. 1a). Three of these boys then stimulated their muscles at 8 Hz for three hours each day for a further eight weeks. This resulted in an increase of the MVC in all three subjects, with an increase in the mean from 1.38 kg to 2.04 kg (Fig. 1b). This suggests that only low frequency stimulation of the muscle is beneficial in DMD.

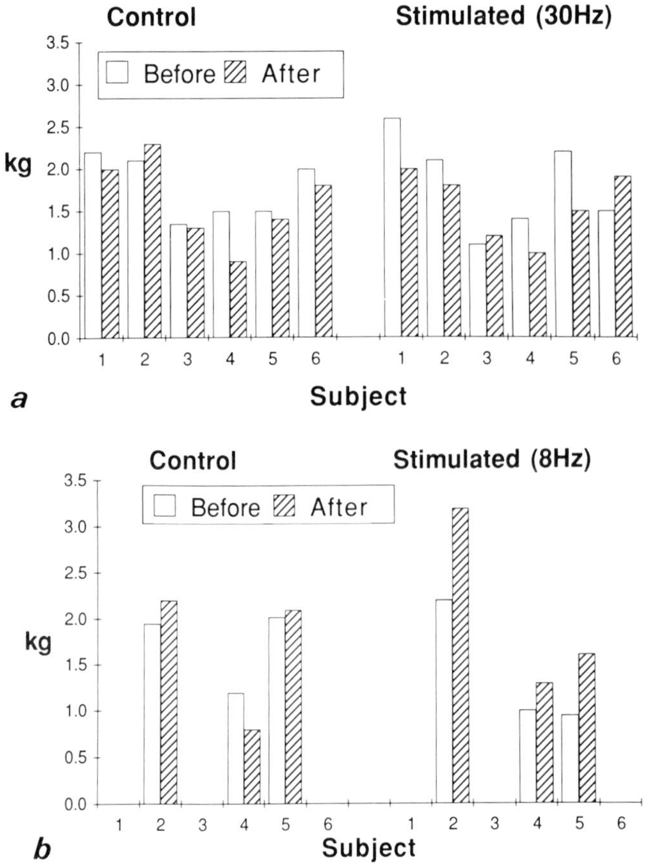

FIG. 1. Histogram showing force of maximum voluntary contraction (MVC) before and after a period of chronic stimulation at 30 Hz in six boys with Duchenne muscular dystrophy, in the stimulated and the unstimulated (control) leg. (Courtesy of Dr Oona Scott.) (b) Histogram showing force of maximum voluntary contraction of three of the subjects in (a) after a further period of stimulation at 8 Hz.

The properties of normal or diseased muscle may also be influenced by procedures such as tenotomy, which change the load on the muscle, or procedures such as denervation or immobilization, which change the activity pattern of the muscle. Karpati et al (1982) observed that early denervation of hindlimb muscles in the dystrophic hamster resulted in the absence of necrosis and central nucleation during the period of denervation, but that the dystrophic changes recurred at a rate comparable to the natural history of the disease once the electrical and mechanical activity of the muscle was resumed. Similar protective effects on dystrophic muscle were produced in the dystrophic mouse by denervation (Jaros & Johnstone 1983), by tenotomy (Wirtz and Loermans 1983), and by immobilization of the hindlimb during early postnatal development (Loermans & Wirtz 1983, Wirtz et al 1986).

Ashmore (1982) obtained a protective effect on the muscle of the dystrophic chicken by immobilizing the wing in a stretched position. The treatment had to be started early, before the first dystrophic symptoms were apparent, and the effects were ascribed to the stretching rather than to the immobilization.

The damaging effect of eccentric contraction (that is, contraction with muscle under stretch) has been demonstrated in the mouse (Faulkner et al 1985) and also in human muscle (Newham et al 1983). The effects of exercise on dystrophic muscle have been studied in the hamster (Howels & Goldspink 1974), in the mouse (Soltan 1962), in the chick (Hudecki et al 1978) and in the human (Vignos & Watkins 1966, Scott et al 1982).

These experimental studies form a useful backdrop to the interesting programme of therapeutic intervention that has been developed in recent years by Rideau in an effort to stabilize the muscle function in Duchenne muscular dystrophy and arrest the progression of the disease (Rideau et al 1986).

In Duchenne muscular dystrophy, affected boys may seem to be normal in the first one to two years of life, although pathological changes in the muscle, consisting of necrosis and regeneration, are already present at birth, as is the leakage of enzymes such as creatine kinase from the muscle into the serum. About half the affected boys are delayed in walking, and do not walk independently until 18 months or later (normal mean is about 13 months and 97th percentile about 17 months). After that there is a gradual progression in disability and weakness with progressive difficulty in going up stairs and in getting up from a supine position. This is achieved by the classical Gowers manoeuvre, in which the children literally climb up their own legs to get to an upright position. By timing the Gowers manoeuvre one can get some indication of the progression of the disease. Normal children will get up from the floor in about one second or less, early Duchenne patients will usually manage in about two seconds, and this gradually increases with age, so that by 8–10 years it is often in the region of 15–20 seconds or more and goes up precipitately until the child is unable to rise unaided (Fig. 2).

Children with Duchenne muscular dystrophy are never able to run in a normal fashion and when they try to run, it accentuates their waddling gait. They are also unable to jump, or to hop on one leg.

Rideau has developed a series of clinical interventions aimed at trying to maximize the available muscle function in boys with DMD. He started by releasing contractures (fixed shortening) of muscles such as the tensor fascia lata, hamstrings and tendo achilles in late cases, and gradually intervened in younger cases. His current programme is to release several of the 'tight' muscles/tendons at the hips, knees and ankles in the early stages of the disease, optimally between four and a half and five years of age, and in

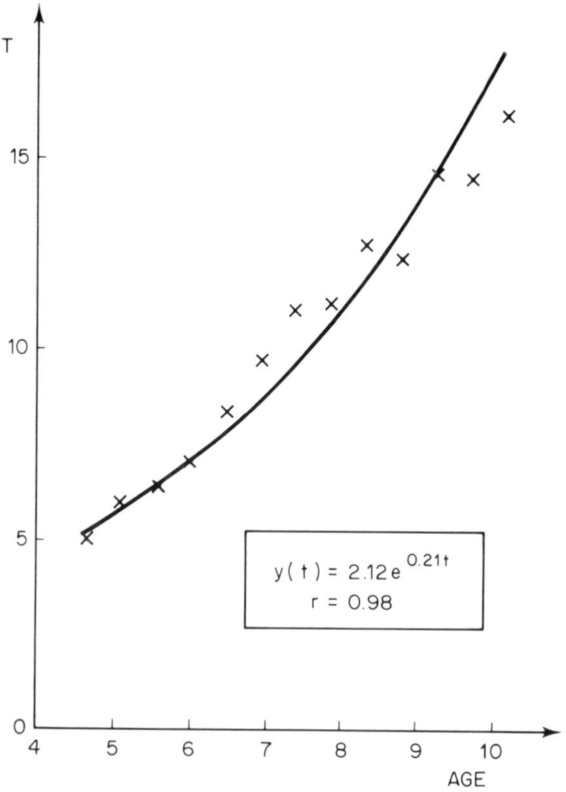

FIG. 2. Graph illustrating the change in the time (T) taken for Gowers manoeuvre with advancing age in Duchenne muscular dystrophy. (Courtesy of Professor Y. Rideau.)

addition has recommended releasing or actually removing the whole fascia lata from the lateral side of the thigh, through a long incision going almost from hip to knee.

These procedures have had a remarkable effect in these boys, which I have recently (November 1987) had an opportunity of assessing at Professor Rideau's Unit in Poitiers. The most striking effects, which I have never seen before in boys with DMD (at any stage), include:

(1) An ability to run in a normal fashion.
(2) An ability to get up from the floor within about a second without a Gowers manoeuvre.
(3) The ability to jump.
(4) The ability to hop on one leg.

In addition, some boys who were operated on about five years previously

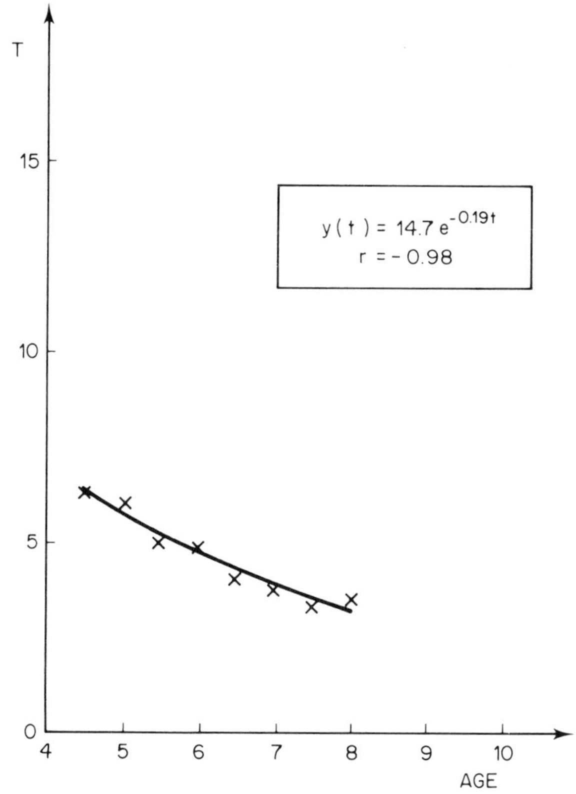

FIG. 3. Graph showing the change in the time (T) taken for Gowers manoeuvre in boys with Duchenne muscular dystrophy after operative intervention. (Courtesy Professor Y. Rideau.)

have not had a recurrence of any contractures, have not needed passive stretching procedures, and have remained in an apparently stable, non-deteriorating state. They have continued to be able to get up from the floor to standing in around two seconds with no increase in the time for the Gowers manoeuvre over the years (Fig. 3). This is quite remarkable, and totally out of keeping with the normal course of Duchenne muscular dystrophy.

After a few years, some of the operated boys have reached a point of rapid loss of function, with increasing Gowers time over a few months and loss of ability to walk, as if the muscle continued at a uniform level and then suddenly became exhausted. In the boys still ambulant at 10 years I have also been struck by the remarkable degree of residual power in the hip abductors and hip extensors, which are usually very weak and have lost anti-gravity power by that stage. There is also a striking prominence and apparent hypertrophy of the vastus lateralis/rectus femoris muscles along the line of

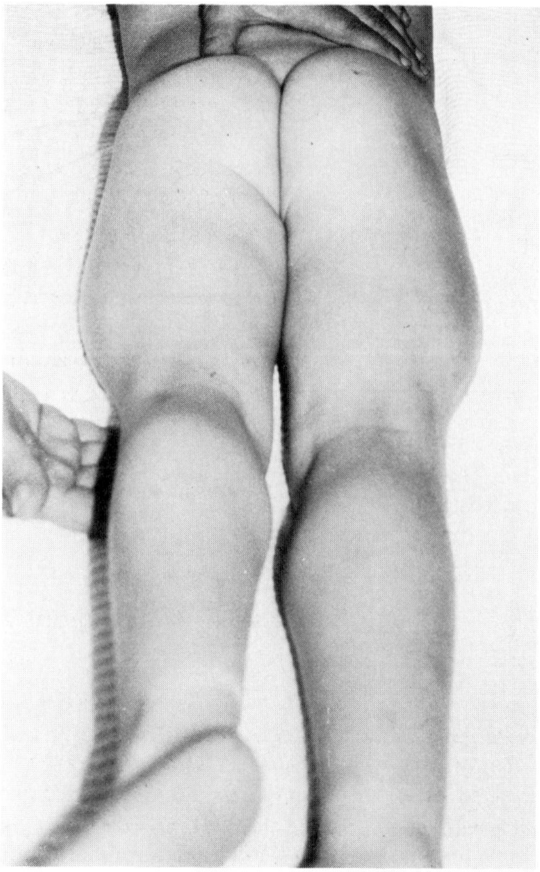

FIG. 4. A boy with Duchenne muscular dystrophy, showing striking focal hypertrophy of the vastus lateralis muscle along the line of excision of the fascia lata.

incision where the fascia lata has been removed (Fig. 4). It gives the impression of previous restraint of the muscle by the fascia lata, like a compartment syndrome, with its release allowing the muscle to hypertrophy in a normal fashion or perhaps even excessively.

This clinical experiment has raised numerous questions about the disease process and the influence of extraneous factors on its progression.

(1) Could the change in length of the muscle have a dramatic effect on its type of contraction and possibly influence whether it contracts concentrically or eccentrically under work load? Eccentric contraction (that is, contraction of a muscle when under a load) may have a damaging effect on the normal as well as the diseased muscle. With the fixed equinus posture produced by the shortening of the gastrocnemius/tendo achilles in Duchenne dystrophy, the tibialis anterior may be overloaded when contracting, as it cannot shorten and

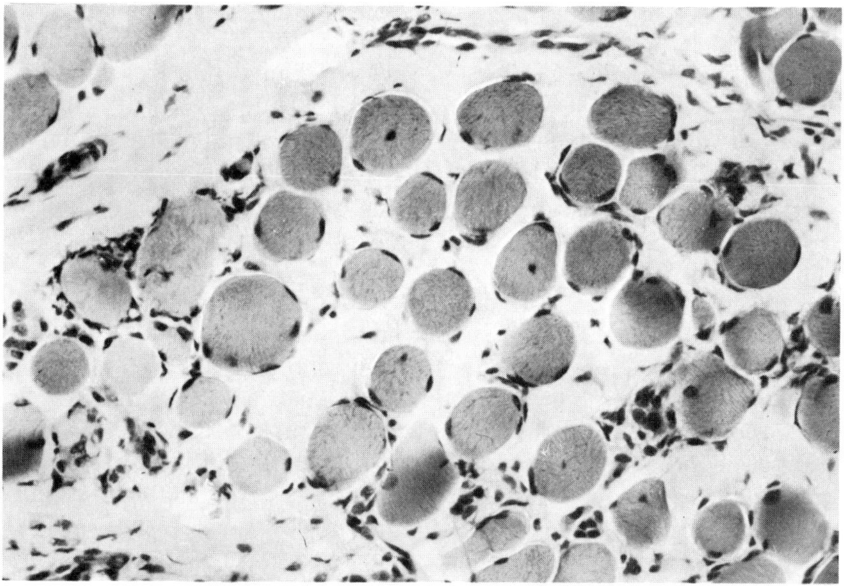

FIG. 5. Cross-section of vastus lateralis muscle of a six-year-old boy with Duchenne muscular dystrophy showing the separation of individual fibres by proliferating endomysial connective tissue.

remains in an extended position. On the other hand, the gastrocnemius may be functioning suboptimally, as it is in a permanently shortened and contracted position. This could also be a factor in the prominence ('pseudohypertrophy') of the calf muscles. One would expect the fast (high ATPase) fibres to be preferentially affected in the tibialis anterior and the slow fibres in the gastrocnemius.

(2) Could the release procedures be restoring the balance between the agonist and antagonist muscles about the hips, knees and ankles, and thus enabling these muscles to function optimally within the constraints imposed by the underlying disease process itself?

(3) Could there be some change in the feedback mechanism from the muscle by which the nerve input is changed and influences the function of the muscle?

(4) Could the fascia lata be having a mechanical constraining effect on the muscle, or be causing occlusion of the vascular supply and focal necrosis of fibres as a result, and possibly wider loss of function of fibres without overt necrosis?

There is a general proliferation of connective tissue in the muscles in DMD from a very early stage, both in the perimysium surrounding the muscle bundles and in the endomysium surrounding the individual fibres (Fig. 5). This may possibly be of primary importance in the pathogenesis of the disease process, as has been suggested from time to time since the writings of Duchenne himself.

The clinical interventions of Rideau suggest that there may be environmental and other factors which are very important in the rate of progression of the disease, rather than the inevitable progression of the underlying pathological process itself. This raises the hope that further means of intervening, at different stages of the disease, may help to prolong not only the ambulation, but also the subsequent functional status and survival, of these afflicted children.

Recent advances in the molecular biology of the DMD have resulted in the gene being cloned and sequenced and the protein encoded by it identified as a 400 kDa protein named 'dystrophin' (Hoffman et al 1987a). This protein is thought to be localized to the microsomal fraction on sucrose gradient separation and to be situated in the triads of the T-tubular system (Hoffman et al 1987b). The protein is absent from the muscle of Duchenne dystrophy patients and also in the X-linked mouse dystrophy (*mdx*) and in an X-linked dystrophy recently discovered in a strain of retriever dogs. Surprisingly the *mdx* mouse has a very benign clinical course with no appreciable weakness, functional deficit or progression, and a normal lifespan. The muscle shows dystrophic features of necrosis plus regeneration but does not have the connective tissue proliferation seen in Duchenne dystrophy. The retriever's dystrophy, on the other hand, is very similar to the human disease, with early onset in the puppies, associated with progressive fixed deformity and a limited range of movement of various joints including the jaw, and a histological pattern in the muscle indistinguishable from Duchenne dystrophy. This further raises the question of how two different animal models with the same biochemical defect could have such divergent clinical syndromes, and suggests that secondary factors, such as load on the muscle, are important in the severity and progression of the disease, independent of the fundamental defect itself.

Could the connective tissue proliferation be a key factor in the progression of the disease and in turn be interrelated with the inability of the dystrophic muscle to mount an adequate regenerative and compensatory response?

Finally, I wish to throw in for discussion, and (I hope) for suggestions for resolution, another common clinical neuromuscular problem, particularly of infancy and childhood, namely spinal muscular atrophy (SMA). This is inherited through an autosomal recessive mechanism and is characterized by symmetrical weakness of skeletal muscles, affecting proximal muscles more than distal, the legs more than the arms, and also affecting the trunk, neck and intercostals, but sparing the diaphragm and the facial and ocular muscles (Dubowitz 1978). The bulbar muscles may be involved and the tongue often shows fasciculation. The onset may be prenatal, with paralysis already present at birth, or postnatal with the infant being completely normal at birth and the paralysis coming on at several weeks of age. The onset often appears to be fairly sudden and the weakness reaches a maximum fairly rapidly, then levels off and may remain static for a considerable time.

The earlier the onset of SMA, the greater the weakness and disability. Infants with the severe form usually die within the first year; children with intermediate severity achieve the ability to sit unsupported and are apparently normal in motor development until the latter half of the first year, when they develop weakness of the legs severe enough to prevent them standing or walking. Children or adults with the mild form have a later onset, after learning to walk.

The histological picture in the muscle is remarkably constant, irrespective of clinical severity. It is characterized by the uniform atrophy of large groups (often whole bundles) of fibres, which are usually of mixed fibre type, plus variably sized clusters of normal or enlarged fibres, generally comprising type I fibres only, suggesting a reinnervation by the sprouting of surviving neurons selectively of one type.

I should be interested to hear from Ron Oppenheim whether one could explain this disease process on the basis of his neurotrophic factor in developing muscle, which normally prevents neuronal death, and how the varying age of onset could fit in with this; and from Chris Henderson whether his concept of a factor in the muscle in spinal muscular atrophy which inhibits the normal neonatal, neuronal, growth-promoting activity could explain the disease, with its varying age of onset and severity, and comparatively stable course.

Acknowledgements

I am grateful to Professor Y. Rideau for permission to quote his data and to include Figs. 2 and 3.

References

Ashmore CR 1982 Stretch-induced growth in chicken wing muscles: effects on hereditary muscular dystrophy. Am J Physiol 242:178–183

Brown MD, Cotter MA, Hudlická O, Vrbová G 1976 The effects of different patterns of muscle activity on capillary density, mechanical properties and structure of slow and fast rabbit muscles. Pfluegers Arch Eur J Physiol 361:241–250

Buller AJ, Eccles JC, Eccles RM 1960 Interactions between motoneurones and muscles in respect of the characteristic speeds of their response. J Physiol (Lond) 150:417–439

Dubowitz V 1967 Cross-innervated mammalian skeletal muscle: histochemical, physiological and biochemical observations. J Physiol (Lond) 193:481–496

Dubowitz V 1978 Muscle disorders in childhood. Saunders, London, Philadelphia & Toronto

Faulkner JA, Jones DA, Round JM 1985 Injury to skeletal muscle of mice by lengthening contractions. J Physiol (Lond) 365:75P

Hoffman EP, Brown RH Jr, Kunkel LM 1987a Dystrophin: the protein product of the Duchenne muscular dystrophy locus. Cell 51:919–928

Hoffman EP, Knudson CM, Campbell KP, Kunkel LM 1987b Subcellular fractionation of dystrophin to the triads of skeletal muscle. Nature (Lond) 330:754–758

Howels KF, Goldspink G 1974 The effect of exercise on the progression of the myopathy in dystrophic hamster muscle fibres. J Anat 117:385–396

Hudecki MS, Pollina C, Bhargava AK, Fitzpatrick JE, Privitera CA, Schmidt D 1978 Effect of exercise on chickens with hereditary muscular dystrophy. Exp Neurol 61:65–73

Jaros E, Johnstone D 1983 Effect of denervation upon muscle fibre number in normal and dystrophic (*dy*/*dy*) mice. J Physiol (Lond) 343:104P–105P

Karpati G, Carpenter S, Prescott S 1982 Prevention of skeletal muscle fibre necrosis in hamster dystrophy. Muscle & Nerve 5:369–372

Loermans H, Wirtz P 1983 Inhibition of the expression of pathology in dystrophic mouse leg muscles by immobilization. Br J Exp Pathol 64:225–230

Luthert P, Vrbová G, Ward KM 1980 Effects of low frequency electrical stimulation on fast muscles of dystrophic mice. J Neurol Neurosurg Psychiatr 43:803–809

Newham DJ, Jones DA, Edwards RHT 1983 Large delayed plasma creatine kinase changes after stepping exercise. Muscle & Nerve 6:380–385

Rideau Y, Duport G, Delaubier A 1986 Premières rémissions reproductibles dans l'évolution de la dystrophie musculaire de Duchenne. Bull Acad Nat Med (Paris) 170:605–610

Salmons S, Vrbová 1969 The influence of activity on some contractile characteristics of mammalian fast and slow muscles. J Physiol (Lond) 201:535–549

Scott OM, Hyde SA, Goddard C, Dubowitz V 1982 Quantitation of muscle function in children: a prospective study in Duchenne muscular dystrophy. Muscle & Nerve 5:291–301

Scott OM, Vrbová G, Hyde SA, Dubowitz V 1985 Effects of chronic low frequency electrical stimulation on normal human tibialis anterior muscle. J Neurol Neurosurg Psychiatr 48:774–781

Scott OM, Vrbová G, Hyde SA, Dubowitz V 1986 Responses of muscles of patients with Duchenne muscular dystrophy to chronic electrical stimulation. J Neurol Neurosurg Psychiatr 49:1427–1434

Soltan HC 1962 Swimming stress and adaptation by dystrophic and normal mice. Am J Physiol 203:91–94

Vignos PJ, Watkins MP 1966 The effect of exercise in muscular dystrophy. JAMA (J Am Med Assoc) 197:121–126

Wirtz P, Loermans H 1983 Immobilization of dystrophic mouse muscle prevents pathology and necrosis of muscle fibres. Muscle & Nerve 6:234–235

Wirtz P, Loermans H, Wallinga-de Jonge W 1986 Long term functional improvement of dystrophic mouse leg muscles upon early immobilization. Br J Exp Pathol 67:201–208

DISCUSSION

Kernell: There is an interesting difference between the way in which muscles from Duchenne patients react to different pulse rates of long-term electrical stimulation and the reactions seen in normal fast cat muscles. In the latter, high pulse rates help to preserve the maximum force as well as the diameter of the muscle fibres (Kernell et al 1987b). Furthermore, as has been observed by us as well as by other groups working on normal animal muscles, the pulse rate of stimulation is fairly unimportant for long-term effects on muscle fatiguability. Provided that the daily time-amount of stimulation is great enough, one can achieve a markedly increased fatigue resistance with high as well as with low pulse rates (e.g., Hudlická et al 1982, Kernell et al 1987a).

Vrbová: Our experience with normal human subjects is exactly the same as that of Dr Kernell. In the trial with Victor Dubowitz (Scott et al 1985) we stimulated at low frequencies, and in the trial we did with Dr Dimitrijevic (Vrbová et al 1987) we stimulated at high frequency. The result as far as fatiguability is concerned is that in normal people or in paraplegics, unlike the Duchenne patient it does not matter what frequency is used.

Dubowitz: In Duchenne muscular dystrophy it is normally a slower muscle that is more fatigue resistant than in control children of the same age. And histochemically, it has always been thought that there is some change towards typeI fibres, so it appears more like slow muscle and there is more slow myosin present on immunocytochemical analysis.

Pette: We have examined the dystrophic mouse (C57 BL/6J) which has a pronounced disturbance of its calcium-sequestering system. We found not only a very low parvalbumin content, but also a reduced calcium uptake capacity (Leberer et al 1988). This reduction could be ascribed to alterations in the properties of the sarcoplasmic reticulum (SR) Ca^{2+}-ATPase, particularly to an increase in inactive enzyme molecules. This disturbance in Ca^{2+} sequestration could mean that stimulation, at least in the mouse, might be harmful to the dystrophic fibres. In collaboration with Gerta Vrbová we have investigated some effects of low frequency stimulation in dystrophic mice of the same strain. An increase in tetanic strength correlated with an increase in glycolytic enzyme activities (Reichmann et al 1981). An increase in mitochondrial enzyme activities was observed only with prolonged stimulation. These observations could indicate that stimulation brings about maturation in the dystrophic muscle, especially of the fast-twitch glycolytic fibres. It is this fibre population which has been shown, in previous studies (Bass et al 1970), to mature most extensively during postnatal growth in normal muscle.

Dubowitz: Because of the difference between animal models in the type of dystrophy, it might be interesting to repeat a similar study in relation to the *mdx* mouse. One of the theories is that dystrophin works via the T-tubule and calcium regulation in the SR.

Vrbová: The *mdx* mouse is quite an athletic-looking animal, however! I wonder whether you need a certain load on the muscle to require a dystrophin type of molecule? If so, Rideau's experiments may be explained by the effect of removing this load. A dystrophic mouse, or any mouse, is very light-footed, so perhaps the load is low per cross-sectional area and this is why the *mdx* mouse can cope without dystrophin, whereas a dog or human, which has a bigger load, needs this molecule much more. Perhaps the size of an animal and the load it has to move are important.

Van Essen: To test the idea of an inadequate load for the *mdx* mice, is it feasible to rear them in a high gravity environment, to see whether that would precipitate the disease?

Vrbová: It hasn't been tried, but you could load them quite easily.

Henderson: Michel Fardeau and I have been developing a model for the pathogenesis of the spinal muscular atrophies (SMA) which started from our results *in vitro* (Henderson et al 1987). Basically, we thought, as have others, that the selective motor neuron loss in SMA might be explained by a defect in the trophic support system. As I said earlier, muscle extracts from spinal muscular atrophy patients and not from a series of other neurological-disease controls and morphologically normal controls inhibited 'neonatal' neurite-promoting activity *in vitro*. If this is at all representative of what is happening *in vivo* in the SMA patient, it might explain some of the surprising features of SMA, including the sudden onset around birth, and variations in the speed and evolution of the disease.

If we imagine that human motoneurons depend first on an embryonic factor and then on a neonatal factor, and that a switch occurs around the time of birth, then even if inhibitors are present that block the neonatal activity, development could occur normally up to just before or after birth. If the inhibitor's effects became apparent at this stage, one could have two extreme situations: either (a) concentrations of inhibitory activity that are so high that neurons are killed at once and have no chance to respond to muscle-derived signals; this situation could correspond to the very rapidly evolving Werdnig–Hoffman form of the disease; or (b) lower amounts of inhibitor might cause withdrawal of motoneurons from the muscle but not lead to outright cell death. The muscle would subsequently respond by excess production of trophic factors, leading to a dynamic equilibrium between forces that are regressive and forces that try to re-form the neuromuscular junction. One could imagine reaching a stable situation that would lead to cycles of denervation and reinnervation, and to the fibre-type grouping that is characteristic of the Kugelberg–Welander disease.

Carlson: Professor Dubowitz, I was struck by the spinal muscular atrophy that you described and the difference in pathology between what we see in regenerated muscles and what you see in your SMA patients, where the atrophic fibres have apparently never been reinnervated and keep the same histochemically defined muscle fibre types. In patients with the early onset form of SMA, how early do the affected fibres appear in human muscles prenatally and how long does it take to develop the degree of atrophy seen at birth? I am wondering how long it takes for the histochemical or molecular pattern of protein to be set in these muscles fibres before birth.

Dubowitz: Differentiation of human muscle starts at about 12 weeks' gestation with the appearance of a population of myotubes that immunocytochemically express slow myosin. Perhaps differentiation goes on after birth, but by 30–36 weeks of gestation you see a mixed pattern.

The speed with which SMA shows the classical pattern is difficult to ascertain; occasionally one sees a patient who is weak from birth, or has an onset within the first few weeks of life. A muscle biopsy shows what we call the pre-pathological stage, with a lot of atrophy, but we don't see reinnervated fibres yet. At a later stage, that child's muscle will show the classical pattern of denervation/reinnervation; even the severe cases show a certain amount of large fibres (compensatory fibres) and reinnervated fibres. The disease process can manifest early in pregnancy, and that might be something that has to be explained in relation to an inhibitor. The earliest I have seen is about 30 weeks' gestation, with the mother being aware of loss of fetal movement by then, and the baby was born prematurely, and affected.

Carlson: Does the pathological picture give any clue to how long it needs for muscle to be innervated before these patterns of metabolic activity are set, or patterns of proteins are fixed?

Dubowitz: We have been trying to address that question with immunocytochemical studies, looking at embryonic, fetal and other myosins, to see the state of maturity of the small fibres and trying to make sure that they are all denervated; but the results are very variable.

Pette: I would like to ask Professor Dubowitz whether some of the histochemically classified type I fibres, especially the large ones, could have been type IIC or type IC fibres.

Dubowitz: They are hypertrophic and in relation to the age of the infant they are relatively big. I don't think they are IIC fibres because they are low in ATPase at pH 9.4, so they are by definition type I rather than type II, but they could be type IC. In addition, they uniformly express slow myosin only, on immunocytochemical study.

Pette: In that case they would co-express at least myosin heavy chains HCIIa and HCI (Staron & Pette 1987) and probably also the neonatal and embryonic myosin heavy chains (A. Maier & D. Pette 1987, unpublished observations).

Dubowitz: No; the large fibres uniformly contain only slow myosin and do not co-express any of the other myosin heavy chains that we have studied (embryonic, neonatal and mature fast).

Hoffman: When you look at the dystrophic muscles in the spinal muscular atrophies, the assumption is that they are denervated because of loss of neurons. How close is the correlation between loss of motoneurons and denervation-like changes? Could there be an initial loss of some trophic effects of the neuron on the muscle, rather than a loss of motoneurons, perhaps followed later by loss of neurons? What is the time course of those events?

Dubowitz: One problem clinically is the lack of correlation between what is seen on muscle biopsy and what is happening in the SMA patient. You often see almost the same degree of atrophy of the muscle and amount of reinnervation in a biopsy of the same type of muscle in very varying clinical cases, one being fairly mild and another much more severe. It is therefore difficult to relate the histological pattern to the functional state or to know how long the disease has been going on, or when it started. But the loss of a trophic effect on the muscle and a physical denervation could, presumably, have the same result, so long as it stimulated the same sprouting and reinnervation, because there is evidence that a lot of sprouting goes on in compensation in spinal muscular atrophy.

Hoffman: Has anyone looked to see whether motoneurons are absolutely lost at the time of denervation?

Dubowitz: The anterior horn cells have been examined histologically. There has been a converse opinion that a lot of the apparently atrophic muscles fibres are not denervated but immature-fetal or arrested muscle fibres. Dr A. Fidzianska has supported this hypothesis with electron microscopic studies of the small muscle fibres, trying to relate them to the early fetal myotube stage of development. Recent immunocytochemical studies in my Unit by Dr Caroline Sewry have also suggested that these fibres may be immature. The acute onset of the weakness, however, suggests that it could be either an acute denervation or possibly an exhaustion of a trophic influence.

Kernell: I had the impression from your slides that a large percentage of the cross-sectional area of the muscles in your SMA patients was taken up by very small atrophic fibres which you assume were denervated. Have you calculated whether the remaining force of those muscles could have been produced by the few remaining large-sized fibres, or would you need the rest of the cross-sectional surface area to produce the force?

Dubowitz: We haven't done that. We have taken a very small selective amount of muscle, and when we originally did open biopsies we sometimes found in one muscle a large area where almost all the fibres were atrophic and an adjacent region with a large area of re-innervated fibres, and then something mixed, between. One almost needs a total cross-section of the muscle to assess the relative distribution of these atrophic fibres.

Vrbová: We are trying to measure the forces developed by these muscles, but it is difficult to correlate them with the biopsies from these children with SMA. The remaining motoneurons in SMA patients are also abnormal: Hausmanowa-Petrusewicz (1985) has shown that there is continuous firing of these surviving motoneurons; they fire even in sleep, all the time. This is an indication that there is a central component to SMA, which could be a result of the target interaction; but it is not a simple situation.

Crowder: Spinal atrophy is an autosomal recessive disease, but different patients have different degrees of expression. Are the family trees extensive enough to say that the phenotype segregates as a single locus, or is it possible that different phenotypes are actually caused by mutations of different genes?

Dubowitz: P. E. Becker suggested that two genes could be involved in SMA and that the combinations of the two give the degrees of severity. This is based partly on the fact that although SMA is usually concordant within a family, with the same degree of severity, and the heterozygotes tend to be free of symptoms, there are sibships where one child has a severe form of the disease and a sib might have the milder form with later onset and longer survival. This would be best explained by more than one gene, and genic interactions.

References

Bass A, Lusch G, Pette D 1970 Postnatal differentiation of the enzyme activity pattern of energy-supplying metabolism in slow (red) and fast (white) muscles of chicken. Eur J Biochem 13:289–292

Hausmanowa-Petrusewicz I 1985 Changes of motor units in neuromuscular diseases. In: Dimitrijevic MR, Kakulas BA, Vrbová G (eds) Recent achievements in restorative neurology. 2. Progressive neuromuscular diseases. Karger, Basle, p 139–151

Henderson CE, Hauser SL, Huchet M et al 1987 Extracts of muscle biopsies from patients with spinal muscular atrophies inhibit neurite outgrowth from spinal neurons. Neurology 37:1361–1364

Hudlická O, Tyler KR, Srihari T, Heilig A, Pette D 1982 The effect of different patterns of long-term stimulation on contractile properties and myosin light chains in rabbit fast muscles. Pfluegers Arch Eur J Physiol 393:164–170

Kernell D, Donselaar Y, Eerbeek O 1987a Effects of physiological amounts of high- and low-rate chronic stimulation on fast-twitch muscle of the cat hindlimb. II. Endurance-related properties. J Neurophysiol 58:614–627

Kernell D, Eerbeek O, Verhey BA, Donselaar Y 1987 Effects of physiological amounts of high- and low-rate chronic stimulation on fast-twitch muscle of the cat hindlimb. I. Speed- and force-related properties. J Neurophysiol 58:598–613

Leberer E, Härtner KT, Pette D 1988 Postnatal development of Ca^{2+}-sequestration by the sarcoplasmic reticulum of fast and slow muscles in normal and dystrophic mice. Eur J Biochem 174:247–253

Reichmann H, Pette D, Vrbová G 1981 Effects of low frequency electrical stimulation on enzyme and isozyme patterns of dystrophic mouse muscle. FEBS (Fed Eur Biochem Soc) Lett 128:55–58

Scott OM, Vrbová G, Hyde SA, Dubowitz V 1985 Effects of acute and chronic electrical stimulation on normal human tibialis anterior muscle. J Neurol Neurosurg Psychiatry 48:774–781

Staron RS, Pette D 1987 The multiplicity of myosin light and heavy chain combinations in histochemically typed single fibres of rabbit soleus muscle. Biochem J 243:687–693

Vrbová G, Dimitrijevic MM, Partridge M, Halter J, Verhagen Metman L 1987 Reversal of increased muscle fatigue in paraplegic patients by electrical stimulation. Neurosci Soc Abstr 17:470–472

Final general discussion

Buller: In this concluding discussion, may we consider the problems and questions that we should now be attacking? I should like to hear the views of those who have not presented formal papers at the symposium.

Kernell: A largely unresolved question concerns the circumstances in which, when one changes the amount and conditions of muscle usage, there are long-term effects on the motoneurons as well as on the muscles themselves. It appears from a number of experiments that properties of motoneurons may indeed be influenced by the muscle they are innervating (e.g. Czéh et al 1978, Foehring et al 1987). Hence, mutual interactions are possible. Yet, in many situations, the motoneurons seem to be rather resistant to change. We have seen evidence of such a stability of certain motoneuronal properties (e.g. soma size) in experiments with heavy chronic stimulation of their axons (Donselaar et al 1986).

Nadal-Ginard: One thing that needs to be done is to dissect the system with which we are working. The tools are becoming available for doing that now. Genetic tools are particularly important here. We need animal models—especially mutant forms—for the different phenotypes. We want to create mutants where we can delete genes one at a time and then study the physiology when one gene is lacking. We can now do site-directed gene insertion in animal systems, and this has to be exploited. It is also important to isolate gene promoters specific for given cell types. That will allow us to do experiments in which the promoter can be linked to a toxic substance or protein that will kill selected groups of cells, leaving the animal intact. These are some of the experiments that we should be doing in the next decade.

Thesleff: One point that has not been touched upon is that when you denervate a mammalian skeletal muscle it becomes spontaneously active and starts to fibrillate. The mechanism seems to be twofold: one aspect is a cyclic change in membrane potential; the other consists of spontaneous action potentials within the T-tubular systems which may trigger an action potential in the sarcolemma. That, in combination with the altered kinetics of the sodium channel, produces fibrillation potentials. The evolutionary value of this kind of change, together with that of the increase of acetylcholine receptors, is to maintain the muscle fibre in good condition so that it is ready to be reinnervated by regenerating or sprouting nerves. The activity will maintain a balance, preventing too much atrophy in the fibre and still keeping it sensitive enough to be reinnervated. This is an adaptive change of the muscle to its denervated condition.

Rieger: So far we have only evoked the specificity of the molecular mechanisms of plasticity of the neuromuscular system in terms of cell–cell interactions. An important set of cell surface macromolecules which are clearly involved in developmental and regenerative events of nerve and muscle are the cell adhesion molecules (CAMs). Substrate adhesion molecules (SAMs) also play a critical role. These adhesion molecules are currently being studied intensively in a number of laboratories, and Dr G.M. Edelman's group at the Rockefeller University has played a pioneering role in determining their critical function in morphogenetic events during embyrogenesis and morphogenesis (for a review see Edelman 1986).

The neural cell adhesion molecule (N-CAM), the neuron-glia cell adhesion molecule (Ng-CAM) and cytotactin are present in the neuromuscular system, and modulation of their distribution, amounts and chemical properties has been demonstrated during development and regeneration (Rieger et al 1985, 1986, Daniloff et al 1986). Localization studies have shown accumulations of these molecules in the developing neuromuscular system at critical sites of cell–cell interactions, namely the neuromuscular junction, the node of Ranvier, and the myotendinous junction. N-CAM and cytotactin seem to play an important role at the neuromuscular junction. N-CAM, for example, is involved in axonal regeneration and muscle reinnervation in the frog. At the node of Ranvier there is a unique accumulation of all three molecules, and antibody perturbation experiments are under way to try to elucidate their respective roles in node of Ranvier formation. Moreover, we have shown (Daniloff et al 1986) that after nerve section, when Schwann cells are activated, N-CAM and Ng-CAM increase dramatically; they return to normal levels once axonal regeneration is completed. All these results suggest that histogenetic and regenerative events in the neuromuscular system are accompanied and may be regulated by subtle modulations of the distribution and physicochemical properties of adhesion molecules.

Van Essen: I would expect that in the next decade or so there will be a substantial increase in the number of candidate substances, molecules and factors that are present at particular stages of development. The challenge will be to transcend simply finding a factor present that might be involved. We need to establish its role in normal development. Much of this will not be explicable just in terms of the presence or absence of a factor; we shall need to understand the kinetics and dynamics of an exquisitely complicated system.

Walsh: A number of biochemical challenges face us with respect to nerve–muscle interactions and also the trophic factor system. As Dr Rieger said, it is clear that when we are thinking of biochemical cues there are now a large number of candidate molecules that are potentially involved in nerve–muscle interactions. It seems quite likely that there is a hierarchy of interactions, rather than one recognition molecule which is particularly dominant in the system.

What are the candidate molecules? From work mostly in *in vitro* model systems it has been shown that the calcium-dependent adhesion molecule, N-cadherin (Takeichi 1987), and the calcium-independent adhesion molecules such as the N-CAMs (Cunningham et al 1987, Dickson et al 1987), which also happen to be members of the immunoglobulin superfamily, and in addition other superfamily members, such as the Thy-1 antigen (Williams & Gagnon 1982), may be important in recognition events. At least some of these recognition molecules have been implicated by Bixby et al (1987). We also have molecules such as fibronectin and laminin and their receptors, which have been found to be important (Bixby et al 1987).

Whereas previously we thought of molecules such as N-CAM as single molecular species or only having a few forms, it now seems likely that there are many forms of such molecules. Alternative mRNA splicing and alternative polyadenylation are mechanisms which have been shown to generate multiple N-CAM forms in brain and muscle (Cunningham et al 1987, Dickson et al 1987). We don't yet know how these individual forms are involved in specific recognition events, but this, plus specific post-translational modifications, is clearly going to be important. Also, as well as the membrane-associated forms of recognition molecules such as N-CAM, there is the exciting possibility of naturally occurring soluble forms (Bock et al 1987, F.S. Walsh, unpublished work), again generated by specific alternative splicing pathways. We therefore have a large range of molecules, both in specific gene families and across gene families, that are likely to play a part in nerve–muscle interactions.

As for the future, the strategy that Bernardo Nadal-Ginard mentioned, namely the reductionist approach, is the only way forward. This should eventually tell us how these exciting molecules are working, by taking the system apart and using transfection-based model systems to dissect specific interactions and to identify important regions of these molecules.

With respect to trophic factors, the experiments that Ron Oppenheim described have for the first time shown that *in vivo* cell survival effects can be obtained by injecting muscle cell extracts and semi-pure factors. A problem is, however, in deciding whether we are dealing with one factor or many different factors. The strategy that Chris Henderson put forward is the one most likely to succeed here. Workers in this area are at present unfortunately using rather loose terms—embryonic factor, neonatal factor, factors expressed on denervation, and so on. It is not yet clear whether we are really dealing with a family of survival factors or one factor. When the factors are purified and tested *in vivo* we shall resolve this important point.

In addition to survival factors there are probably important factors that are involved in, for example, the induction of choline acetyltransferase activity, and choline uptake; how the whole system operates will not be known until each of these molecules is purified by the methods that Dr Henderson discussed.

Anne Mudge made a point earlier about what happens when cells die as a result of loss of trophic support. The NGF research area has given us a good idea of what happens when this trophic factor interacts with its receptor. NGF has been shown to bind to specific receptor in the cell membrane; this in turn activates a number of second messenger systems in NGF-responsive cells. A comparable set of mechanisms will surely be found for the motoneuron system, with a receptor for the growth factors and signalling mechanisms that are subsequently activated. Presumably the decision as to whether a motoneuron dies or lives will depend on which genes and gene families are activated as a result of transduction and signalling processes.

With respect to the challenge that Victor Dubowitz has made regarding agents that cause neurons to survive in the spinal muscular atrophies, it may be worth pointing out that it is now known that NGF-sensitive neurons exist in the CNS and that populations of these may disappear in certain disease states, such as Alzheimer's disease. We may be able to think of ways of altering the signalling after NGF has bound by manipulation of the second messenger systems pharmacologically, thereby causing cells which would otherwise die to survive. This approach could also be considered in the neuromuscular field. We may not need to introduce a pure trophic factor to obtain motoneuron survival. We might be able to bypass the trophic factor by pharmacological means.

Zak: I would like to ask Professor Dubowitz what he thinks will be the impact of the availability of probes for dystrophin, either for the gene or for the protein, on the clinical problems of muscular dystrophy.

Dubowitz: The first stage in trying to categorize the dystrophies is to see what happens in forms of dystrophy other than Duchenne muscular dystrophy, using the dystrophin antibody to divide them into dystrophin-present and dystrophin-absent forms, and also trying to gain a better understanding of the variation within DMD itself, from the severe type through to the milder Becker type of dystrophy.

The major problem is to understand how the lack of this low abundance protein produces the disease; the search for mechanisms will go on from that. Whether we shall achieve this understanding from the molecular biology alone, or whether we shall need other approaches, awaits resolution. There may possibly be a central role for the connective tissues in the pathogenesis of the disease process, and the proliferation of connective tissue may be directly linked with the absence of dystrophin.

Zak: The *dmx* mouse is challenging in relation to these studies. Apparently, from the pictures that Gerta Vrbová has shown me, this mouse mutant undergoes devastating yet transient changes in the musculature.

Vrbová: Yes. Each skeletal muscle undergoes more or less complete destruction at 2–4 weeks after birth, but the pathology resembles polymyositis rather than Duchenne dystrophy. Then very rapidly, at 4–5 weeks after birth, the muscles recover, and by 5–6 weeks of age the muscle strength and morphology are normal. The muscle fibres have centrally placed nuclei and the mice have high levels of serum creatine kinase. It is very challenging as to why this disease is not progressive.

Buller: About three years ago, Dr Vrbová did her histological study of the *dmx* mouse and from these studies it was concluded that this was not a model of Duchenne dystrophy. Now this mouse strain has been shown to lack dystrophin, which is apparently the gene product absent in DMD. So the *dmx* mouse is again considered to be a model of Duchenne muscular dystrophy.

Vrbová: It is not a model of the disease, because DMD is a progressive disease, whereas in the *dmx* mouse the muscles recover.

Buller: How then do you define a disease? If we are thinking in reductionist terms, we must define DMD in terms of the basic genetic defect and the lack of the gene product. As Victor Dubowitz says, we must define the dystrophin-deficient forms of muscular dystrophy. As yet we have only three examples: human Duchenne, the golden retriever model, and the *dmx* mouse. It is a fascinating problem why the phenotypes of these three deficiencies should be different, but we must define diseases in terms of their primary causation, not of the phenotype that results.

Dubowitz: We can also define the regenerative capacity of muscle. We have been comparing regenerating fibres in Duchenne dystrophy with normal regeneration in the model of necrosis and regeneration following eccentric exercise in healthy volunteers. There are distinct differences. Regenerating fibres are believed to arise from the fusion of satellite cells within the basal lamina of necrotic fibres. Human muscle damaged by eccentric exercise shows focal necrosis, and desmin antibodies identify the cuff of new myocytes that form at the periphery of the necrotic fibre. Once myotubes are formed, lower levels of β-spectrin expression are seen as well as traces of slow myosin and sometimes fast myosin. Embryonic myosin does not appear to be present except at very early stages. In DMD, in contrast, the desmin-positive peripheral myocytes are not found, despite the abundance of histologically apparent regenerating basophilic fibres. Basophilic fibres also differ in DMD, in that several (but not all) basophilic fibres show no detectable β-spectrin and several contain embryonic myosin. Groups of basophilic fibres in DMD often show no remnants of necrotic fibres.

These results suggest: (1) an alternative origin of regenerating fibres in DMD, and (2) that regeneration in DMD is not normal. Perhaps the loss of regenerative capacity in dystrophic muscle is the key to the pathogenesis of the disease process and its relentlessly progressive course. It will be interesting to compare the regenerative potential of the dystrophic muscle in the *mdx* mouse and the retriever dog with that of Duchenne dystrophy.

Buller: The phenotype is similar but the disease process may be totally different. The history of progress in clinical medicine consists of dissecting out phenotypes into different primary causations. For DMD we now have the genetic defect identified, and also the gene product. From there on, other factors may cause the phenotypes to vary.

Holder: It seems to me that an important phase in the development of the neuromuscular system is the stage before the nerves and muscles interact. In the early stages of embryonic development, nerve and muscle develop independently, and study of that phase may prove most instructive. For example, the pattern of where each of the muscles forms, which is crucial for subsequent development, is established quite separately from that of the nervous system (see, for example, Chevallier et al 1977). Also, the initial muscle fibre type pattern develops independently of the nervous system (Laing & Lamb 1983, Butler et al 1982). Equally, the motor nerves arise in the neural tube and grow towards the muscles, and yet their axons reach the right places and generate the correct pattern of connections, independently of their targets. I'm sure that in the next few years we shall dissect some of these independent processes out, using the methods of molecular biology.

Finally, the challenge is to understand these early processes so that we can discover how the system repairs itself and can, therefore, persuade the nerves not only to grow but to go to the right places. For that we need to know how they initially got there.

Lowrie: In our discussion of specificity and plasticity the consensus seems to be that we are seeing increasing possibilities of specificity, and that this is at the expense of plasticity, or adaptation, if we prefer the term. I look at it more optimistically and feel that, given the increase in specificity, it is remarkable how much adaptation is possible. I would endorse the point made earlier, that adaptation may appear limited because as yet we do not know how to switch on the appropriate gene. I would also say that although the adaptive response to, say, the disruption of nerve–muscle interactions during a critical period of development, before the motoneuron and muscle have both matured, is clearly less able to restore the system to normality than in the adult, I would nevertheless not describe it as restricted adaptation—rather as just a different type of plasticity.

Zak: It seems to me that at this meeting we have started to modify the concept of muscle plasticity. In the past we had been impressed by the extent of a muscle's promiscuity. Now we begin to see that there are constraints on the adaptive options of a muscle. The gradual restriction of differentiation programmes initiated upon commitment of stem cells is not an all-or-none phenomenon but continues through overt muscle cell differentiation and later during the maturation and diversification of myocytes. The challenge of future research is to delineate the adaptive options available to a given stage of muscle differentiation.

My second impression of the symposium is that so far the responses of individual gene families to the perturbation of muscle function or its metabolic environment do not follow a common theme. Consequently, defining an alteration in muscular phenotype on the basis of one gene family only is undoubtedly an oversimplification. To answer these questions in a more meaningful way will require further understanding of the functional correlates of the multiple variants of muscle constituents.

* * * *

Buller: I find it difficult to add anything useful to the discussions that have gone on at the symposium, and I do not intend to summarize the views expressed. However, as the oldest participant, who has been to many meetings on this and related subjects, I can say that this is the first meeting I have attended in which there has been such marked emphasis on recombinant DNA technology, and a clear belief in its potential value for future progress in the field. That does not mean to say that other types of work should not continue. But, in terms of moving forward in the immediate future, this new technology seems to most of us to be the most powerful tool available. Indeed, I would expect its jargon to dominate the next symposium on the plasticity of the neuromuscular system!

Meetings of this quality do not just happen. While all of you, the participants, have tempered your (formal) discussion to an admirable extent—perhaps a little more public belligerence would have been helpful to readers of the volume—we have been given adequate opportunity to argue (and sometimes agree) outside the formal sessions. To the Ciba Foundation, and to those who helped to gather us together—and I would especially mention Professor Gerta Vrbová and Professor Dirk Pette—I offer our most sincere thanks. We separate determined to be better informed when next we meet, and with gratitude for all that has been done to make this meeting not only productive, but also enjoyable.

References

Bixby JL, Pratt RS, Lilien J, Reichardt LF 1987 Neurite outgrowth on muscle cell surfaces involves extracellular matrix receptors as well as Ca^{++}-dependent and -independent cell adhesion molecules. Proc Natl Acad Sci USA 84:2555–2559

Bock E, Edvardsen K, Gibson A, Linnemann D, Lyles JM, Nybroe O 1987 Characterization of soluble forms of NCAM. FEBS (Fed Eur Biochem Soc) Lett 225:33–36

Butler J, Cosmos E, Brierley J 1982 Differentiation of muscle fiber types in aneurogenic brachial muscles of the chick embryo. J Exp Zool 224:65–80

Chevallier A, Kieny M, Mauger A 1977 Limb somite relationship: origin of the limb musculature. J Embryol Exp Morphol 41:245–258

Cunningham BA, Hemperly JJ, Murray BA, Prediger EA, Brackenbury R, Edelman GM 1987 Neural cell adhesion molecule: structure, immunoglobulin-like domains, cell surface modulation, and alternative RNA splicing. Science (Wash DC) 236:799–806

Czéh G, Gallego R, Kudo N, Kuno M 1978 Evidence for the maintenance of motoneuron properties by muscle activity. J Physiol (Lond) 281:239–252

Daniloff JK, Levi G, Grumet M, Rieger F, Edelman GM 1986 Altered expression of neuronal cell adhesion molecules induced by nerve injury and repair. J Cell Biol 103:924–945

Dickson G, Gower HJ, Barton CH et al 1987 Human muscle neural cell adhesion molecule (N-CAM): identification of a muscle specific sequence in the extracellular domain. Cell 50:1119–1130

Donselaar Y, Kernell D, Eerbeek O 1986 Soma size and oxidative enzyme activity in normal and chronically stimulated motoneurons of the cat's spinal cord. Brain Res 385:22–29

Edelman GM 1986 Cell adhesion molecules in the regulation of animal form and tissue pattern. Annu Rev Cell Biol 2:81–116

Foehring RC, Sypert GW, Munson JB 1987 Motor-unit properties following cross-reinnervation of cat lateral gastrocnemius and soleus muscles with medial gastrocnemius nerve. II. Influence of muscle on motoneurons. J Neurophysiol 57:1227–1245

Laing NG, Lamb AH 1983 The distribution of muscle fibre types in chick embryo wings transplanted to the pelvic region is normal. J Embryol Exp Morphol 78:67–82

Rieger F, Grumet M, Edelman GM 1985 N-CAM at the vertebrate neuromuscular junction. J Cell Biol 101:285–293

Rieger F, Daniloff JK, Pinçon-Raymond M, Crossin KL, Grumet M, Edelman GM 1986 Neuronal cell adhesion molecules and cytotactin are colocalized at the node of Ranvier. J Cell Biol 103:379–391

Takeichi M 1987 Cadherins: a molecular family essential for selective cell–cell adhesion and animal morphogenesis. Trends Genet 3:213–217

Williams AF, Gagnon J 1982 Neuronal cell Thy-1 glycoprotein: homology with immunoglobulin. Science (Wash DC) 216:696–703

Index of contributors

Non-participating co-authors are indicated by asterisks. Entries in bold type indicate papers; other entries refer to discussion contributions

Indexes compiled by John Rivers

Subject index